Modern Birkhäuser Classics

Many of the original research and survey monographs in pure and applied mathematics published by Birkhäuser in recent decades have been groundbreaking and have come to be regarded as foundational to the subject. Through the MBC Series, a select number of these modern classics, entirely uncorrected, are being re-released in paperback (and as eBooks) to ensure that these treasures remain accessible to new generations of students, scholars, and researchers.

Robust Nonlinear Control Design

State-Space and Lyapunov Techniques

Randy A. Freeman
Petar V. Kokotović

Reprint of the 1996 Edition

Birkhäuser
Boston • Basel • Berlin

Randy A. Freeman
Department of Electrical
and Computer Engineering
Northwestern University
Evanston, IL 60208
U.S.A.

Petar Kokotović
Department of Electrical
and Computer Engineering
University of California
Santa Barbara, CA 93106
U.S.A.

Originally published in the series *Systems & Control: Foundations & Applications*

ISBN-13: 978-0-8176-4758-2 e-ISBN-13: 978-0-8176-4759-9
DOI: 10.1007/978-0-8176-4759-9

Library of Congress Control Number: 2007940262

Cover design by Alex Gerasev.

Printed on acid-free paper.

9 8 7 6 5 4 3 2 1

www.birkhauser.com

Randy A. Freeman
Petar V. Kokotović

Robust Nonlinear Control Design

State-Space and Lyapunov Techniques

Birkhäuser 1996
Boston • Basel • Berlin

Randy A. Freeman
Department of Electrical and Computer Engineering
Northwestern University
Evanston, IL 60208

Petar V. Kokotović
Department of Electrical and Computer Engineering
University of California
Santa Barbara, CA 93106

Printed on acid-free paper

Birkhäuser 

ISBN 0-8176-3930-6
ISBN 3-7643-3930-6
Typeset by the authors in TEX
Printed and bound by Edwards Brothers, Ann Arbor, MI
Printed in the United States of America

9 8 7 6 5 4 3 2 1

Preface

This is the first book entirely dedicated to the design of robust nonlinear control systems. We believe that every effort in this direction is timely and will be highly rewarding in both theoretical and practical results.

Although the problem of achieving robustness with respect to disturbances and model uncertainty is as old as feedback control itself, effective systematic methods for the robust design of linear systems have been developed only recently. That such methods are already being successfully applied by a large community of practicing engineers testifies to a vital technological need.

Limitations of a popular methodology have always been among the factors stimulating new research. Such is the case with the inability of robust linear control to cope with nonlinear phenomena which become dominant when commands or disturbances cause the system to cover wide regions of its state space. In this situation it is natural to turn to nonlinear approaches to robust control design.

There are obvious reasons why robustness studies of nonlinear systems have been incomparably less numerous than their luckier linear cousins. The complexity of nonlinear phenomena is daunting even in the absence of disturbances and other uncertainties. It is not surprising that it has taken some time for a "clean" theory to discover classes of nonlinear systems with tractable analytic and geometric properties. During the last ten years, much progress has been made in this direction by nonlinear differential-geometric control theory. Most recently, a merger of this theory with classical Lyapunov stability theory led to the systematic adaptive "backstepping" design of nonlinear control systems with unknown constant parameters. However, the adaptive control paradigm is not suitable for handling fast time-varying and functional uncertainties which are the main topic of this book.

Wide operating regimes involving large magnitudes of state and control variables, such as torques, pressures, velocities, and accelerations, are becoming increasingly common in modern aircraft, automotive systems, and industrial processes. In these regimes, nonlinearities which are not confined to "linear sectors" (namely those which exhibit super-linear growth) often cause severe, or even catastrophic, forms of instability. For this reason, our theory and design methods take such critical nonlinearities into account and focus on large-signal (global) behavior rather than small-signal (local) behavior. While not restricting nonlinear growth, we do consider systems with a particular structure.

Often a control design is performed on a model having no uncertainties. The robustness of the resulting system is then analyzed, possibly followed by a redesign to improve robustness. In contrast, our approach is to explicitly include uncertainties in the design model, taking them into account during the design itself. We therefore extend the theory behind Lyapunov design to include uncertainties by introducing the *robust control Lyapunov function* (rclf). Just as the existence of a control Lyapunov function is equivalent to the nonlinear stabilizability of systems without uncertainties, the existence of our rclf is equivalent to the nonlinear robust stabilizability of systems with uncertainties. The task of constructing an rclf thereby becomes a crucial step in robust nonlinear control design.

Our recursive methods for constructing rclf's remove the "matching condition" constraint which severely limited the applicability of early robust Lyapunov designs. Already these designs exploited a worst-case differential game formulation, and we adopt a similar viewpoint in our approach to robust control design. Our solution of an inverse optimal robust stabilization problem shows that every rclf is the value function associated with a meaningful game. The resulting inverse optimal designs prevent the wasteful cancellation of nonlinearities which are beneficial in achieving the control objective, and they also inherit the desirable stability margins guaranteed by optimality.

The theoretical foundation of the entire book is established in Chapter 3 where we develop the rclf framework. Chapter 4 contains new results in inverse optimality and relates them to crucial issues in control design and performance. The bulk of the design content of this book appears in Chapters 5–8. In Chapter 5 we present the recursive Lyapunov design procedure we call *robust backstepping.* This design procedure is modified to accommodate measurement disturbances in Chapter 6. A dynamic feedback version of backstepping is developed in Chapter 7. In Chapter 8 we combine these robust and dynamic backstepping methods to obtain a robust nonlinear version of classical proportional/integral (PI) control. Illustrative examples appear throughout the book, while Chapters 7 and 8 include detailed design examples.

This book is intended for graduate students and researchers in control theory, serving as both a summary of recent results and a source of new research problems. We assume the reader has a basic knowledge of nonlinear analysis and design tools, including Lyapunov stability theory, input/output linearization, and optimal control. For those readers not familiar with elementary concepts from set-valued analysis, we provide a review of set-valued maps in Chapter 2.

* * *

We thank Tamer Başar for helping to direct our path, especially as we developed the inverse optimality results in Chapter 4. Also, we benefited greatly from frequent discussions with Miroslav Krstić and Ioannis Kanellakopoulos, whose contributions in adaptive nonlinear control directly inspired the dynamic backstepping methods in Chapters 7 and 8. We are grateful for the insights we gained from these colleagues. We thank Mohammed Dahleh, Laurent Praly, and Eduardo Sontag for sharing with us their technical expertise which helped shape many of our results. We are grateful to John Cheng of Rockwell International for providing us with physical examples motivating the material in Chapter 8. Many other researchers and educators influenced the content of this book, including Mrdjan Janković, Art Krener, Philippe Martin, Rodolphe Sepulchre, Stephen Simons, and Mark Spong.

Finally, this work would not have been possible without the patient support of our wives, Lisa and Anna—it is *analisa* that lies behind each of our control designs.

The research presented in this book was supported in part by the National Science Foundation under Grant ECS-9203491 and by the Air Force Office of Scientific Research under Grant F49620-92-J-0495, both through the University of California at Santa Barbara, and by the U.S. Department of Energy under Grant DE-FG-02-88-ER-13939 through the University of Illinois at Urbana-Champaign.

Randy Freeman
Evanston, Illinois

Petar Kokotović
Santa Barbara, California

March 1996

Contents

Chapter 1

Introduction

The main purpose of every feedback loop, created by nature or designed by engineers, is to reduce the effect of uncertainty on vital system functions. Indeed, feedback as a design paradigm for dynamic systems has the potential to counteract uncertainty. However, dynamic systems with feedback (closed-loop systems) are often more complex than systems without feedback (open-loop systems), and the design of feedback controllers involves certain risks. Feedback can be used for stabilization, but inappropriately designed feedback controllers may reduce, rather than enlarge, regions of stability. A feedback controller that performs well on a linearized model may in fact drastically reduce the stability region of the actual nonlinear system.

Broadly speaking, robustness is a property which guarantees that essential functions of the designed system are maintained under adverse conditions in which the model no longer accurately reflects reality. In modeling for robust control design, an exactly known *nominal plant* is accompanied by a description of plant uncertainty, that is, a characterization of how the "true" plant might differ from the nominal one. This uncertainty is then taken into account during the design process.

A popular version of this robust control paradigm, depicted in Figure 1.1, involves the interconnection of a nominal plant G, a controller K, and an uncertainty Δ. The precisely known nominal plant G may be some generalized plant which includes design artifacts such as frequency-dependent weights on the uncertainty Δ. Once G is determined, the robust control problem is to construct a controller K which guarantees

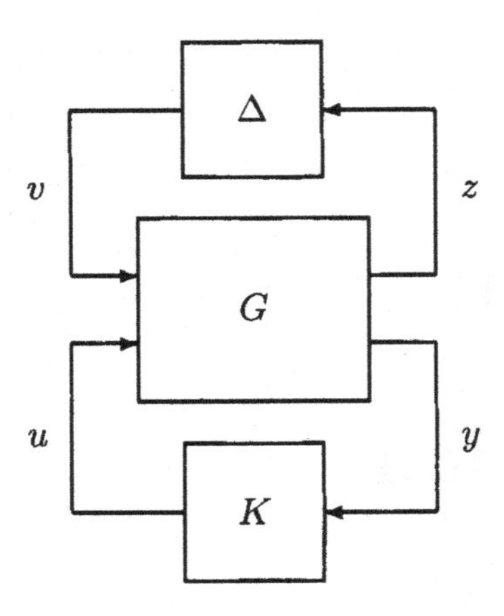

Figure 1.1: A robust control paradigm.

closed-loop stability and performance in the presence of every Δ belonging to a given family $\mathcal{F}_\Delta$ of admissible uncertain systems. This is a *worst-case* or *deterministic* paradigm for robust control because all uncertainties in $\mathcal{F}_\Delta$ are assumed to be "equally likely." The task of choosing an appropriate family $\mathcal{F}_\Delta$ is a crucial ingredient in robust control design.

Much of current robust control theory is *linear*, that is, its results have been obtained under the assumption that the nominal plant G is linear (usually also finite-dimensional and time-invariant). A wide variety of families $\mathcal{F}_\Delta$ of admissible uncertain systems have been considered in this context, including families of structured and unstructured uncertainties, memoryless (real) and dynamic (complex) uncertainties, linear and nonlinear uncertainties, time-invariant and time-varying uncertainties, etc. Moreover, different measures for the "size" of an uncertainty have lead to different frameworks for robust control (H_∞, L_1, etc.), and the most common choice for the family $\mathcal{F}_\Delta$ is a set of uncertainties satisfying some norm bound. When the actual (physical) system exhibits nonlinear behavior, the family $\mathcal{F}_\Delta$ must be chosen large enough to encompass the nonlinear phenomena because G is restricted to be linear. A disadvantage of this approach is that it ignores available information about existing nonlinearities, and the resulting controllers may be too conservative (especially when the nonlinearities are significant).

A natural attempt to overcome this drawback of robust linear linear control is to allow the nominal plant G to be *nonlinear* and thereby pursue *robust nonlinear control design.* This is the main purpose of this book.

1.1 A Lyapunov framework for robust control

While frequency domain methods have been instrumental in advancing the theory of robust linear control, they are conspicuously absent from the nonlinear arsenal.[1] On the other hand, state space methods have been, and continue to be, rigorously developed for general nonlinear systems. Nonlinear input/output methods have a long history in determining the stability of feedback interconnections of separate subsystems based on their individual input/output properties [126, 159, 160, 25, 125, 142], though the characterization of these properties is usually accomplished with state space methods rooted in Lyapunov stability theory [98, 158, 49, 77, 152]. Connections between input/output and state space stability for nonlinear systems have been established by the theory of dissipativity [155, 56, 57, 55] in which a Lyapunov-like *storage function* is used to monitor the "energy" flowing in and out of a system.

One of the earliest frameworks for robust nonlinear control is the *guaranteed stability* or *Lyapunov min-max* approach, developed in [89, 48, 90, 23, 11]. This framework includes as a special case the *quadratic stability* approach for uncertain linear systems, surveyed in [22]. More recently, game theory [9] and the theory of dissipativity have led to the so-called "nonlinear H_∞" approach to robust nonlinear control [8, 150, 151, 61, 10, 83, 65]. Finally, the concept of *input-to-state stability* (ISS) introduced in [131] has led to an input/output methodology intimately related to state space Lyapunov stability [133, 134, 99, 95, 119, 143, 141, 67, 135, 137, 138, 66].

In Chapter 3, we develop a general state space Lyapunov framework

[1] A noteworthy exception is the describing function method (method of harmonic balance) [105, 77] for the analysis of periodic solutions to systems with sector nonlinearities.

for robust nonlinear control which encompasses both the guaranteed stability and the ISS frameworks. In our problem formulation in Section 3.1, we use set-valued maps (reviewed in Chapter 2) to describe constraints on uncertainties, controls, and measurements. The use of set-valued maps in control theory dates back to the early sixties [27, 153], and it continues in the current study of robust and optimal control, especially in the contexts of viability theory for differential inclusions [127, 128, 5, 79] and nonsmooth analysis [21, 107, 108]. In Section 3.2, we show that the ISS disturbance attenuation problem, which is a type of "nonlinear L_1 control" problem important for the application of recent small gain theorems [67, 66, 99, 141], is included in our formulation.

Having formulated a robust control problem in the state space, we are motivated to find necessary and sufficient conditions for its solution. The most important necessary and sufficient condition for the stability of a nonlinear system is the existence of a Lyapunov function. Even though this condition is not computable in general, Lyapunov functions have been used extensively in nonlinear stability analysis. Lyapunov theory was developed for systems without inputs and has therefore traditionally been applied only to closed-loop control systems, that is, systems for which a feedback control has already been selected.

However, candidate Lyapunov functions can be useful design tools: they provide guidelines for choosing feedback controls as one may impose the constraint of "making the Lyapunov derivative negative" in the construction of the feedback. This idea is not new [71, 82, 63, 68], but it has been made explicit only relatively recently with the introduction of the *control Lyapunov function* (clf) for systems with control inputs [3, 130]. In short, a clf for a nonlinear control system of the form

$$\dot{x} \quad = \quad f(x, u) \tag{1.1}$$

is a candidate Lyapunov function $V(x)$ with the property that for every fixed $x \neq 0$ there exists an admissible value u for the control such that $\nabla V(x) \cdot f(x, u) < 0$. In other words, a clf is simply a candidate Lyapunov function whose derivative can be made negative *pointwise* by the choice of control values. Clearly, if f is continuous and there exists a continuous state feedback for (3.18) such that the point $x = 0$ is a globally asymptotically stable equilibrium of the closed-loop system, then by standard converse Lyapunov theorems [87, 104] there must exist a clf for the system (1.1). If f is affine in the control variable, then the existence of a clf for (3.18) is also sufficient for stabilizability via continuous[2] state feedback [3, 132].

From the above discussion, the following parallelism is clear: just as the existence of a Lyapunov function is necessary and sufficient for the stability of a system without inputs, the existence of a clf is necessary and sufficient for the ***stabilizability*** of a system with a control input. However, neither the Lyapunov function nor the clf is adequate for our purposes because our nominal plant G has *two* different inputs, one from the controller K and one from the uncertainty Δ in Figure 1.1. Furthermore, the clf methodology applies only to state feedback, and our problem formulation includes general types of measurement feedback.

We therefore introduce the *robust control Lyapunov function* (rclf) in Section 3.3. Not only does the rclf generalize the clf to systems with both control and uncertainty inputs, it also generalizes the "output clf" defined in [148, 149, 147]. In Sections 3.4–3.5, we show that the existence of an rclf is necessary and sufficient for the solvability of our robust control problem.

The robust control Lyapunov function characterizes the solvability of our problem, and at the same time it raises two important design issues:

1. How does one construct an rclf for an uncertain nonlinear system? Are there significant classes of systems for which a systematic construction is possible?
2. Once an rclf has been found, how does one construct the robust controller?

Although the second design issue is more easily resolved than the first, it is of no less importance in achieving desirable closed-loop behavior. We address the second issue in Chapter 4, and then we return to the first issue in Chapters 5–8.

---

[2]Continuity of the feedback at the point $x = 0$ requires an extra condition.

1.2 Inverse optimality in robust stabilization

In Chapter 4 we show how one might construct a robust controller once a robust control Lyapunov function has been found. We now present an elementary example that will illustrate the main point. Suppose we wish to robustly stabilize the system

$$\dot{x} \;=\; -x^3 + u + wx \tag{1.2}$$

where u is an unconstrained control input. Robustness is to be achieved with respect to a disturbance w known to take values in the interval $[-1, 1]$. It is clear by inspection that a robustly stabilizing state feedback control law for this system is

$$u \;=\; x^3 - 2x \tag{1.3}$$

This control law, which cancels the nonlinearity $-x^3$, is a result of feedback linearization [60, 111]. However, it is an absurd choice because the control term x^3 wastefully cancels a beneficial nonlinearity. Moreover, this term is actually positive feedback which increases the risk of instability due to other uncertainties not taken into account in the design.

Although it is easy to find a better control law for this simple system, what we seek is a *systematic* method for choosing a reasonable control law for a general system for which an rclf is known. One approach would be to formulate and solve an optimal robust stabilization problem with a cost functional which penalizes control effort. For the system (1.2), the cost functional

$$J \;=\; \int_0^\infty \left[x^2 + u^2\right] dt \tag{1.4}$$

is minimized (in the worst case) by the optimal feedback law

$$u \;=\; x^3 - x - x\sqrt{x^4 - 2x^2 + 2} \tag{1.5}$$

The control laws (1.3) and (1.5) are plotted in Figure 1.2. The optimal control law (1.5) recognizes the benefit of the nonlinearity $-x^3$ and accordingly produces little control effort for large x; moreover, this optimal

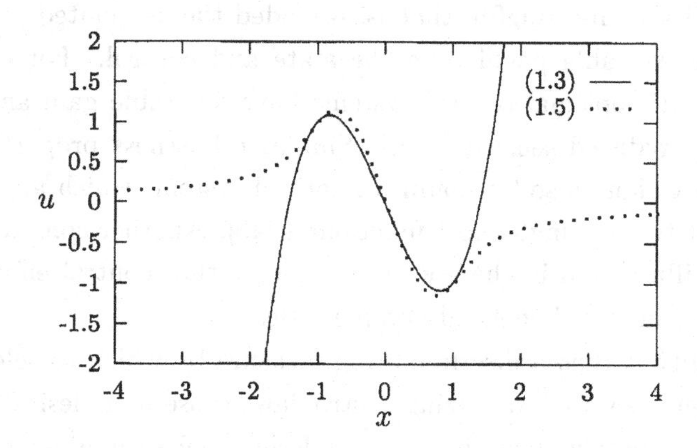

Figure 1.2: A comparison between the control laws (1.3) and (1.5)

control law never generates positive feedback. However, such superiority comes at the price of solving a steady-state Hamilton-Jacobi-Isaacs (HJI) partial differential equation, a task feasible only for the simplest of nonlinear systems. Indeed, for a general system and cost functional

$$\dot{x} \;=\; f(x,u,w)\,, \qquad\qquad J \;=\; \int_0^\infty L(x,u)\,dt \tag{1.6}$$

the steady-state HJI equation is

$$0 \;=\; \min_u \max_w \Big[L(x,u) \;+\; \nabla V(x) \cdot f(x,u,w) \Big] \tag{1.7}$$

where the *value function* $V(x)$ is the unknown. For an appropriate choice of the function $L(x,u)$ in (1.6), a smooth positive definite solution $V(x)$ to this equation (1.7) will lead to a continuous state feedback control $u(x)$ which provides optimality, stability, and robustness with respect to the disturbance w. However, such smooth solutions may not exist or may be extremely difficult to compute.

We will show that a known rclf for a system can be used to construct an optimal control law directly and explicitly, without recourse to the HJI equation (1.7). This will be accomplished by solving an *inverse optimal robust stabilization problem.*

The relationship between stability and optimality has been a central issue in the optimal stabilization problem ever since the advent of the steady-state Hamilton-Jacobi-Bellman (HJB) equation. Optimal feedback systems enjoy many desirable properties beyond stability, provided the optimality is meaningful, that is, provided the associated cost functional places suitable penalty on the state and control. For example, linear-quadratic optimal control systems have favorable gain and phase margins and reduced sensitivity [2]. Similar robustness properties have been shown to hold also for nonlinear control systems which are optimal with respect to meaningful cost functionals [46]. Another consequence of optimality, illustrated in the above example, is that control effort is not wasted to counteract beneficial nonlinearities.

Optimality is thus a discriminating measure by which to select from among the entire set of stabilizing control laws those with desirable properties. Unfortunately, its usefulness as a design tool for nonlinear systems is hampered by the computational burden associated with the HJB and HJI equations. Suppose, however, that we have found an rclf for our system, perhaps through the recursive construction of Chapter 5. If we can find a meaningful cost functional such that the given rclf is the corresponding value function, then we will have indirectly obtained a solution to the HJI equation and we can therefore compute the optimal control law. As long as the cost functional belongs to a meaningful class, the resulting control law will inherit all of the benefits of optimality listed above. Motivated by such reasoning, we pose the *inverse* optimal robust stabilization problem of finding a meaningful cost functional such that a given rclf is the corresponding value function.

Inverse problems in optimal control have a long history [70, 2, 106, 110, 109, 63, 50, 64]. The first inverse problem to be formulated and solved was for linear time-invariant systems [70, 2, 106]. These results provided a characterization of those stabilizing control gain matrices that were also optimal with respect to some quadratic cost. Inverse problems for nonlinear systems have since been considered, but with less success; some solutions for open-loop stable nonlinear systems are given in [63, 64, 46], and homogeneous systems are discussed in [53, 54]. The results presented in Chapter 4 extend the existing results in two significant directions. First, we pose and solve the inverse problem in the setting of a two-person zero-sum differential game [9], the opposing players being the control and the disturbance. Our inverse problem thus takes system uncertainties into account as we consider *robust* stabilization (cf. Section 3.1). Second, our results are valid for all robustly stabilizable systems, including open-loop unstable systems.

We show that every rclf solves the steady-state HJI equation associated with a meaningful game. As a consequence of this result, if an rclf is known, we can construct a feedback law which is optimal with respect to a meaningful cost functional. Moreover, we can accomplish this *without* solving the HJI equation for the value function. In fact, we do not even need to construct the cost functional because we can calculate the optimal feedback directly from the rclf without recourse to the HJI equation. Indeed, we provide a formula which generates a class of such optimal control laws and which involves only the rclf, the system equations, and design parameters.

The control laws given by our formula are called *pointwise min-norm* control laws, and each one inherits the desirable properties of optimality. For example, the simplest pointwise min-norm control law for the system (1.2) is

$$u = \begin{cases} x^3 - 2x & \text{when } x^2 < 2 \\ 0 & \text{when } x^2 \geq 2 \end{cases} \tag{1.8}$$

This control law is compared with the optimal control law (1.5) in Figure 1.3. We see that these two control laws, both optimal with respect to a meaningful cost functional, are qualitatively the same. They both recognize the benefit of the nonlinearity $-x^3$ in (1.2) and accordingly expend little control effort for large signals; moreover, these control laws never generate positive feedback. The main difference between them lies in their design: the pointwise min-norm control law (1.8) came from the simple formula we provide in Section 4.2, while the control law (1.5) required the solution of an HJI equation. In general, the pointwise min-norm calculation is feasible but the HJI calculation is not.

1.3 Recursive Lyapunov design

In Chapters 5–8 we return to the main robust control design issue, namely, how to construct a robust control Lyapunov function for an uncertain nonlinear system. We begin with a review of the method of *Lyapunov re-*

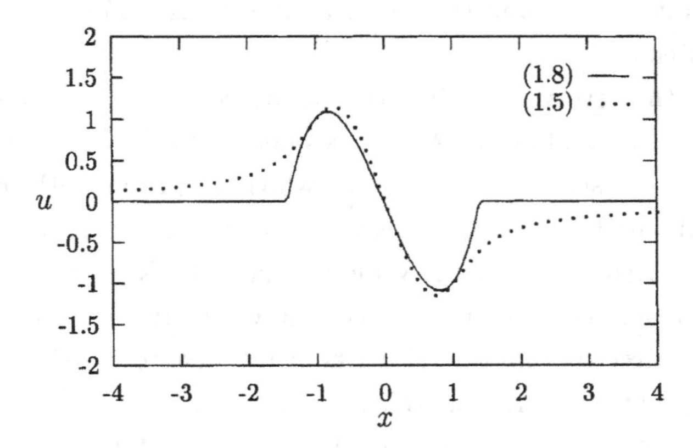

Figure 1.3: A comparison between the control laws (1.8) and (1.5)

design or *min-max design*, clearly presented in [77, 139]. In this method, developed in [89, 48, 90, 23, 11], one begins with a control Lyapunov function for the nominal system (the system without uncertainties) and then attempts to use this clf as an rclf for the uncertain system. The success of this method is guaranteed under a restrictive *matching condition* which requires all uncertainties to enter the system through the same channels as the control variables. Much effort has been devoted to weakening the matching condition [144, 12, 19, 154, 120, 18, 31, 101, 129, 121, 34]. Early results for nonlinear systems [12, 19] required that the unmatched uncertainties be sufficiently small. Greater success was achieved for the special case of quadratic stability for uncertain linear systems: a generalized matching condition was proposed in [144], and the antisymmetric stepwise uncertainty structure developed in [154] was shown to be not only sufficient but also *necessary* for robust (quadratic) stabilizability.[3]

Progress for nonlinear systems was slow until the breakthrough [72] in adaptive nonlinear control. This breakthrough was based on the nonlinear stabilization technique of "adding an integrator" introduced in [136, 17, 146]. The recursive application of this technique [81], which eventually led to the adaptive control results of [72], is known as *integrator backstepping* or simply *backstepping* [73, 80, 85].

[3]Necessity is proved under the assumption that the uncertainties are independent in a certain set of standard coordinates.

For the nonlinear robust control problem formulated in Chapter 3, backstepping led to the discovery of a structural *strict feedback condition* (much weaker than the matching condition) under which the systematic construction of an rclf is always possible. These *robust backstepping* results first appeared in [31] and were obtained independently in [101, 129, 121].

In its simplest form, robust backstepping leads to an rclf which is quadratic in a set of transformed coordinates [31, 101, 102, 129, 121]. However, this type of rclf can generate unnecessarily "hard" robust control laws, that is, control laws with unnecessarily high local gains in some regions of the state space. These high gains can cause excessive control effort such as high-magnitude chattering in the control signal. Moreover, this hardening property is propagated and amplified through each step of the recursive backstepping design. To overcome this drawback, we introduce a *flattened* rclf which leads to a dramatic reduction in hardening and thereby requires much less control effort with no sacrifice in performance. This can be seen in Figure 1.4, which compares simulation results for a second-order uncertain system under robust feedback control laws designed using the quadratic and the flattened rclf. Only the control law designed using the quadratic rclf exhibits high-magnitude chattering. This chattering is wasteful because the trajectories in the state space (not shown) are nearly identical for the two control schemes.

Smoothness assumptions are common in recursive backstepping designs because of the need to calculate derivatives of various functions during the construction of the control law and Lyapunov function. For some nonlinear systems, however, such assumptions cannot be satisfied. We show how to relax the smoothness assumptions for the case in which nonlinearities are locally Lipschitz continuous. We use our robust design methods along with the tools developed in [20] to accommodate this nonsmoothness, which we view as "uncertainty" in the derivative.

The recursive backstepping design presented in Chapter 5 leads to the systematic construction of rclf's for strict feedback systems under the assumption of perfect state feedback. In Chapter 6, we show that such systems admit rclf's (and are thus robustly stabilizable) even when the state measurement is corrupted by disturbances (such as sensor noise).

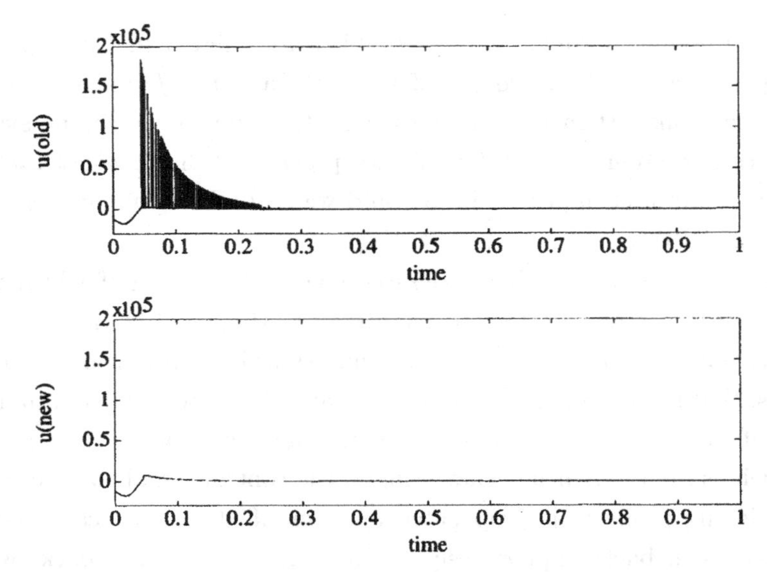

Figure 1.4: Comparison of control signals generated by quadratic and flattened rclf. *Source:* Freeman, R. A. and Kokotović, P. V. 1993. Design of 'softer' robust nonlinear control laws. *Automatica* **29**(6), 1425–1437. With permission.

In this case, a modified recursive construction of the rclf incorporates the flattening technique of Chapter 5 as well as key mathematical inequalities introduced in the Appendix. This result for strict feedback systems is significant because not all nonlinear systems which are stabilizable under perfect state feedback can be made (globally) robust to state measurement disturbances (for the class of memoryless time-invariant control laws). We illustrate this fundamental observation with a counterexample: we construct a second-order single-input system which is globally exponentially stabilizable via perfect state feedback, but for which *no* (memoryless time-invariant) control law can prevent finite escape times in the presence of small measurement disturbances.

In Chapter 7 we present some results on global stabilization and tracking via dynamic partial state feedback. Although the design of partial state feedback tracking controllers for nonlinear systems is of great practical interest, systematic design procedures for broad classes of systems are yet to be developed. For the case where the reference signals to be tracked are generated by autonomous exosystems, local results have been obtained in [62, 58] in the context of the nonlinear regulator problem. Semi-global extensions of these results for a class of uncertain feedback linearizable systems are provided in [76].

Our goal is to solve the *global* tracking problem for reference signals not necessarily generated by autonomous exosystems. This problem has already been solved for a class of systems which can be transformed into minimum-phase linear systems perturbed by output nonlinearities [100, 102, 73]. We introduce a broader class of *extended strict feedback systems* in which unmeasured states enter in an affine manner, and we present a dynamic backstepping controller design for this class of systems. We make the common assumption of global input/output linearizability (via full state feedback), and we assume that the inverse dynamics are globally bounded-input/bounded-state stable. This assumption is weaker than the minimum-phase assumption in [100, 102, 73] because it allows part of the zero dynamics to be Lyapunov stable *but not necessarily asymptotically stable.* We can therefore include exogenerators for unmeasured disturbances as in the output regulation problem in [62]. For the special case in which such unmeasured disturbances are constant and there are no other unmeasured states, our design reduces to the adaptive control design in [84]. In this manner we extend the applicability of the tuning function approach in [84] beyond the adaptive control problem. We illustrate our design on a nonlinear arm/rotor/platform system.

In Chapter 8 we combine the design techniques of Chapters 5 and 7 to obtain a robust nonlinear state feedback version of classical proportional-integral (PI) control. We solve the global set-point tracking problem for a class of nonlinear strict feedback systems having arbitrarily fast time-varying uncertainties as well as time-invariant uncertain nonlinear functions. For the special case in which the uncertain function is simply an unknown constant parameter, our results reduce to a robust extension of the tuning function adaptive control design in [84]. In this special case, we are no longer restricted to tracking constant reference signals. We illustrate our method by designing a speed controller for a fan having an uncertain drag nonlinearity.

Chapter 2

Set-Valued Maps

In robust control theory, an uncertain dynamical system is described by a set of models rather than a single model. For example, a system with an unknown parameter generates a set of models, one for each possible value of the parameter; likewise for a system with an unknown disturbance (which can be a function of time as well as state variables and control inputs). As a result, any map one might define for a single model becomes a *set-valued map*. Such is the case with an input/output map, a map from initial states to final states, or a map from disturbances to values of cost functionals. It is therefore natural that, in our study of robust nonlinear control, we use the language and mathematical apparatus of set-valued maps. In doing so, we follow the tradition started in the optimal control literature in the early sixties [27, 153] and continued in the control-related fields of nonsmooth analysis, game theory, differential inclusions, and viability theory [21, 127, 128, 5, 79].

We will use set-valued maps to formulate a general robust nonlinear control problem in Chapter 3 and then to develop an associated Lyapunov stability theory. Set-valued maps will also be instrumental in solving the inverse optimal robust control problem of Chapter 4. To prepare for these applications, we review some elementary results from set-valued analysis in this chapter. This review is based on material in [6, 7, 79].

The Lyapunov stability theory we develop in Chapter 3 will enable us in Chapter 4 to reduce the problem of constructing an optimal robust feedback control law to that of solving a parameter-dependent constrained optimization problem in a finite-dimensional space. Set-valued maps occur naturally in the study of such mathematical programming problems. In mathematical programming [96], it is often the case that the objective function $\phi(x, z)$ depends on a parameter x as well as the unknown variable z, and the problem is to maximize this function subject to some constraint on z. The constraint can be written $z \in F(x)$, where the constraint set F is also allowed to depend on the parameter x. The solution to this problem, defined as $f(x) := \sup\{\phi(x, z) \,:\, z \in F(x)\}$, is called a *marginal function* of the parameter x. It is natural to investigate the continuity properties of f, as this will provide information on the sensitivity of the optimization problem to variations in x. Such continuity properties will depend on how both the objective function ϕ and the constraint set F vary with the parameter x. This motivates the study of the continuity of set-valued maps such as the map F from the parameter space to the subsets of the z-space. We present common notions of set-valued continuity in Section 2.1 below, and in Section 2.2 we give the basic results on the continuity of marginal functions. Multiple constraints in optimization problems give rise to set-valued maps defined by intersections, and the continuity properties of such maps are described in Section 2.3.

A *selection* of a set-valued map F is a single-valued map f such that $f(x) \in F(x)$ for all x. The axiom of choice postulates the existence of selections for set-valued maps having nonempty values, but we need to know which additional conditions guarantee the existence of *continuous* selections. We present some continuous selection theorems in Section 2.4, including a well-known theorem due to Michael. As is done in viability theory [5], we will use these theorems in Chapters 3 and 4 to prove the existence of continuous feedback control laws having desired properties (in our case robust stability and optimality). Finally, in Section 2.5 we give some results on parameterizations of set-valued maps.

Terminology in this chapter is from the real analysis textbook [123]. Throughout, we let $I\!R$ and $I\!R_+$ denote the sets of real and nonnegative real numbers, respectively. We adopt the conventions $\sup \varnothing = -\infty$ and $\inf \varnothing = +\infty$, where $\varnothing$ denotes the empty set.

2.1 Continuity of set-valued maps

A *set-valued map*, also called a *multifunction*, is a map from one nonempty set X to the subsets of another nonempty set Z. We will write $F : X \rightsquigarrow Z$ for a set-valued map F from X to the subsets of Z; other existing notation includes $F : X \rightrightarrows Z$. Because one can regard a set-valued map $F : X \rightsquigarrow Z$ as a single-valued map $F : X \to 2^Z$ (where 2^Z denotes the power set of all subsets of Z), a separate theory for set-valued maps may appear redundant. However, we find such a theory extremely useful because of its concise language and clear visualization of concepts.

We define the *domain*, *image*, and *graph* of $F : X \rightsquigarrow Z$ as follows:

$$\mathrm{Dom}(F) := \big\{x \in X : F(x) \neq \varnothing\big\} \tag{2.1}$$

$$\mathrm{Im}(F) := \bigcup_{x \in X} F(x) \tag{2.2}$$

$$\mathrm{Graph}(F) := \big\{(x,z) \in X \times Z : z \in F(x)\big\} \tag{2.3}$$

We see from the definition of the graph of F that there is a one-to-one correspondence between set-valued maps from X to Z and subsets of the product space $X \times Z$ (such subsets are also known as *relations*). The domain and image of F are simply the projections of the graph of F onto X and Z, respectively.

For subsets $K \subset X$ and $L \subset Z$, we define

$$F(K) := \bigcup_{x \in K} F(x) \tag{2.4}$$

$$F^{-1}(L) := \big\{x \in X : F(x) \cap L \neq \varnothing\big\} \tag{2.5}$$

and adopt the convention $F(\varnothing) = \varnothing$. We see that $F(X) = \mathrm{Im}(F)$ and $F^{-1}(Z) = \mathrm{Dom}(F)$, and that (2.5) defines the *inverse map* $F^{-1} : Z \rightsquigarrow X$ with $F^{-1}(z) := F^{-1}(\{z\})$. Note that the inverse of a set-valued map is always well-defined, unlike the inverse of a single-valued map.

2.1.1 Upper and lower semicontinuity

When X and Z are topological spaces, we can define the continuity of a set-valued map $F : X \rightsquigarrow Z$. There are two common ways to generalize the definition of the continuity of a single-valued function to that of a set-valued map:

Definition 2.1 *A set-valued map $F : X \rightsquigarrow Z$ is* **upper semicontinuous (usc) at** $x \in X$ *when for every open set $U \subset Z$ such that $F(x) \subset U$ there exists a neighborhood V of x such that $F(V) \subset U$. We say that F is* **upper semicontinuous (usc)** *when it is usc at every point in X.*

Definition 2.2 *A set-valued map $F : X \rightsquigarrow Z$ is* **lower semicontinuous (lsc) at** $x \in X$ *when for every open set $U \subset Z$ such that $x \in F^{-1}(U)$ there exists a neighborhood V of x such that $V \subset F^{-1}(U)$. We say that F is* **lower semicontinuous (lsc)** *when it is lsc at every point in X.*

When F is single-valued, both of these definitions reduce to the usual definition of continuity. The reader should take note, however, that these semicontinuity definitions for set-valued maps are different from the standard semicontinuity definitions for (extended) real-valued functions. For example, an upper semicontinuous real-valued function $f : X \to \mathbb{R}$ need not be either usc or lsc when regarded as a set-valued map $x \mapsto \{f(x)\}$. Although the identical terminology for the two different concepts can be justified, we will not do so here. Finally, we say that a set-valued map is *continuous* when it is both usc and lsc.

As is the case for functions, the continuity of set-valued maps can be characterized in terms of inverse images:

Proposition 2.3 *A set-valued map $F : X \rightsquigarrow Z$ is usc if and only if $F^{-1}(K)$ is closed for every closed set $K \subset Z$. Also, $F : X \rightsquigarrow Z$ is lsc if and only if $F^{-1}(L)$ is open for every open set $L \subset Z$.*

Proof: Let F be usc, let $K \subset Z$ be closed, and define $M := \{x \in X : F(x) \subset (Z \backslash K)\}$. It follows from Definition 2.1 that M is open, which means $F^{-1}(K) = X \backslash M$ is closed. Conversely, suppose $F^{-1}(K)$ is closed for every closed set $K \subset Z$, and suppose $F(x) \subset U$ for some $x \in X$ and some open set $U \subset Z$. Then the set $V := X \backslash F^{-1}(Z \backslash U)$ is open and satisfies $x \in V$ and $F(V) \subset U$, which means F is usc at x. The corresponding lsc result is immediate from Definition 2.2. ∎

Because the domain of a set-valued map is the inverse image of a topological space (which is both open and closed), we have the following:

Corollary 2.4 *The domain of an usc set-valued map is closed, and the domain of a lsc set-valued map is open.*

We shall later need the following two results:

Proposition 2.5 *If Z is regular and if $F : X \rightsquigarrow Z$ is usc with closed values, then* Graph(F) *is closed.*

Proof: Let $(x,z) \in X \times Z$ be such that $z \notin F(x)$. Because Z is regular and $F(x)$ is closed, there exist disjoint open sets $U_1 \subset Z$ and $U_2 \subset Z$ such that $z \in U_1$ and $F(x) \subset U_2$. It follows from Definition 2.1 that there exists an open neighborhood V of x such that $F(V) \subset U_2$. We conclude that $(V \times U_1) \cap \text{Graph}(F) = \varnothing$, and it follows that Graph(F) is closed. ∎

Proposition 2.6 *If $F : X \rightsquigarrow Z$ is usc with compact values, then $F(K)$ is compact for every compact set $K \subset X$.*

Proof: Let $\{U_\lambda\}_{\lambda \in \Lambda}$ be an open cover of $F(K)$. Then for each $x \in K$, the open cover $\{U_\lambda\}$ of the compact set $F(x)$ has a finite subcover $\{U_\lambda\}_{\lambda \in \Lambda_x}$. For each $x \in K$, define the open set $U_x := \bigcup_{\lambda \in \Lambda_x} U_\lambda$. It follows from Proposition 2.3 that $\{X \backslash F^{-1}(Z \backslash U_x)\}_{x \in K}$ is an open cover of K and thus has a finite subcover $\{X \backslash F^{-1}(Z \backslash U_{x_i})\}$, $1 \le i \le n$. Therefore $\{U_\lambda\}_{\lambda \in \Lambda_{x_1} \cup \ldots \cup \Lambda_{x_n}}$ is a finite subcover of $F(K)$. ∎

2.1.2 Lipschitz and Hausdorff continuity

In nonlinear control theory, we often seek locally Lipschitz continuous feedback control laws (rather than merely continuous ones) because then the ordinary differential equations of closed-loop systems admit unique solutions. This motivates us to define Lipschitz continuity for set-valued maps between metric spaces. Let X and Z be metric spaces,[1] and consider a set-valued map $F : X \rightsquigarrow Z$.

Definition 2.7 *A set-valued map $F : X \rightsquigarrow Z$ is* **Lipschitz continuous on** $K \subset X$ *when there exists $k \in I\!R_+$ such that for all $x, y \in K$,*

$$F(x) \;\subset\; \Big\{ z \in Z \;:\; d(z, F(y)) \;\le\; k\, d(x,y) \Big\} \tag{2.6}$$

We say that F is **locally Lipschitz continuous** *when it is Lipschitz continuous on a neighborhood of every point in X.*

[1] Throughout this chapter, we let d denote both the metric and the point-to-set distance function $d(x,K) := \inf\{d(x,y) \,:\, y \in K\}$ with the convention $d(x,\varnothing) = \infty$. The underlying space will always be clear from context.

We will see below (Proposition 2.8) that a locally Lipschitz set-valued map is lsc, and that a locally Lipschitz set-valued map with compact values is continuous.

Alternative characterizations of continuity can be obtained by assigning a metric to the power set 2^Z of all subsets of Z. Given two subsets K and L of the metric space Z, we define the *Hausdorff distance* $\mathcal{H}(K, L)$ between K and L as follows:

$$\mathcal{H}(K,L) \;:=\; \begin{cases} \max\left\{\sup\limits_{z\in K} d(z,L),\ \ \sup\limits_{z\in L} d(z,K)\right\} & \text{when } K \cup L \neq \varnothing \\ 0 & \text{when } K = L = \varnothing \end{cases}$$

One can verify that $\mathcal{H}$ is an extended pseudometric for the set 2^Z, and we can therefore use it to define continuity properties of a set-valued map $F : X \rightsquigarrow Z$ regarded as a single-valued function $F : X \to 2^Z$. The relationships between such Hausdorff continuity properties and the continuity properties described above are summarized as follows:

Proposition 2.8 *Let X and Z be metric spaces, and consider a set-valued map $F : X \rightsquigarrow Z$. If F is Hausdorff continuous, then F is lsc. Also, if F has compact values, then F is Hausdorff continuous if and only if F is continuous. Finally, F is Hausdorff locally Lipschitz continuous if and only if F is locally Lipschitz continuous.*

Proof: Suppose F is continuous with respect to the Hausdorff distance. Fix $x \in X$, and let $U \subset Z$ be open and such that $F(x) \cap U \neq \varnothing$. Let $z \in F(x) \cap U$; then $z \in U$ and so there exists $\varepsilon > 0$ such that $d(z, w) < \varepsilon$ implies $w \in U$. Let $\delta > 0$ be such that $d(x, y) < \delta$ implies $\mathcal{H}(F(x), F(y)) < \varepsilon$. Now $z \in F(x)$, and thus $d(x, y) < \delta$ implies $d(z, F(y)) < \varepsilon$ which in turn implies $F(y) \cap U \neq \varnothing$. We conclude that F is lsc at x. Next suppose $F(x)$ is compact, and let $U \subset Z$ be open and such that $F(x) \subset U$. It follows that there exists $\varepsilon > 0$ such that $d(z, F(x)) < \varepsilon$ implies $z \in U$. Let $\delta > 0$ be such that $d(x, y) < \delta$ implies $\mathcal{H}(F(x), F(y)) < \varepsilon$. Then $d(x, y) < \delta$ and $z \in F(y)$ imply $d(z, F(x)) < \varepsilon$ which in turn implies $z \in U$. Thus $d(x, y) < \delta$ implies $F(y) \subset U$, and we conclude that F is usc at x.

Suppose next that F is continuous with compact values, fix $x \in X$, and let $\varepsilon > 0$. It follows from Definition 2.1 that there exists $\delta_0 > 0$ such that $d(x,y) < \delta_0$ and $z \in F(y)$ imply $d(z, F(x)) < \varepsilon$. Thus $d(x,y) < \delta_0$ implies $\sup\{d(z,F(x)) : z \in F(y)\} \leq \varepsilon$. Now because $F(x)$ is compact, there exist a finite number of points $z_i \in F(x)$, $1 \leq i \leq n$, such that $d(z,\{z_i\}) < \frac{1}{2}\varepsilon$ for all $z \in F(x)$. It follows from Definition 2.2 that for each $i \in \{1,\ldots,n\}$ there exists $\delta_i > 0$ such that $d(x,y) < \delta_i$ implies $d(z_i, F(y)) < \frac{1}{2}\varepsilon$. Let $\delta = \min\{\delta_0,\ldots,\delta_n\}$ and suppose $d(x,y) < \delta$. For each $z \in F(x)$ we have $d(z,z_i) < \frac{1}{2}\varepsilon$ for some i, and it follows that $d(z,F(y)) \leq d(z,z_i) + d(z_i,F(y)) < \varepsilon$. Therefore $d(x,y) < \delta$ implies both $\sup\{d(z,F(y)) : z \in F(x)\} \leq \varepsilon$ and $\sup\{d(z,F(x)) : z \in F(y)\} \leq \varepsilon$ and thus $\mathcal{H}(F(x),F(y)) \leq \varepsilon$. We conclude that F is continuous at x with respect to the Hausdorff distance.

Finally, note that (2.6) is equivalent to $\sup\{d(z,F(y)) : z \in F(x)\} \leq k\, d(x,y)$. By interchanging the roles of x and y in (2.6), we see that (2.6) is true for all $x,y \in K$ if and only if $\mathcal{H}(F(x),F(y)) \leq k\, d(x,y)$ for all $x,y \in K$. ∎

2.2 Marginal functions

We next investigate the continuity properties of parameter-dependent solutions of constrained optimization problems. Let X and Z be topological spaces in which the parameter x and the unknown z take their values (respectively). We consider an objective function $\phi : X \times Z \to I\!R \cup \{\pm\infty\}$ and a set-valued constraint $F : X \rightsquigarrow Z$, and we define the optimal solution $f : X \to I\!R \cup \{\pm\infty\}$ to be

$$f(x) \quad := \quad \sup_{z \in F(x)} \phi(x,z) \tag{2.7}$$

This solution f is called a *marginal function* of the parameter x. The next two propositions describe conditions under which the marginal function f inherits the continuity properties of the objective function ϕ and the set-valued constraint map F.

Proposition 2.9 *If F and ϕ are lsc, then f is lsc. If F and ϕ are usc and F has compact values, then f is usc.*

Proof: Suppose F and ϕ are lsc, and let $x \in X$ be such that $f(x) > c$ for some $c \in I\!R \cup \{-\infty\}$. Then there exists $z \in F(x)$ such that $\phi(x,z) > c$.

Because ϕ is lsc, there exist open neighborhoods V_1 of x and U of z such that $\phi > c$ on $V_1 \times U$. It follows from Definition 2.2 that there exists a neighborhood $V \subset V_1$ of x such that $F(y) \cap U \neq \emptyset$ for all $y \in V$. Thus for each $y \in V$ there exists $z \in F(y)$ such that $\phi(y, z) > c$. Therefore $y \in V$ implies $f(y) > c$, and we conclude that f is lsc.

Next suppose F and ϕ are usc, and suppose F has compact values. Let $x \in X$ be such that $f(x) < c$ for some $c \in \mathbb{R} \cup \{\infty\}$. Then $z \in F(x)$ implies $\phi(x, z) < c$, and because ϕ is usc there exist $\varepsilon_z > 0$ and open neighborhoods V_z of x and U_z of z such that $\phi < c - \varepsilon_z$ on $V_z \times U_z$. Because $\{x\} \times F(x)$ is compact, its open cover $\{V_z \times U_z\}$ has a finite subcover $\{V_i \times U_i\}$, $1 \leq i \leq n$, and for each i there exists $\varepsilon_i > 0$ such that $\phi < c - \varepsilon_i$ on $V_i \times U_i$. It follows from Definition 2.1 that there exists a neighborhood $V \subset \bigcap_i V_i$ of x such that $F(y) \subset \bigcup_i U_i$ for all $y \in V$. Thus if $y \in V$ and $z \in F(y)$, then $(y, z) \in V_i \times U_i$ for some i which means $\phi(y, z) < c - \min\{\varepsilon_i\}$. Therefore $y \in V$ implies $f(y) < c$, and we conclude that f is usc. ∎

Proposition 2.10 *Suppose X and Z are metric spaces with X locally compact. If F and ϕ are locally Lipschitz continuous and F has nonempty compact values, then f is locally Lipschitz continuous.*

Proof: Fix $x \in X$; then by the local compactness of X and the local Lipschitz continuity of F there exist a compact neighborhood V of x and a constant $L_1 \in \mathbb{R}_+$ such that $\mathcal{H}(F(x_1), F(x_2)) \leq L_1 d(x_1, x_2)$ for all $x_1, x_2 \in V$. It follows from Proposition 2.6 that $V \times F(V)$ is compact, and we conclude from Lemma A.12 that there exists a constant $L_2 \in \mathbb{R}_+$ such that $|\phi(x_1, z_1) - \phi(x_2, z_2)| \leq L_2\, d(x_1, x_2) + L_2\, d(z_1, z_2)$ for all $x_1, x_2 \in V$ and all $z_1, z_2 \in F(V)$. Let $x_1, x_2 \in V$. Because ϕ is continuous and $F(x_1)$ is compact, there exists $z_1 \in F(x_1)$ such that $f(x_1) = \phi(x_1, z_1)$. Because $\mathcal{H}(F(x_1), F(x_2)) \leq L_1 d(x_1, x_2)$ we have $d(z_1, F(x_2)) \leq L_1 d(x_1, x_2)$, which means there exists $z_2 \in F(x_2)$ such that $d(z_1, z_2) \leq L_1 d(x_1, x_2)$. It follows that $f(x_1) - f(x_2) \leq \phi(x_1, z_1) - \phi(x_2, z_2) \leq L_2\, d(x_1, x_2) + L_2\, d(z_1, z_2) \leq (1 + L_1) L_2\, d(x_1, x_2)$. By symmetry we obtain $|f(x_1) - f(x_2)| \leq (1 + L_1) L_2\, d(x_1, x_2)$, and we conclude that f is Lipschitz continuous on V and thus locally Lipschitz continuous. ∎

2.3 Intersections

In our study of robust nonlinear control, we will encounter parameter-dependent optimization problems with multiple constraints. Such problems give rise to set-valued constraint maps defined by intersections. It is therefore of interest to determine when such intersections inherit the continuity properties of the individual maps. We will first investigate general continuity properties in Section 2.3.1 and then specialize to Lipschitz continuity properties in Section 2.3.2.

2.3.1 Continuity of intersections

We begin by investigating the continuity of maps whose graphs are subsets of the graphs of continuous maps:

Proposition 2.11 *Let set-valued maps $F, G : X \rightsquigarrow Z$ be such that* $\mathrm{Graph}(F) \subset \mathrm{Graph}(G)$. *If G is lsc and* $\mathrm{Graph}(F)$ *is open relative to* $\mathrm{Graph}(G)$, *then F is lsc. If G is usc with compact values and* $\mathrm{Graph}(F)$ *is closed relative to* $\mathrm{Graph}(G)$, *then F is usc.*

Proof: Suppose G is lsc and $\mathrm{Graph}(F)$ is open relative to $\mathrm{Graph}(G)$. Fix $x \in X$ and let $U \subset Z$ be open and such that $F(x) \cap U \neq \varnothing$. Then there exists $z \in G(x) \cap U$ such that $(x, z) \in \mathrm{Graph}(F)$. Because $\mathrm{Graph}(F)$ is open relative to $\mathrm{Graph}(G)$, there exist open neighborhoods V_1 of x and U_1 of z such that $(V_1 \times U_1) \cap \mathrm{Graph}(G) \subset \mathrm{Graph}(F)$. Thus $G(x) \cap U_1 \cap U \neq \varnothing$, and it follows from Definition 2.2 that there exists a neighborhood $V \subset V_1$ of x such that $G(y) \cap U_1 \cap U \neq \varnothing$ for all $y \in V$. Hence $F(y) \cap U \neq \varnothing$ for all $y \in V$, and we conclude that F is lsc at x.

Next suppose G is usc with compact values and $\mathrm{Graph}(F)$ is closed relative to $\mathrm{Graph}(G)$. Let $K \subset Z$ be closed; it follows from Proposition 2.3 that we need to show that $F^{-1}(K)$ is closed. Fix $x \in X \backslash F^{-1}(K)$ and define the set $M := \mathrm{Graph}(F) \cap (X \times K)$. Then for every $z \in G(x)$ we have $(x, z) \notin M$, and because M is closed relative to $\mathrm{Graph}(G)$ there exist open neighborhoods V_z of x and U_z of z such that $(V_z \times U_z) \cap M = \varnothing$. Because $\{x\} \times G(x)$ is compact, its open cover $\{V_z \times U_z\}$ has a finite subcover $\{V_i \times U_i\}$, $1 \leq i \leq n$. It follows from Definition 2.1 that there exists a neighborhood $V \subset \bigcap_i V_i$ of x such that $G(V) \subset \bigcup_i U_i$. Thus if $y \in V$

and $z \in G(y)$, then $(y,z) \in V_i \times U_i$ for some i which means $z \notin F(y) \cap K$. Hence $y \in V$ implies $F(y) \cap K = \varnothing$, and it follows that $V \subset X \backslash F^{-1}(K)$. We conclude that $X \backslash F^{-1}(K)$ is open, that is, $F^{-1}(K)$ is closed. ■

From this result we deduce the following properties of maps defined by intersections:

Corollary 2.12 *Let $G : X \rightsquigarrow Z$ and $H : X \rightsquigarrow Z$ be given, and define $F : X \rightsquigarrow Z$ by $F(x) := G(x) \cap H(x)$. If G is lsc and* Graph(H) *is open, then F is lsc. If G is usc with compact values and* Graph(H) *is closed, then F is usc.*

Corollary 2.13 *Let $G : X \rightsquigarrow Z$ be lsc and $\phi : X \times Z \to I\!R \cup \{\pm\infty\}$ be usc. Then $F : X \rightsquigarrow Z$ defined by $F(x) := \{z \in G(x) : \phi(x,z) < 0\}$ is also lsc.*

2.3.2 Lipschitz continuity of intersections

The proofs of the Lipschitz versions of the results in the previous section are considerably more technical. We begin with a Lipschitz version of Corollary 2.13:

Proposition 2.14 *Suppose X is a locally compact metric space and Z is a normed space. Let $G : X \rightsquigarrow Z$ be locally Lipschitz continuous with nonempty compact convex values. Let $\phi : X \times Z \to I\!R$ be locally Lipschitz continuous and such that the mapping $z \mapsto \phi(x,z)$ is convex for each fixed $x \in X$. Then $F : X \rightsquigarrow Z$ defined by $F(x) := \{z \in G(x) : \phi(x,z) < 0\}$ is locally Lipschitz continuous on* Dom(F).

Proof: It follows from Proposition 2.8, Corollary 2.13, and Corollary 2.4 that Dom(F) is open. Fix $x \in$ Dom(F); then by the local compactness of X and the local Lipschitz continuity of G there exist a compact neighborhood $V \subset$ Dom(F) of x and a constant $L_1 \in I\!R_+$ such that $\mathcal{H}(G(x_1), G(x_2)) \leq L_1 d(x_1, x_2)$ for all $x_1, x_2 \in V$. Because F has bounded values, we can define $\beta : V \to I\!R$ by $\beta(y) := \inf\{\phi(y,z) : z \in F(y)\}$. It follows from Proposition 2.8, Corollary 2.13, and Proposition 2.9 that β is usc. Also, we have $\beta(y) < 0$ for all $y \in V$, and so $\alpha := \max\{\beta(y) : y \in V\}$ satisfies $\alpha < 0$. It follows from Proposition 2.6 that $V \times G(V)$ is compact, and we conclude from Lemma A.12 that there exists a constant $L_2 \in I\!R_+$ such that $|\phi(x_1, z_1) - \phi(x_2, z_2)| \leq L_2 \, d(x_1, x_2) + L_2 \, \|z_1 - z_2\|$ for all $x_1, x_2 \in V$ and all $z_1, z_2 \in G(V)$. Fix $x_1, x_2 \in V$ and let $z \in F(x_1)$. Because $G(x_2)$ is compact, there exists $z_1 \in G(x_2)$ such that $\|z - z_1\| = d(z, G(x_2))$. Now $z \in G(x_1)$ which means $d(z, G(x_2)) \leq L_1 \, d(x_1, x_2)$, thus

$$\begin{aligned} d(z, F(x_2)) &\leq \|z - z_1\| + d(z_1, F(x_2)) \\ &\leq L_1 \, d(x_1, x_2) + d(z_1, F(x_2)) \end{aligned} \tag{2.8}$$

Also, $\phi(x_2, z_1) \leq \phi(x_1, z) + L_2 \, d(x_1, x_2) + L_2 \|z - z_1\|$, and because $\phi(x_1, z) < 0$ we have $\phi(x_2, z_1) < L \, d(x_1, x_2)$ where $L := (1 + L_1) L_2$. Suppose first that $\phi(x_2, z_1) \geq 0$, and let $\varepsilon \in (0, \frac{1}{2})$. Then $\beta(x_2) < (1 - \varepsilon)\alpha$, which means there exists $z_2 \in F(x_2)$ such that $\phi(x_2, z_2) \leq (1 - \varepsilon)\alpha < \varepsilon\alpha$. From the intermediate value theorem there exists $\theta \in (0, 1)$ such that $z_3 := \theta z_1 + (1 - \theta) z_2$ satisfies $\phi(x_2, z_3) = \varepsilon\alpha$. Because $G(x_2)$ is convex we have $z_3 \in G(x_2)$, and it follows that $z_3 \in F(x_2)$. From the convexity of $\phi(x_2, \cdot)$ we have

$$\begin{aligned} \varepsilon\alpha &\leq \theta\phi(x_2, z_1) + (1 - \theta)\phi(x_2, z_2) \\ &\leq L \, d(x_1, x_2) + (1 - \theta)(1 - \varepsilon)\alpha \end{aligned} \tag{2.9}$$

Also, we have $d(z_1, F(x_2)) \leq \|z_1 - z_3\| \leq (1 - \theta)\|z_1 - z_2\| \leq (1 - \theta)D$ where D is the finite constant $D := \operatorname{diam}(G(V))$. Substituting for $(1 - \theta)$ from (2.9), we obtain

$$d(z_1, F(x_2)) \leq \frac{D}{(1 - \varepsilon)|\alpha|} \left[L \, d(x_1, x_2) + \varepsilon |\alpha| \right] \tag{2.10}$$

This is true for every $\varepsilon \in (0, \frac{1}{2})$, and it follows that $d(z_1, F(x_2)) \leq \frac{1}{|\alpha|} DL \, d(x_1, x_2)$. Suppose next that $\phi(x_2, z_1) < 0$; then $z_1 \in F(x_2)$, and in both cases the inequality (2.8) becomes $d(z, F(x_2)) \leq (L_1 + \frac{1}{|\alpha|} DL) \, d(x_1, x_2)$. This is true for every $z \in F(x_1)$, and it follows that $\sup\{d(z, F(x_2)) : z \in F(x_1)\} \leq (L_1 + \frac{1}{|\alpha|} DL) \, d(x_1, x_2)$. By symmetry we have $\mathcal{H}(F(x_1), F(x_2)) \leq (L_1 + \frac{1}{|\alpha|} DL) \, d(x_1, x_2)$ for all $x_1, x_2 \in V$, and we conclude that F is locally Lipschitz continuous on $\operatorname{Dom}(F)$. ■

When Z is an inner product space, we can relax the requirement in Proposition 2.14 that the set-valued map G have compact values. We first prove a technical lemma based on [7, Lemma 9.4.2] and then present our results.

Lemma 2.15 *Let Z be an inner product space with closed unit ball B, and let $\alpha \in (0,1)$. Then for every pair of convex sets $F, G \subset Z$ and every $r, s \in I\!R_+$ such that $F \cap \alpha r B \neq \varnothing$ and $G \cap \alpha s B \neq \varnothing$, we have $\mathcal{H}(F \cap rB,\, G \cap sB) \leq (1 + \frac{1}{\sqrt{1-\alpha^2}})\,[\mathcal{H}(F,G) + |r-s|]$.*

Proof: Define $L \in (2,\infty)$ and $h \in [0,\infty]$ by $L := 1 + \frac{1}{\sqrt{1-\alpha^2}}$ and $h := \mathcal{H}(F,G) + |r-s|$, respectively. Let $z \in F \cap rB$ and let $\varepsilon > 0$; then there exists $z_1 \in G$ such that $\|z - z_1\| \leq \mathcal{H}(F,G) + \varepsilon$. First suppose $\|z_1\| \leq s$; then $z_1 \in G \cap sB$ which means $d(z, G \cap sB) \leq \|z - z_1\| \leq \mathcal{H}(F,G) + \varepsilon \leq h + \varepsilon$, and because $L > 1$ we have

$$d(z, G \cap sB) \;\leq\; L(h+\varepsilon) \tag{2.11}$$

Next suppose $\|z_1\| > s$; then because $G \cap \alpha s B \neq \varnothing$ and G is convex, there exists $z_2 \in G$ such that $\|z_2\| = \alpha s$. If $z_2 = 0$ then $s = 0$ which means $d(z, G \cap sB) = \|z\| \leq r \leq h$, and we again satisfy the inequality (2.11). Otherwise we have $\|z_2\| = \alpha s > 0$. It follows from the convexity of G that there exists $\theta \in (0,1)$ such that $z_3 := \theta z_1 + (1-\theta) z_2$ satisfies $z_3 \in G$ and $\|z_3\| = s$. Define $e_1 := (z_1 - z_3)/\|z_1 - z_3\|$ and $z_4 := z_3 - \langle z_3, e_1 \rangle\, e_1$. Now $z_4 \perp (z_4 - z_2)$ which means $\|z_4\| \leq \|z_2\| = \alpha s$, and because $z_4 \perp (z_4 - z_3)$ we have $\|z_4 - z_3\|^2 = \|z_3\|^2 - \|z_4\|^2 \geq (1-\alpha^2)s^2$. Define $e_2 := z_3/s$ and $z_5 := \langle z_1, e_2 \rangle\, e_2$. We then have $s\|z_5 - z_3\| = s\|\langle z_1, e_2\rangle\, e_2 - s e_2\| = |\langle z_1, z_3 \rangle - s^2| = |\langle z_1 - z_3, z_3 \rangle| = \|z_4 - z_3\| \cdot \|z_1 - z_3\| \geq s\sqrt{1-\alpha^2}\,\|z_1 - z_3\|$, and it follows that

$$\|z_1 - z_3\| \;\leq\; (L-1)\|z_5 - z_3\| \tag{2.12}$$

Now $\langle z_2, e_2 \rangle \leq \|z_2\| \leq s$ which means $\langle z_2 - z_3, e_2 \rangle \leq 0$, and it follows that $\langle z_5 - z_3, e_2 \rangle = \langle z_1 - z_3, e_2 \rangle = \frac{\theta - 1}{\theta} \langle z_2 - z_3, e_2 \rangle \geq 0$. Therefore $\|z_5 - z_3\| = \|z_5 - z_3\| \cdot \|e_2\| = \langle z_5 - z_3, e_2 \rangle = \langle z_1 - z_3, e_2 \rangle \leq \|z_1\| - s \leq \|z_1 - z\| + \|z\| - s \leq \mathcal{H}(F,G) + |r-s| + \varepsilon = h + \varepsilon$. It follows from (2.12) that

$$\begin{aligned} d(z, G \cap sB) &\leq \|z - z_3\| \;\leq\; \|z - z_1\| + \|z_1 - z_3\| \\ &\leq (h+\varepsilon) + (L-1)\|z_5 - z_3\| \\ &\leq L(h+\varepsilon) \end{aligned} \tag{2.13}$$

The inequality (2.11) is therefore satisfied in every case, and by taking the supremum over $z \in F \cap rB$ and letting $\varepsilon \to 0$ we obtain $\sup\{d(z, G \cap sB)$:

$z \in F \cap rB\} \leq L h$. From symmetry we obtain $\mathcal{H}(F \cap rB, G \cap sB) \leq L[\mathcal{H}(F,G) + |r-s|]$ as desired. ■

Proposition 2.16 *Suppose X is a locally compact metric space and Z is an inner product space with closed unit ball B. Let $G : X \rightsquigarrow Z$ be locally Lipschitz continuous with convex values. Let $r : X \to \mathbb{R}_+$ and $\varepsilon : X \to (0,1)$ be locally Lipschitz continuous, and suppose we have $G(x) \cap \varepsilon(x)r(x)B \neq \varnothing$ for all $x \in X$. Then $F : X \rightsquigarrow Z$ defined by $F(x) := G(x) \cap r(x)B$ is locally Lipschitz continuous.*

Proof: Fix $x \in X$; then by the local compactness of X and the local Lipschitz continuity of G and r there exist a compact neighborhood $V \subset U$ of x and a constant $L \in \mathbb{R}_+$ such that $\mathcal{H}(G(x_1), G(x_2)) + |r(x_1) - r(x_2)| \leq L\, d(x_1, x_2)$ for all $x_1, x_2 \in V$. We define $\alpha := \max\{\varepsilon(x) : x \in V\}$. Fix $x_1, x_2 \in V$. Then $G(x_1) \cap \alpha r(x_1)B \neq \varnothing$ and $G(x_2) \cap \alpha r(x_2)B \neq \varnothing$, and it follows from Lemma 2.15 that

$$\begin{aligned} \mathcal{H}(F(x_1), F(x_2)) &\leq \left(1 + \tfrac{1}{\sqrt{1-\alpha^2}}\right)\left[\mathcal{H}(G(x_1), G(x_2)) + |r(x_1) - r(x_2)|\right] \\ &\leq \left(1 + \tfrac{1}{\sqrt{1-\alpha^2}}\right) L\, d(x_1, x_2) \qquad (2.14) \end{aligned}$$

We conclude that F is locally Lipschitz continuous. ■

Proposition 2.17 *Suppose X is a locally compact metric space and Z is an inner product space with closed unit ball B. Let $G : X \rightsquigarrow Z$ be locally Lipschitz continuous with nonempty closed convex values. Let $\phi : X \times Z \to \mathbb{R}$ be locally Lipschitz continuous and such that the mapping $z \mapsto \phi(x,z)$ is convex for each fixed $x \in X$. Define the set-valued map $F : X \rightsquigarrow Z$ by $F(x) := \{z \in G(x) : \phi(x,z) < 0\}$. Then there exists a locally Lipschitz continuous function $r : \mathrm{Dom}(F) \to \mathbb{R}_+$ such that the set-valued map $x \rightsquigarrow F(x) \cap r(x)B$ is locally Lipschitz continuous with nonempty values on* $\mathrm{Dom}(F)$.

Proof: It follows from Proposition 2.8, Corollary 2.13, and Proposition 2.9 that the mapping $x \mapsto \inf\{\|z\| : z \in F(x)\}$ is usc. Thus for $\varepsilon \in (0,1)$ there exists a locally Lipschitz function $r : \mathrm{Dom}(F) \to \mathbb{R}_+$ such that $F(x) \cap \varepsilon r(x)B \neq \varnothing$ for all $x \in \mathrm{Dom}(F)$. It follows from Proposition 2.16 that the set-valued map $G_r : X \rightsquigarrow Z$ defined by $G_r(x) := G(x) \cap$

$r(x)B$ is locally Lipschitz continuous with nonempty compact convex values on $\mathrm{Dom}(F)$. It then follows from Proposition 2.14 that the set-valued map $F_r : X \rightsquigarrow Z$ defined by $F_r(x) := \{z \in G_r(x) : \phi(x,z) < 0\}$ is locally Lipschitz continuous with nonempty values on $\mathrm{Dom}(F)$. ∎

2.4 Selection theorems

We next investigate conditions under which a set-valued map $F : X \rightsquigarrow Z$ admits a *continuous selection*, that is, a continuous (single-valued) function $f : X \to Z$ such that $f(x) \in F(x)$ for all $x \in X$. The selection theorems presented below will be used in Chapters 3 and 4 to prove the existence of continuous feedback control laws having robust stability and optimality properties.

2.4.1 Michael's theorem

The most widely known selection theorem is due to Michael:

Theorem 2.18 *Suppose X is a metric space and Z is a Banach space. Let $F : X \rightsquigarrow Z$ be lsc with nonempty closed convex values. Then there exists a continuous function $f : X \to Z$ such that $f(x) \in F(x)$ for all $x \in X$.*

The proof of this theorem, which can be found for example in [79, p. 58], is not constructive: it is based on partitions of unity and the uniform convergence of a sequence of continuous functions from X to Z.

2.4.2 Minimal selections

If Z is a Hilbert space and $F : X \rightsquigarrow Z$ has nonempty closed convex values, then we can define a function $m : X \to Z$ by

$$m(x) \quad := \quad \arg\min\Big\{\|z\| \ : \ z \in F(x)\Big\} \tag{2.15}$$

This function m is called the *minimal selection of F*.

Proposition 2.19 *Suppose Z is a Hilbert space, and let $F : X \rightsquigarrow Z$ be lsc with closed graph and nonempty closed convex values. Then m is* *locally bounded and* Graph(m) *is closed. If furthermore the dimension of* Z *is finite, then* m *is continuous.*

Proof: Because F is lsc, it follows from Proposition 2.9 that the mapping $x \mapsto \|m(x)\|$ is usc. Thus m is locally bounded and the set-valued mapping $x \rightsquigarrow \|m(x)\|B$ has closed graph, where B denotes the closed unit ball of Z. Because Graph(F) is closed and $\{m(x)\} = F(x) \cap \|m(x)\|B$, we conclude that Graph(m) is closed. Suppose now that the dimension of Z is finite. Because m is locally bounded, it takes its values locally in a compact subset of Z. Now a function from a topological space to a compact space is continuous if its graph is closed, and we conclude that m is continuous. ∎

2.4.3 Lipschitz selections

Let $K \subset \mathbb{R}^n$ be nonempty, compact, and convex. Recall that the support function of K is the convex function $\sigma_K : \mathbb{R}^n \to \mathbb{R}$ given by $\sigma_K(x) := \max\{\langle x, k\rangle : k \in K\}$. Because σ_K is locally Lipschitz continuous, it follows from Rademacher's theorem that the gradient $\nabla\sigma_K$ exists almost everywhere and is locally bounded. As a result, the following integral is well-defined:

$$\mathrm{St}(K) \quad := \quad \frac{1}{\int_B d\mu} \int_B \nabla\sigma_K \, d\mu \tag{2.16}$$

where B is the unit ball in $\mathbb{R}^n$ and μ is the Lebesgue measure. This point $\mathrm{St}(K) \in \mathbb{R}^n$ is called the *Steiner point* (or *curvature centroid*) of K. It follows from [122, Corollary 23.5.3] that whenever $\nabla\sigma_K(x)$ exists we have $\nabla\sigma_K(x) = \arg\max\{\langle x, k\rangle : k \in K\}$, that is, $\nabla\sigma_K(x) \in K$. Therefore the point $\mathrm{St}(K)$, being the weighted average of points in K, belongs to K. The significance of the Steiner point is the following Lipschitz property [7, Theorem 9.4.1]: if K and L are nonempty compact convex subsets of $\mathbb{R}^n$, then

$$\|\mathrm{St}(K) - \mathrm{St}(L)\| \quad \leq \quad n\,\mathcal{H}(K, L) \tag{2.17}$$

We can therefore use the Steiner point to obtain Lipschitz selections of Lipschitz maps:

Proposition 2.20 *Suppose X is a locally compact metric space, and let $F : X \rightsquigarrow I\!R^n$ be locally Lipschitz continuous with nonempty closed convex values. Then there exists a locally Lipschitz continuous function $f : X \to I\!R^n$ such that $f(x) \in F(x)$ for all $x \in X$.*

Proof: Let $m : X \to I\!R^n$ be the minimal selection of F. It follows from Propositions 2.8 and 2.9 that the mapping $x \mapsto \|m(x)\|$ is usc, and so for $\varepsilon \in (0,1)$ there exists a locally Lipschitz function $r : X \to I\!R_+$ such that $\|m(x)\| \leq \varepsilon r(x)$ for all $x \in X$. It then follows from Proposition 2.16 that the set-valued map $G : X \rightsquigarrow I\!R^n$ defined by $G(x) := F(x) \cap r(x)B$ is locally Lipschitz continuous with nonempty compact convex values. If we define $f(x) := \mathrm{St}(G(x))$, we see that f is a locally Lipschitz continuous selection of G and thus of F. ∎

2.5 Parameterized maps

Let X, Z, and W be topological spaces, let $G : X \rightsquigarrow W$ be a given set-valued map, and let $f : \mathrm{Graph}(G) \to Z$ be a given function. We define a *parameterized* set-valued map $F : X \rightsquigarrow Z$ by $F(x) := f(x, G(x))$.

Proposition 2.21 *Suppose f is continuous. If G is lsc, then F is lsc. If G is usc with compact values, then F is usc.*

Proof: Suppose G is lsc, and let $L \subset Z$ be open. It follows from Corollary 2.12 and the continuity of f that $H : X \rightsquigarrow W$ defined by $H(x) := \{w \in G(x) : f(x,w) \in L\}$ is lsc. Because $F^{-1}(L) = \mathrm{Dom}(H)$, it follows from Corollary 2.4 that $F^{-1}(L)$ is open, and we conclude from Proposition 2.3 that F is lsc. Next suppose G is usc with compact values, and let $K \subset Z$ be closed. It follows from Corollary 2.12 and the continuity of f that $H : X \rightsquigarrow W$ defined by $H(x) := \{w \in G(x) : f(x,w) \in K\}$ is usc. Because $F^{-1}(K) = \mathrm{Dom}(H)$, it follows from Corollary 2.4 that $F^{-1}(K)$ is closed, and we conclude from Proposition 2.3 that F is usc. ∎

Conversely, a *parameterization* of a set-valued map $F : X \rightsquigarrow Z$ is a function $f : X \times W \to Z$ such that $F(x) = f(x, W)$ for all $x \in X$. Continuous set-valued maps with nonempty compact convex values in $I\!R^n$ admit continuous parameterizations:

Proposition 2.22 *Let X be a metric space, and let B denote the closed unit ball of $\mathbb{R}^n$. If $F : X \rightsquigarrow \mathbb{R}^n$ is continuous with nonempty compact convex values, then there exists a continuous function $f : X \times \mathbb{R}^n \to \mathbb{R}^n$ such that $F(x) = f(x, B)$ for all $x \in X$. Furthermore, if X is locally compact and F is locally Lipschitz continuous, then f can be chosen to be locally Lipschitz continuous.*

Proof: We define a function $h : X \times \mathbb{R}^n \to \mathbb{R}^n$ by

$$h(x, y) \quad := \quad y \cdot \max_{z \in F(x)} \|z\| \tag{2.18}$$

It follows from Proposition 2.9 that h is continuous. We next define a set-valued map $G : X \times \mathbb{R}^n \rightsquigarrow \mathbb{R}^n$ with bounded convex values by

$$G(x, y) \quad := \quad \Big\{ z \in F(x) : \|z - h(x, y)\| < 2d(h(x, y), F(x)) \Big\} \tag{2.19}$$

The mapping $(x, y) \mapsto d(h(x, y), F(x))$ is continuous by Proposition 2.9, and it follows from Corollary 2.13 that G is lsc. The lsc property is not affected by taking closures, and it follows from Theorem 2.18 that there exists a continuous function $g : \mathrm{Dom}(G) \to \mathbb{R}^n$ such that $g(x, y) \in \overline{G(x, y)}$ for all $(x, y) \in \mathrm{Dom}(G)$. From (2.19) we have $G(x, y) = \varnothing$ if and only if $d(h(x, y), F(x)) = 0$, so we can define $f : X \times \mathbb{R}^n \to \mathbb{R}^n$ as follows:

$$f(x, y) \quad := \quad \begin{cases} g(x, y) & \text{when } d(h(x, y), F(x)) \neq 0 \\ h(x, y) & \text{when } d(h(x, y), F(x)) = 0 \end{cases} \tag{2.20}$$

For all $(x, y) \in X \times \mathbb{R}^n$ we have $\|f(x, y) - h(x, y)\| \leq 2d(h(x, y), F(x))$, and it follows that f is continuous. Fix $x \in X$. Clearly $f(x, y) \in F(x)$ for all $y \in \mathbb{R}^n$, and thus we have $f(x, B) \subset F(x)$. Conversely, let $z \in F(x)$; then from (2.18) there exists $y \in B$ such that $z = h(x, y) = f(x, y)$, and we conclude that $F(x) \subset f(x, B)$.

Next suppose X is locally compact and F is locally Lipschitz continuous. It follows from Proposition 2.10 that h is locally Lipschitz continuous. We define a set-valued map $H : X \times \mathbb{R}^n \rightsquigarrow \mathbb{R}^n$ with nonempty compact convex values by

$$H(x, y) \quad := \quad F(x) \cap \Big[h(x, y) + 2d(h(x, y), F(x))B \Big] \tag{2.21}$$

The mapping $(x, y) \mapsto d(h(x, y), F(x))$ is locally Lipschitz continuous by Proposition 2.10, and, observing that $H(x, y) - h(x, y) = [F(x) - h(x, y)] \cap$

$2d(h(x,y), F(x))B$, it follows from Proposition 2.16 that H is locally Lipschitz continuous. From Proposition 2.20 there exists a locally Lipschitz continuous function $f : X \times I\!R^n \to I\!R^n$ such that $f(x,y) \in H(x,y)$ for all $(x,y) \in X \times I\!R^n$, and as above we have $F(x) = f(x,B)$. ∎

2.6 Summary

This review of basic definitions and results in set-valued analysis prepares us for the formulation of a general robust nonlinear control problem in Chapter 3. Marginal functions will arise naturally as we define worst-case quantities. Selection theorems will allow us to conclude the existence of feedback control laws with desired properties. Parameterizations will enable us to simplify problem statements without loss of generality.

It would be possible to cast many of the results in the following chapters in the language of viability theory for differential inclusions [5], but for our purposes there is no need to introduce the additional terminology.

Chapter 3

Robust Control Lyapunov Functions

Significant advances in the theory of linear robust control in recent years have led to powerful new tools for the design and analysis of control systems. A popular paradigm for such theory is depicted in Figure 3.1, which shows the interconnection of three system blocks G, K, and Δ. The plant G relates a control input u and a disturbance input v to a measurement output y and a penalized output z. The control input u is generated from the measured output y by the controller K. All uncertainty is located in the block Δ which generates the disturbance input v from the penalized output z. The plant G, which is assumed to be linear and precisely known, may incorporate some nominal plant as well as frequency-dependent weights on the uncertainty Δ (for this reason G is sometimes called a generalized plant). Once G is determined, the *robust stabilization problem* is to construct a controller K which guarantees closed-loop stability for all systems Δ belonging to a given family of admissible (possibly nonlinear) uncertain systems.

Necessary and sufficient conditions for the solvability of the linear robust stabilization problem have been derived for a variety of families of admissible uncertainties Δ. Many such results rely on small gain theorems which convert the robust stabilization problem into the following *disturbance attenuation problem*: construct a controller K which guarantees that some gain between the input v and the output z is smaller than unity. Examples of such gains include the H_∞ norm, the L_1 norm, and

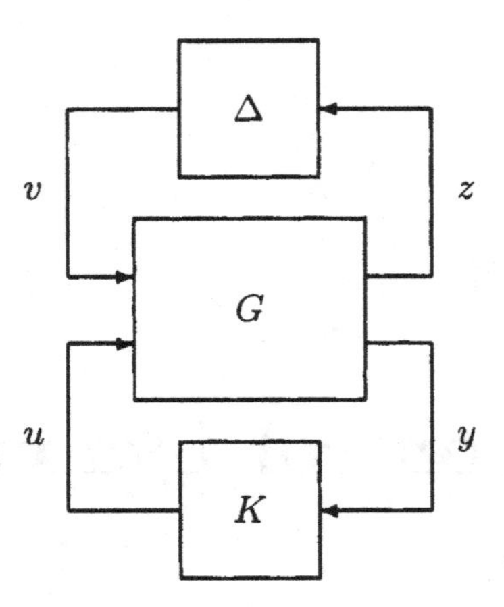

Figure 3.1: Robust control paradigm.

the structured singular value of the linear closed-loop system relating the input v to the output z.

In this chapter we formulate a *nonlinear* robust stabilization problem in which both G and Δ are nonlinear (Section 3.1), we investigate its relationship to a nonlinear disturbance attenuation problem (Section 3.2), and we provide necessary and sufficient conditions for its solvability (Sections 3.3–3.5).

Input/output frequency domain methods, which are effective for linear systems, are less suitable for nonlinear systems; the Lyapunov function remains the only general tool for nonlinear stability analysis. We will therefore pose our nonlinear robust stabilization problem in Section 3.1 in a form conducive to the application of Lyapunov stability theory, that is, in some underlying state space. To formulate Lyapunov conditions for robust stabilizability, we will require that this state space remain the same for the families of admissible controllers K and admissible uncertain systems Δ; in other words, we will only consider controllers and uncertain systems of *fixed* dynamic order. More precisely, given a plant G and some nonnegative integers p and q, we will derive necessary and sufficient conditions for the existence of an admissible controller K of dynamic order p which guarantees closed-loop stability for all admissible uncertain systems Δ of dynamic order q. We will show in Section 3.2 that the dynamic order restriction on the uncertainty Δ disappears when we interpret our conditions in the context of nonlinear disturbance attenuation; in this case, stability will be guaranteed for uncertainties of *any* dynamic order provided that they satisfy an appropriate small gain condition.

Once we fix the dynamic orders of the blocks K and Δ, we can augment the plant G with an appropriate number of integrators and obtain an equivalent formulation with *memoryless* blocks K and Δ. For example, if we are considering a class of third-order controllers K, we can add three states to the plant G, redefine the signals u and y, and reduce the original class of controllers to a class of memoryless controllers. Without loss of generality, therefore, we will assume that K and Δ are memoryless blocks; the details of the problem reformulation required by such an assumption will be given at the end of Section 3.1.1.

We will make specific choices for the signals v and z in Figure 3.1. The penalized output z will have two components, the underlying state x and the control input u. Likewise, the disturbance input v will have two components, the measured output y and a new disturbance input w. This configuration will allow us to model internal plant uncertainty as well as uncertainty in the control input and measured output (see Figure 3.2).

Throughout this chapter we will make extensive use of the background material on set-valued maps provided in Chapter 2. Although we could present most of our results in the language of viability theory for differential inclusions [5], we choose not to introduce the additional terminology.

3.1 Nonlinear robust stabilization

3.1.1 System description

We consider four finite-dimensional Euclidean spaces: the *state space* $\mathcal{X}$, the *control space* $\mathcal{U}$, the *disturbance space* $\mathcal{W}$, and the *measurement space* $\mathcal{Y}$. Given a continuous function $f : \mathcal{X} \times \mathcal{U} \times \mathcal{W} \times I\!R \to \mathcal{X}$, we form a differential equation

$$\dot{x} \;=\; f(x, u, w, t) \tag{3.1}$$

where $x \in \mathcal{X}$ is the state variable, $u \in \mathcal{U}$ is the control input, $w \in \mathcal{W}$ is the disturbance input, and $t \in I\!R$ is the independent (time) variable.

Associated with this differential equation are admissible measurements, admissible disturbances, and admissible controls, each characterized by a set-valued constraint.

A *measurement* for the equation (3.1) is a function $y : \mathcal{X} \times I\!R \to \mathcal{Y}$ such that $y(\cdot\,, t)$ is continuous for each fixed $t \in I\!R$ and $y(x, \cdot)$ is locally L_∞ for each fixed $x \in \mathcal{X}$.[1] Given a *measurement constraint* $Y : \mathcal{X} \times I\!R \rightsquigarrow \mathcal{Y}$, we say a measurement $y(x,t)$ is *admissible* when $y(x,t) \in Y(x,t)$ for all $(x,t) \in \mathcal{X} \times I\!R$. Thus Y defines an output inclusion

$$y \;\in\; Y(x,t) \tag{3.2}$$

which characterizes all admissible measurements. This is clearly a generalization of the output equation $y = h(x,t)$ usually associated with the differential equation (3.1). Such a generalization allows for measurement uncertainty due to imperfect sensors because there may be several different admissible measurement trajectories associated with a single state trajectory. For example, suppose we have a sensor which provides a measurement of the state x with an error of up to ten percent of its magnitude, plus some noise with unity maximum amplitude. In this case our measurement constraint would be $Y(x,t) = x + 0.1|x|B + B$ where B denotes the closed unit ball.

A *disturbance* for the equation (3.1) is a function $w : \mathcal{X} \times \mathcal{U} \times I\!R \to \mathcal{W}$ such that $w(\cdot\,,\cdot\,,t)$ is continuous for each fixed $t \in I\!R$ and $w(x,u,\cdot)$ is locally L_∞ for each fixed $(x,u) \in \mathcal{X} \times \mathcal{U}$. Given a *disturbance constraint* $W : \mathcal{X} \times \mathcal{U} \times I\!R \rightsquigarrow \mathcal{W}$, we say a disturbance $w(x,u,t)$ is *admissible* when $w(x,u,t) \in W(x,u,t)$ for all $(x,u,t) \in \mathcal{X} \times \mathcal{U} \times I\!R$. Admissible disturbances include both exogenous disturbances and feedback disturbances, and they therefore encompass a large class of memoryless plant and input uncertainties. Such memoryless uncertainties form the basis of the *guaranteed stability* framework for robust nonlinear control developed in [89, 48, 90, 23, 11], which includes as a special case the *quadratic stability* framework for linear systems (see the survey paper [22]).

A *control* for (3.1) is a function $u : \mathcal{Y} \times I\!R \to \mathcal{U}$ such that $u(\cdot\,,t)$ is continuous for each fixed $t \in I\!R$ and $u(y,\cdot)$ is locally L_∞ for each fixed

[1] We say a function is locally L_∞ when it is (essentially) bounded on a neighborhood of every point.

$y \in \mathcal{Y}$. Given a *control constraint* $U : \mathcal{Y} \times \mathbb{R} \rightsquigarrow \mathcal{U}$, we say a control $u(y,t)$ is *admissible* when $u(y,t) \in U(y,t)$ for all $(y,t) \in \mathcal{Y} \times \mathbb{R}$ and furthermore $u(y,t)$ is jointly continuous in (y,t). One might expect a *constant* control constraint $U(y,t) \equiv U_0$ to be sufficient for our purposes, but there are good reasons for allowing this constraint to depend on the measurement y. For example, one may want to deactivate some expensive component of the control u when the measurement y lies within a normal operating region; in this case, the control constraint for normal values of y would be a strict subset of the control constraint for extreme values of y. Moreover, one may need to redefine the control variable to meet the assumption below that the system (3.1) is affine in u, in which case a constant control constraint would likely be converted into one which depends on y.

The function f together with the three set-valued constraints U, W, and Y comprise a *system* $\Sigma = (f, U, W, Y)$. By a solution to Σ we mean a solution $x(t)$ to the initial value problem

$$\dot{x} = f\big(x,\, u(y(x,t),t),\, w(x,u(y(x,t),t),t),\, t\big) \qquad x(t_0) = x_0 \tag{3.3}$$

given a measurement $y(x,t)$, a disturbance $w(x,u,t)$, a control $u(y,t)$, and an initial condition $(x_0,t_0) \in \mathcal{X} \times \mathbb{R}$. Our definitions guarantee that the right-hand side of (3.3) is continuous in x and locally L_∞ in t, and it follows from classical existence theorems that solutions to Σ always exist (locally in t) but need not be unique. Figure 3.2 shows the signal flow diagram of the system; note that the dotted blocks relate this diagram to the robust control paradigm in Figure 3.1. The W and Y components of the uncertainty Δ generate multiple solutions to Σ for a given control and initial condition.

We say a system $\Sigma = (f, U, W, Y)$ is *time-invariant* when the mappings f, U, W, and Y are all independent of t. In this case we abuse notation and write $f(x,u,w)$, $U(y)$, $W(x,u)$, and $Y(x)$. Likewise, we say a control is time-invariant when $u(y,t) = u(y)$.

Although we have included only memoryless uncertainties and controls in our formulation, we can accommodate *fixed-order* dynamics by redefining the system Σ. For example, to allow dynamic feedback of order q in the controller, we extend the state, control, and measurement

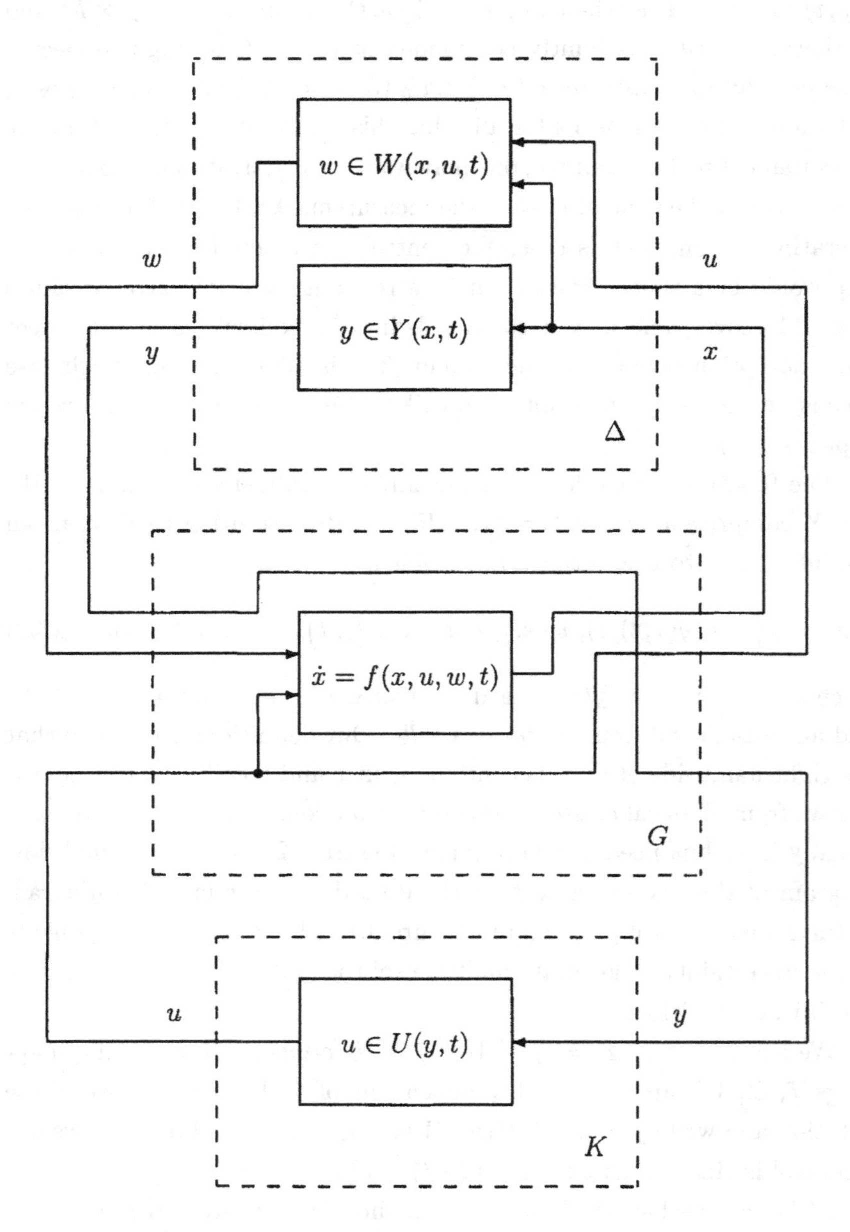

Figure 3.2: Signal flow diagram of a system Σ.

spaces by introducing an auxiliary state variable $x_a \in \mathbb{R}^q$, an auxiliary control variable $u_a \in \mathbb{R}^q$, and an auxiliary measurement variable $y_a \in \mathbb{R}^q$. The augmented equations are

$$\dot{x} = f(x, u, w, t) \tag{3.4}$$

$$\dot{x}_a = u_a \tag{3.5}$$

$$y \in Y(x, t) \tag{3.6}$$

$$y_a \in \{x_a\} \tag{3.7}$$

which represent an augmented system $\Sigma_a = (f_a, U_a, W_a, Y_a)$. It is clear that a (memoryless) control for the augmented system Σ_a corresponds to a qth-order dynamic control for the original system Σ.

3.1.2 Problem statement

Given a control for a system Σ, we can examine the behavior of the set of solutions to Σ for all initial conditions, admissible measurements, and admissible disturbances. When every such solution converges to a compact *residual set* $\Omega \subset \mathcal{X}$ containing a desired operating point (for convenience taken to be the origin $0 \in \mathcal{X}$), we will say that these solutions are robustly stable. To give a more precise definition, we first recall from Lyapunov stability theory that a continuous function $\beta : \mathbb{R}_+ \times \mathbb{R}_+ \to \mathbb{R}_+$ belongs to class $\mathcal{KL}$ when $\beta(\cdot, s) \in \mathcal{K}$ for each fixed $s \in \mathbb{R}_+$ and $\beta(r, s)$ decreases to zero as $r \in \mathbb{R}_+$ remains constant and $s \to \infty$.

Definition 3.1 *Fix a control for Σ, and let $\Omega \subset \mathcal{X}$ be a compact set containing $0 \in \mathcal{X}$. The solutions to Σ are* **robustly globally uniformly asymptotically stable with respect to Ω (RGUAS-Ω)** *when there exists $\beta \in \mathcal{KL}$ such that for all admissible measurements, admissible disturbances, and initial conditions $(x_0, t_0) \in \mathcal{X} \times \mathbb{R}$, all solutions $x(t)$ exist for all $t \geq t_0$ and satisfy*[2]

$$|x(t)|_\Omega \leq \beta(|x_0|_\Omega, t - t_0) \tag{3.8}$$

for all $t \geq t_0$. Also, the solutions to Σ are **RGUAS** *when they are* RGUAS-$\{0\}$.

[2]The notation $|\cdot|_\Omega$ represents the Euclidean point-to-set distance function, that is, $|\cdot|_\Omega := d(\cdot, \Omega)$.

Note that RGUAS-Ω implies that the residual set Ω is (robustly) positively invariant. In particular, RGUAS (with $\Omega = \{0\}$) implies that $0 \in \mathcal{X}$ is a equilibrium point (in forward time) for every admissible measurement and admissible disturbance.

The robust stabilization problem for a system Σ which results from Definition 3.1 is to construct an admissible control such that the solutions to Σ are RGUAS-Ω for some residual set Ω. We define three types of robust stabilizability according to how small we can make this set Ω:

Definition 3.2 *The system* Σ *is* **robustly asymptotically stabilizable** *when there exists an admissible control such that the solutions to* Σ *are* RGUAS. *The system* Σ *is* **robustly practically stabilizable** *when for every* $\varepsilon > 0$ *there exists an admissible control and a compact set* $\Omega \subset \mathcal{X}$ *satisfying* $0 \in \Omega \subset \varepsilon B$ *such that the solutions to* Σ *are* RGUAS-Ω. *The system* Σ *is* **robustly stabilizable** *when there exists an admissible control and a compact set* $\Omega \subset \mathcal{X}$ *satisfying* $0 \in \Omega$ *such that the solutions to* Σ *are* RGUAS-Ω.

Clearly robust asymptotic stabilizability implies robust practical stabilizability, which in turn implies robust stabilizability. The difference between practical stabilizability and mere stabilizability is that the the residual set Ω of a practically stabilizable system can be made arbitrarily small by choice of the control. If the system is not practically stabilizable, we might examine whether or not we can make Ω arbitrarily small in some desired directions by allowing it to grow in others.

3.2 Nonlinear disturbance attenuation

In Section 3.1 we formulated our robust stabilization problem in an underlying state space $\mathcal{X}$ of fixed dimension, and in doing so we apparently restricted the admissible uncertainties to those of a fixed dynamic order. In this section, we formulate a nonlinear disturbance attenuation problem which, in conjunction with recently developed small gain theorems, leads to a new robust stabilization problem with no restrictions on the dynamic order of the admissible uncertainties. We then show that this new problem formulation is actually a special case of the one given in Section 3.1.

3.2.1 Input-to-state stability

To formulate a nonlinear disturbance attenuation problem, we need an appropriate definition for the "gain" of a nonlinear system. One might consider using standard input/output definitions for the gain of an operator between signal spaces (such as L_2 or L_∞), but such definitions may be inadequate for several reasons:

- the gain definition should reflect the fact that, for nonlinear systems, relationships between inputs and outputs depend on signal size,
- the definition should also incorporate the nonlinear effects of initial conditions, and,
- methods should exist for computing the defined quantities.

Motivated by such considerations, Sontag introduced *input-to-state stability* (ISS) in his landmark paper [131].

Input-to-state stability is a property of a nonlinear system

$$\dot{x} = F(x,v) \tag{3.9}$$

where x and v are, respectively, the system state and an external input signal (both living in finite-dimensional Euclidean spaces), and F is continuous with $F(0,0) = 0$.

Definition 3.3 *The system* (3.9) *is* **input-to-state stable (ISS)** *when there exist functions $\beta \in \mathcal{KL}$ and $\gamma \in \mathcal{K}$ such that for every initial condition x_0 and every bounded input $v \in L_\infty$, every solution $x(t)$ of* (3.9) *starting from $t = 0$ satisfies*

$$|x(t)| \leq \beta(|x_0|, t) + \gamma(\|v\|_\infty) \tag{3.10}$$

for all $t \geq 0$.

The function β in this definition describes the decaying effect of the initial condition x_0, while the function γ describes the effect of the input signal v. One can think of this *gain function* γ as a generalization of the gain of an operator between L_∞ spaces. Indeed, suppose the initial condition is $x_0 = 0$ and suppose γ is a linear function $\gamma(r) = \gamma_0 r$ for some constant $\gamma_0 > 0$; then (3.10) reduces to the standard gain definition

$$\|x\|_\infty \;\leq\; \gamma_0 \|v\|_\infty \tag{3.11}$$

If the system relating v to x were linear and time-invariant, the constant γ_0 would be bounded from below by the L_1-norm of the impulse response. For this reason, theory developed using the ISS property could fall under the heading "nonlinear L_1 control."

The ISS definition of stability is rapidly becoming standard in the nonlinear control community, and many alternative stability definitions proposed in the literature have recently been shown to be equivalent to ISS [138]. Moreover, Lyapunov functions, which are ubiquitous in nonlinear stability analysis, are the main tools for computing the functions β and γ in Definition 3.3. Therefore ISS is the framework of choice for the formulation of nonlinear disturbance attenuation problems, especially because recent ISS small gain theorems [67, 99] relate such problems to robust stabilization. The only other framework which has received considerable attention in the literature is the so-called "nonlinear H_∞" framework based on the standard operator gain between L_2 spaces [8, 150, 151, 61, 10, 83, 65]. Both the ISS and the H_∞ frameworks fall under the general theory of dissipative systems developed in [155, 56, 57] (see [135] for a discussion on the connections between ISS and dissipativity).

3.2.2 Nonlinear small gain theorems

Because ISS generalizes finite-gain stability (in L_∞), it is of interest to investigate how standard small gain theorems for interconnected finite-gain subsystems can be extended to interconnected ISS subsystems. As reported in [67, 99], a feedback connection of ISS subsystems retains the ISS property when the composition of the gain functions around the loop is "strictly" less than the identity function (the notion of strictness here is related to the topological separation property of [125]). We will state one of the simpler ISS small gain theorems for systems of the form

$$\dot{x}_1 \;=\; F_1(x_1, x_2) \tag{3.12}$$

$$\dot{x}_2 = F_2(x_2, x_1) \tag{3.13}$$

where F_1 and F_2 are sufficiently smooth with $F_1(0,0)=0$ and $F_2(0,0)=0$.

Theorem 3.4 *Suppose the x_1-subsystem (3.12), with x_2 regarded as the input, is ISS with gain function $\gamma_1 \in \mathcal{K}$. Likewise, suppose the x_2-subsystem (3.13), with x_1 regarded as the input, is ISS with gain function $\gamma_2 \in \mathcal{K}$. Suppose there exist functions $\rho_1, \rho_2 : I\!R_+ \to I\!R_+$ such that the mappings $s \mapsto \rho_1(s) - s$ and $s \mapsto \rho_2(s) - s$ are of class $\mathcal{K}_\infty$ and furthermore*

$$(\rho_1 \circ \gamma_1 \circ \rho_2 \circ \gamma_2)(s) \leq s \tag{3.14}$$

for all $s \geq 0$. Then the zero solution of (3.12)–(3.13) is globally asymptotically stable.

If the two gain functions γ_1 and γ_2 were linear, the small gain condition (3.14) would reduce the standard condition that the product of their slopes be (strictly) less than unity. Theorem 3.4, which is actually a corollary of a more general result in [67], motivates the nonlinear disturbance attenuation problem presented in the next section.

3.2.3 Disturbance attenuation vs. robust stabilization

Let us consider the following nonlinear system with output:

$$\dot{x} = f(x, u, w) \tag{3.15}$$
$$y = h(x, d) \tag{3.16}$$

where x is the state variable, u is the control input, y is the measured output, w and d are disturbance inputs, $f(0,0,0) = 0$, and $h(0,0) = 0$. We pose a nonlinear disturbance attenuation problem as follows: construct an admissible (output feedback) control $u(y)$ for (3.15)–(3.16), with $u(0) = 0$, such that the closed-loop system is ISS with respect to the external inputs w and d. According to ISS small gain theorems (such as Theorem 3.4), the solution of this problem provides robust stabilization with respect to the uncertain system Δ shown in Figure 3.3. Here the system Δ can be of *arbitrary* dynamic order; it need only be ISS with

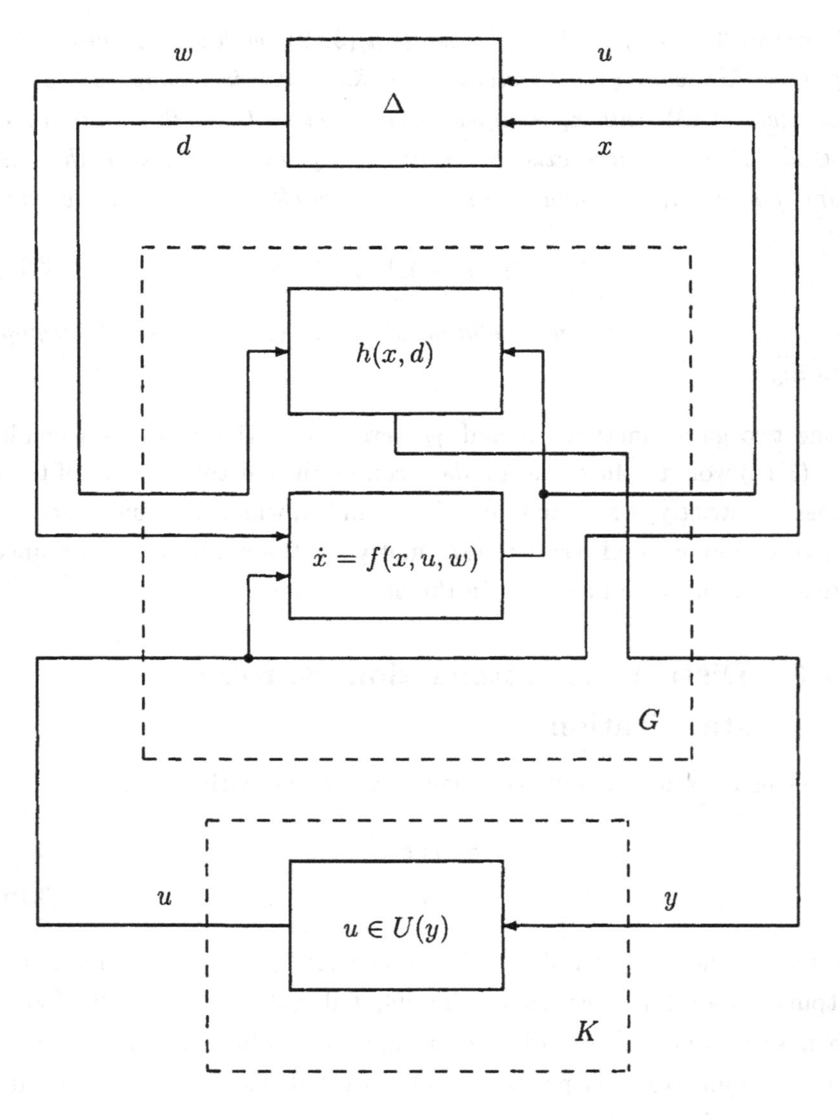

Figure 3.3: A nonlinear disturbance attenuation problem.

a sufficiently "small" gain function γ. Therefore this disturbance attenuation problem seems to reflect the paradigm of linear robust control theory more fully than the robust stabilization problem posed in Section 3.1. However, this new problem is actually a special case of the previous problem as can be determined from the following result [137]:

Proposition 3.5 *The system* (3.9) *is ISS if and only if there exist functions $\beta \in \mathcal{KL}$ and $\rho \in \mathcal{K}_\infty$ such that for every input function $v(x,t)$ which is continuous in x and locally L_∞ in t and which satisfies $|v(x,t)| \le \rho(|x|)$ for all x and t, every solution $x(t)$ of* (3.9) *starting from an initial condition x_0 at time $t = 0$ satisfies*

$$|x(t)| \;\le\; \beta(|x_0|, t) \tag{3.17}$$

for all $t \ge 0$.

Indeed, let us define a disturbance constraint $W(x) := \rho(|x|)B$ and a measurement constraint $Y(x) := h(x,\, \rho(|x|)B)$ for some function $\rho \in \mathcal{K}_\infty$, where B denotes the appropriate closed unit ball. Then it is clear from Proposition 3.5 that the above nonlinear disturbance attenuation problem for the system (3.15)–(3.16) admits a solution if and only if the system $\Sigma = (f, U, W, Y)$ is robustly asymptotically stabilizable (cf. Definition 3.2) for some function $\rho \in \mathcal{K}_\infty$.

We conclude that the robust stabilization problem posed in Section 3.1 is more general than the nonlinear disturbance attenuation problem posed in this section. In the remaining sections of this chapter, we provide necessary and sufficient conditions for the solvability of the general problem.

3.3 Robust control Lyapunov functions

The existence of a Lyapunov function is the most important necessary and sufficient condition for the stability of a nonlinear system. The sufficiency was proved by Lyapunov [98], and the necessity was established half a century later with the appearance of the so-called converse theorems [87, 104]. Having been developed much earlier than control theory, Lyapunov theory deals with dynamical systems without inputs. For this reason, it has traditionally been applied only to *closed-loop* control systems, that is, systems for which the input has been eliminated through the substitution of a predetermined feedback control. However, some authors started using candidate Lyapunov functions in feedback design itself by "making the Lyapunov derivative negative" when choosing the control [71, 82, 63, 68]. Such ideas have been made precise with the introduction of the *control Lyapunov function* for systems with control inputs [3, 130].

3.3.1 Control Lyapunov functions

A control Lyapunov function (clf) for a system of the form

$$\dot{x} \;=\; f(x,u) \tag{3.18}$$

is a C^1, positive definite, radially unbounded function $V(x)$ such that

$$x \neq 0 \quad\Longrightarrow\quad \inf_{u\in U} \nabla V(x)\cdot f(x,u) \;<\; 0 \tag{3.19}$$

where U is a convex set of admissible values of the control variable u. In other words, a clf is simply a candidate Lyapunov function whose derivative can be made negative *pointwise* by the choice of control values. Clearly, if f is continuous and there exists a continuous state feedback for (3.18) such that the point $x = 0$ becomes a globally asymptotically stable equilibrium of the closed-loop system, then by standard converse Lyapunov theorems [87, 104] there must exist a clf for the system (3.18). If f is affine in the control variable, then the existence of a clf for (3.18) is also sufficient for stabilizability via continuous[3] state feedback [3, 132].

To summarize, just as the existence of a Lyapunov function is necessary and sufficient for the stability of a system without inputs, the existence of a clf is necessary and sufficient for the stabilizability of a system with a control input.

Example 3.6 Let us consider the second-order system

$$\dot{x}_1 \;=\; -x_1^3 + x_2\,\phi(x_1,x_2) \tag{3.20}$$

$$\dot{x}_2 \;=\; u + \psi(x_1,x_2) \tag{3.21}$$

[3]Continuity of the feedback at the point $x = 0$ requires an extra condition on the clf known as the *small control property* (see Section 3.4.1).

where ϕ and ψ are continuous functions and u takes values in the set $U = I\!R$. The function $V(x_1, x_2) := \frac{1}{2}x_1^2 + \frac{1}{2}x_2^2$ satisfies

$$\begin{aligned}\inf_{u\in U} \nabla V(x)\cdot f(x,u) &= \inf_{u\in U} \Big[-x_1^4 + x_1x_2\,\phi(x_1,x_2) \\ &\qquad + x_2\,u + x_2\,\psi(x_1,x_2)\Big] \\ &= \begin{cases} -x_1^4 & \text{when } x_2 = 0 \\ -\infty & \text{when } x_2 \neq 0 \end{cases} \end{aligned} \tag{3.22}$$

and is therefore a clf for this system. We conclude that this system is globally asymptotically stabilizable. In this case we could also prove stabilizability by constructing a particular feedback control; indeed, the control $u(x_1, x_2) = -x_2 - \psi(x_1, x_2) - x_1\,\phi(x_1, x_2)$ renders $\dot{V}$ negative definite and thus guarantees global asymptotic stability.

Example 3.7 (feedback linearizable systems) Suppose there exists a diffeomorphism $\xi = \Phi(x)$ with $\Phi(0) = 0$ which transforms the system (3.18) into

$$\dot{\xi} = F\xi + G\big[\ell_0(\xi) + \ell_1(\xi)\,u\big] \tag{3.23}$$

where the matrix pair (F, G) is controllable and the functions ℓ_0 and ℓ_1 are continuous with $\ell_1(\xi)$ nonsingular for all ξ. Suppose $U = \mathcal{U}$, and let P be the symmetric positive definite solution to the Riccati equation

$$F^{\mathrm{T}}P + PF - PGG^{\mathrm{T}}P + I = 0 \tag{3.24}$$

The function $V(x) := \Phi(x)^{\mathrm{T}}P\Phi(x) = \xi^{\mathrm{T}}P\xi$ satisfies

$$\begin{aligned}\inf_{u\in\mathcal{U}} \nabla V(x)\cdot f(x,u) &= \inf_{u\in\mathcal{U}} \Big[\xi^{\mathrm{T}}[F^{\mathrm{T}}P + PF]\xi + 2\xi^{\mathrm{T}}PG\,[\ell_0(\xi) + \ell_1(\xi)\,u]\Big] \\ &= \inf_{u\in\mathcal{U}} \Big[-\xi^{\mathrm{T}}\xi + \xi^{\mathrm{T}}PG\,[G^{\mathrm{T}}P\xi + 2\ell_0(\xi) + 2\ell_1(\xi)\,u]\Big] \\ &= \begin{cases} -\xi^{\mathrm{T}}\xi & \text{when } \xi^{\mathrm{T}}PG = 0 \\ -\infty & \text{when } \xi^{\mathrm{T}}PG \neq 0 \end{cases}\end{aligned} \tag{3.25}$$

and is therefore a clf for this system. As in the previous example, we conclude that this system is globally asymptotically stabilizable. Note that we have reached this conclusion without constructing the linearizing feedback control law.

These two examples illustrate the methodological difference between the clf and the classical Lyapunov function: the clf allows us to conclude the stabilizability of a system with an *undefined* feedback control, whereas the classical Lyapunov function only allows us to conclude the stability of the closed-loop system generated by a *predetermined* feedback control.

In the remaining sections of this chapter, we extend the clf methodology to systems with both control and disturbance inputs: we introduce the *robust control Lyapunov function* (rclf) and show that its existence is necessary and sufficient for robust stabilizability as defined in Section 3.1.

3.3.2 Rclf: general definition

Recall from Section 3.1.1 that a system Σ consists of a continuous function f and three set-valued maps U, W, and Y. We let $\mathcal{V}(\mathcal{X})$ denote the set of all C^1 functions $V : \mathcal{X} \times I\!R \to I\!R_+$ such that there exist class $\mathcal{K}_\infty$ functions α_1 and α_2 satisfying

$$\alpha_1(\|x\|) \leq V(x,t) \leq \alpha_2(\|x\|) \tag{3.26}$$

for all $(x,t) \in \mathcal{X} \times I\!R$. The set $\mathcal{V}(\mathcal{X})$ is the set of all candidate Lyapunov functions for testing the robust stability of a system Σ. This set $\mathcal{V}(\mathcal{X})$ is contained in the broader set $\mathcal{P}(\mathcal{X})$ of functions which need not be differentiable or radially unbounded, that is, the set of all continuous functions $\alpha : \mathcal{X} \times I\!R \to I\!R_+$ such that there exist $\chi_1, \chi_2 \in \mathcal{K}$ satisfying $\chi_1(\|x\|) \leq \alpha(x,t) \leq \chi_2(\|x\|)$ for all $(x,t) \in \mathcal{X} \times I\!R$. Given $V \in \mathcal{V}(\mathcal{X})$ and a system Σ, we define the *Lyapunov derivative* $L_f V : \mathcal{X} \times \mathcal{U} \times \mathcal{W} \times I\!R \to I\!R$ by the equation

$$L_f V(x,u,w,t) \quad := \quad V_t(x,t) + V_x(x,t) \cdot f(x,u,w,t) \tag{3.27}$$

where V_t and V_x denote the respective partial derivatives of V. Clearly $L_f V$ is continuous, and it follows from (3.26) that $L_f V(0,\cdot,\cdot,\cdot) \equiv 0$. We next define a set-valued map $Q : \mathcal{Y} \times I\!R \times I\!R_+ \rightsquigarrow \mathcal{X}$ by

$$Q(y,c,t) \quad := \quad \big\{ x \in \mathcal{X} \;:\; y \in Y(x,t) \text{ and } V(x,t) \geq c \big\} \tag{3.28}$$

Given a measurement $y \in \mathcal{Y}$, a time $t \in I\!R$, and a constant $c \in I\!R_+$, the set $Q(y,c,t)$ is the set of all states $x \in X$ which lie outside the c-level set of V and which are consistent with the output inclusion (3.2).

Definition 3.8 *A function $V \in \mathcal{V}(\mathcal{X})$ is called a* **robust control Lyapunov function (rclf)** *for a system Σ when there exist $c_V \in I\!R_+$ and $\alpha_V \in \mathcal{P}(\mathcal{X})$ such that*

$$\inf_{u \in U(y,t)} \sup_{x \in Q(y,c,t)} \sup_{w \in W(x,u,t)} \Big[L_f V(x,u,w,t) + \alpha_V(x,t) \Big] < 0 \quad (3.29)$$

for all $y \in \mathcal{Y}$, all $t \in I\!R$, and all $c > c_V$.

The rclf generalizes the clf in several directions. First, both control *and* disturbance inputs are represented in the definition. Second, this definition supports measurement feedback as well as state feedback, and it generalizes the "output clf" defined in [148, 149, 147]. Finally, the parameter c_V allows us to consider all three types of stabilizability in Definition 3.2, not just asymptotic stabilizability.

3.3.3 Rclf: state-feedback for time-invariant systems

We obtain a simpler definition of an rclf when we consider the special case of state feedback for time-invariant systems:

Proposition 3.9 *Let Σ be time-invariant, suppose $\mathcal{X} = \mathcal{Y}$, and suppose $Y(x) = \{x\}$ for all $x \in \mathcal{X}$. Then a time-invariant function $V \in \mathcal{V}(\mathcal{X})$ is an rclf for Σ if and only if there exist $c_V \in I\!R_+$ and a time-invariant function $\alpha_V \in \mathcal{P}(\mathcal{X})$ such that*

$$\inf_{u \in U(x)} \sup_{w \in W(x,u)} \Big[L_f V(x,u,w) + \alpha_V(x) \Big] < 0 \quad (3.30)$$

whenever $V(x) > c_V$.

Proof: First note that α_V in Definition 3.8 can always be chosen time-invariant. Suppose first that V satisfies (3.29); in particular we have

$$\inf_{u \in U(y)} \sup_{x \in Q(y,V(y))} \sup_{w \in W(x,u)} \Big[L_f V(x,u,w) + \alpha_V(x) \Big] < 0 \quad (3.31)$$

whenever $V(y) > c_V$. Now $x \in Q(y, V(y))$ if and only if $x = y$; thus

$$\inf_{u \in U(y)} \sup_{w \in W(y,u)} \Big[L_f V(y,u,w) + \alpha_V(y) \Big] < 0 \quad (3.32)$$

whenever $V(y) > c_V$. We therefore obtain (3.30) as desired. Next fix $y \in \mathcal{Y}$ and $c > c_V$. If $V(y) < c$ then $Q(y,c) = \varnothing$, which means

$$\inf_{u \in U(y)} \sup_{x \in Q(y,c)} \sup_{w \in W(x,u)} \Big[L_f V(x,u,w) + \alpha_V(x) \Big] < 0 \tag{3.33}$$

as desired. If $V(y) \geq c$ then $Q(y,c) = \{y\}$, and therefore (3.33) follows from the inequality (3.30). ■

The inequality (3.30) can be interpreted as follows: for every fixed x there exists an admissible value u for the control such that the Lyapunov derivative is negative for any admissible value w for the disturbance. This is a natural generalization of the clf for systems with disturbance inputs.

Example 3.10 (feedback linearizable systems) Let us modify the feedback linearizable system (3.23) from Example 3.7 by adding a disturbance input w as follows:

$$\dot{\xi} = F\xi + G\Big[\ell_0(\xi,w) + \ell_1(\xi)\,u\Big] \tag{3.34}$$

Suppose the measurement constraint is $Y(x) = \{x\}$ for all $x \in \mathcal{X}$ (state feedback), suppose the control constraint is $U(x) \equiv \mathcal{U}$, and suppose the disturbance constraint is such that $W(x, \mathcal{U})$ is bounded for every $x \in \mathcal{X}$. Let P be the symmetric positive definite solution to the matrix Riccati equation (3.24), and consider the function $V(x) := \Phi(x)^{\mathrm{T}} P \Phi(x) = \xi^{\mathrm{T}} P \xi$. If we select $\alpha_V(x) := \frac{1}{2}\xi^{\mathrm{T}}\xi$, then we have

$$\begin{aligned}
&\inf_{u \in \mathcal{U}} \sup_{w \in W(x,u)} \Big[L_f V(x,u,w) + \alpha_V(x) \Big] \\
&\quad = \inf_{u \in \mathcal{U}} \sup_{w \in W(x,u)} \Big[\xi^{\mathrm{T}}[F^{\mathrm{T}}P + PF]\,\xi \\
&\qquad\qquad + 2\xi^{\mathrm{T}}PG\,[\ell_0(\xi,w) + \ell_1(\xi)\,u] + \tfrac{1}{2}\xi^{\mathrm{T}}\xi \Big] \\
&\quad = \inf_{u \in \mathcal{U}} \sup_{w \in W(x,u)} \Big[-\tfrac{1}{2}\xi^{\mathrm{T}}\xi + \xi^{\mathrm{T}}PG\,[G^{\mathrm{T}}P\xi + 2\ell_0(\xi,w) + 2\ell_1(\xi)\,u] \Big] \\
&\quad = \begin{cases} -\tfrac{1}{2}\xi^{\mathrm{T}}\xi & \text{when } \xi^{\mathrm{T}}PG = 0 \\ -\infty & \text{when } \xi^{\mathrm{T}}PG \neq 0 \end{cases}
\end{aligned} \tag{3.35}$$

We conclude from Proposition 3.9 that V is an rclf for the system (3.34) with $c_V = 0$. If we compare this example with Example 3.7, we see that the clf V for the system (3.23) (which had no disturbance input) became an rclf for the system (3.34) (which we obtained by adding a disturbance input w). As we shall see in Chapter 5, not every clf will become an rclf when a disturbance input is added to the system. In general, this occurs only when the disturbance input is *matched* with the control input, that is, when the disturbance w enters the system through the same channel as the control u and therefore could be cancelled by the control if it were a measured signal. This is the case in the above system (3.34) because the function $\ell_1(\xi)$ is nonsingular for all ξ.

3.3.4 Rclf: absence of disturbance input

The rclf is defined for systems much more general than those for which the clf is defined. In particular, the rclf allows for disturbance inputs, measurement feedback, and non-constant control constraints. It is of mathematical interest, however, to determine whether or not the rclf and clf coincide for the narrow class of systems for which we can define a clf. Let us therefore consider a time-invariant system Σ with perfect state measurement ($Y(x) = \{x\}$ for all $x \in \mathcal{X}$), a constant control constraint ($U(x) \equiv U$), and no disturbance input. Recall from (3.19) that a (time-invariant) function $V \in \mathcal{V}(\mathcal{X})$ is a clf for $\dot{x} = f(x,u)$ when it satisfies

$$x \neq 0 \quad \Longrightarrow \quad \inf_{u \in U} L_f V(x,u) \; < \; 0 \tag{3.36}$$

It is clear from (3.30) that every rclf for this system (with $c_V = 0$) is also a clf. We next investigate whether or not every clf is also an rclf in this restricted case.

Suppose V is a clf with the additional property that the left-hand side of the inequality in (3.36) is bounded away from zero for large x:

$$\begin{gathered} \forall c{>}0 \;\; \exists \delta{>}0 \quad \text{such that} \\ V(x) \geq c \;\Longrightarrow\; \inf_{u \in U} L_f V(x,u) \;\leq\; -\delta \end{gathered} \tag{3.37}$$

Every clf with this additional property is an rclf according to (3.30): if (3.37) is true, then we can find a (time-invariant) function $\alpha_V \in \mathcal{P}(\mathcal{X})$ such that the addition of $\alpha_V(x)$ to the left-hand side of (3.36) preserves the inequality. Conversely, if a given clf does not satisfy the additional property (3.37), it will not be an rclf because an appropriate function

$\alpha_V \in \mathcal{P}(\mathcal{X})$ will not exist (α_V must be bounded from below by a class $\mathcal{K}$ function of the norm). However, given a clf V which does not satisfy (3.37), we can always construct another clf Υ which does satisfy (3.37) and which is therefore an rclf. Indeed, let V be a clf, and define the function $\beta : \mathbb{R}_+ \to \mathbb{R}_+ \cup \{\infty\}$ by

$$\beta(r) \quad := \quad - \sup_{\|x\|=r} \ \inf_{u \in U} \ L_f V(x,u) \tag{3.38}$$

It follows from (3.36) and Proposition 2.9 that β is lsc and strictly positive on $(0,\infty)$. Consequently there exists a continuous positive definite function $\rho : \mathbb{R}_+ \to \mathbb{R}_+$ such that $\rho \leq \beta$ on $\mathbb{R}_+$. Let $\alpha_1 \in \mathcal{K}_\infty$ be as in (3.26), and define $\mu : \mathbb{R}_+ \to \mathbb{R}_+$ by

$$\mu(r) \quad := \quad \begin{cases} \rho(\alpha_1^{-1}(1)) & \text{when } 0 \leq r \leq 1 \\ \min\limits_{1 \leq s \leq r} \rho(\alpha_1^{-1}(s)) & \text{when } r > 1 \end{cases} \tag{3.39}$$

Clearly μ is decreasing with strictly positive values, and furthermore $\mu(\alpha_1(r)) \leq \rho(r)$ whenever $\alpha_1(r) \geq 1$. Also, it follows from Proposition 2.9 that μ is continuous. We can therefore define a new candidate clf $\Upsilon \in \mathcal{V}(\mathcal{X})$ by

$$\Upsilon(x) \quad := \quad \int_0^{V(x)} \frac{1}{\mu(r)} \, dr \tag{3.40}$$

We now verify that Υ satisfies property (3.37) (with V replaced by Υ). First observe that

$$\inf_{u \in U} \ L_f \Upsilon(x,u) \quad = \quad \frac{1}{\mu(V(x))} \inf_{u \in U} \ L_f V(x,u) \tag{3.41}$$

for all $x \in \mathcal{X}$. Suppose $\alpha_1(\|x\|) \geq 1$; then because μ is decreasing we have $\mu(V(x)) \leq \mu(\alpha_1(\|x\|)) \leq \rho(\|x\|) \leq \beta(\|x\|)$, and from (3.38) we obtain

$$\mu(V(x)) \quad \leq \quad - \inf_{u \in U} \ L_f V(x,u) \tag{3.42}$$

Combining (3.41) and (3.42), we obtain $\inf_{u \in U} L_f \Upsilon(x,u) \leq -1$ whenever $\alpha_1(\|x\|) \geq 1$. This together with the fact that $\inf_{u \in U} L_f \Upsilon(x,u)$ achieves its maximum on compact sets implies property (3.37) (with V replaced by Υ). As we have seen above, this implies further that Υ is an rclf.

To summarize, we have compared the clf and rclf definitions for the narrow class of systems for which we can define a clf. Although the definitions do not exactly coincide, we have shown that the existence of an rclf is equivalent to the existence of a clf.

3.4 Rclf implies robust stabilizability

We now investigate conditions under which the existence of an rclf for a system Σ implies robust stabilizability per Definition 3.2. We begin by considering the following set of assumptions on the constraint maps U, W, and Y:

C1: the control constraint U is lsc with nonempty closed convex values,
C2: the disturbance constraint W is usc with nonempty compact values, and,
C3: the set-valued map $(y,t) \rightsquigarrow \{x \in \mathcal{X} : y \in Y(x,t)\}$ is usc with compact values.

The continuity requirement in each of these assumptions is natural because we are looking for continuous feedback controls. The requirement in C2 that the disturbance constraint have compact values is our version of "bounded uncertainty." The compactness requirement in C3 is not as natural: for a given time t, the set of states x consistent with the value y of the measurement is assumed to be a bounded set. This is adequate for describing uncertain state feedback where the instantaneous measurement determines the state within a bounded set, but for true output feedback we will likely satisfy only the weaker assumption

C3′: the set-valued map $(y,t) \rightsquigarrow \{x \in \mathcal{X} : y \in Y(x,t)\}$ is usc with closed values.

We will refer to both assumptions C3 and C3′ in our proofs so that it is clear where we use the compactness assumption in C3. In Section 3.4.2 below, we will discuss how we can replace C3 with C3′ in the case of output feedback.

Our second set of assumptions concerns the structure of the function f which determines the dynamics (3.1) of the system Σ, together with the structure of W:

A1: for each fixed $(x, w, t) \in \mathcal{X} \times \mathcal{W} \times I\!R$, the mapping $u \mapsto f(x, u, w, t)$ is affine, and,

A2: the disturbance constraint W is independent of the control input u, that is, $W(x, u, t) = W(x, t)$.

The assumption A1 that f be affine in the control variable is standard in the clf literature. It can be dropped if one considers *relaxed* controls [3], that is, controls defined as mappings from the measurement space $\mathcal{Y}$ to the space of probability distributions over the control space $\mathcal{U}$. We do not pursue such relaxed controls here and thus restrict ourselves to systems which satisfy A1. Assumption A2 eliminates the dependence of the disturbance constraint W on the control, but it does *not* eliminate the dependence of admissible disturbances on the control. This assumption can be relaxed when f is jointly affine in u and w; to be precise, A1 and A2 can be replaced by the following:

A1$'$: for each fixed $(x, t) \in \mathcal{X} \times I\!R$, the mapping $(u, w) \mapsto f(x, u, w, t)$ is affine, and,

A2$'$: the disturbance constraint W is convex in u, that is,

$$W(x,\, \theta u_1 + (1-\theta)u_2,\, t) \;\subset\; \theta\, W(x, u_1, t) \;+\; (1-\theta)\, W(x, u_2, t)$$

for all $x \in \mathcal{X}$, all $t \in I\!R$, all $u_1, u_2 \in \mathcal{U}$, and all $\theta \in [0,1]$.

The convexity property in A2$'$ neither implies nor is implied by the convexity of the graph or the values of W. An example of a disturbance constraint satisfying both C2 and A2$'$ is $W(x, u, t) = \rho(x, u, t)B$ for a continuous positive real-valued function ρ which is convex in u. In this case, a disturbance $w(x, u, t)$ is admissible if and only if it satisfies the bound $\|w(x, u, t)\| \leq \rho(x, u, t)$.

Theorem 3.11 *Let Σ satisfy* C1–C3 *and either* A1–A2 *or* A1$'$–A2$'$. *If there is an rclf V for Σ, then Σ is robustly stabilizable. If furthermore $c_V = 0$, then Σ is robustly practically stabilizable. Also, if Σ and V are* *time-invariant, then the robustly stabilizing admissible control can always be chosen to be time-invariant.*

Proof: We begin by defining $D : \mathcal{Y}\times I\!R_{+}\times \mathcal{U}\times I\!R \to I\!R \cup \{\pm\infty\}$ by

$$D(y,c,u,t) \quad := \quad \sup_{x\in Q(y,c,t)} \; \sup_{w\in W(x,u,t)} \Big[L_f V(x,u,w,t) \,+\, \alpha_V(x,t)\Big] \tag{3.43}$$

We claim that D is usc. Indeed, it follows from C3, Corollary 2.12, and the definition of Q in (3.28) that Q is usc with compact values. It then follows from C2 and Proposition 2.9 that D is usc. Note that $D(y,c,u,t) = -\infty$ whenever $Q(y,c,t)$ is empty. We next show that D is convex in u. The function $L_f V$ is affine in u by A1, and if A2 is true we conclude that the mapping $u \mapsto D(y,c,u,t)$ is convex because it is the pointwise supremum of a collection of affine mappings. If A2 is not true but A1$'$ and A2$'$ are true, we compute as follows for $\theta \in [0,1]$:

$$\begin{aligned}
D(y,c,\,\theta u_1 + (1-\theta)u_2,\,t) \;&=\; \sup_{x\in Q(y,c,t)} \; \sup_{w\in W(x,\,\theta u_1+(1-\theta)u_2,\,t)} \\
&\Big[L_f V(x,\,\theta u_1 + (1-\theta)u_2,\,w,t) \,+\, \alpha_V(x,t)\Big] \\
&\le\; \sup_{x\in Q(y,c,t)} \; \sup_{w\in\theta\, W(x,u_1,t)+(1-\theta)\, W(x,u_2,t)} \\
&\Big[L_f V(x,\,\theta u_1 + (1-\theta)u_2,\,w,t) \,+\, \alpha_V(x,t)\Big] \\
&\le\; \sup_{x\in Q(y,c,t)} \; \sup_{w_1\in W(x,u_1,t)} \; \sup_{w_2\in W(x,u_2,t)} \Big[\alpha_V(x,t) \\
&+\; L_f V(x,\,\theta u_1 + (1-\theta)u_2,\,\theta w_1 + (1-\theta)w_2,\,t)\Big] \\
&\le\; \sup_{x\in Q(y,c,t)} \; \sup_{w_1\in W(x,u_1,t)} \; \sup_{w_2\in W(x,u_2,t)} \Big[\alpha_V(x,t) \\
&+\; \theta\, L_f V(x,u_1,w_1,t) \,+\, (1-\theta)\, L_f V(x,u_2,w_2,t)\Big] \\
&\le\; \theta\, D(y,c,u_1,t) \,+\, (1-\theta)\, D(y,c,u_2,t)
\end{aligned}$$

We again conclude that D is convex in u. We next use D to define a set-valued *regulation map* $K : \mathcal{Y}\times I\!R_{+}\times I\!R \rightsquigarrow \mathcal{U}$ as follows:

$$K(y,c,t) \quad := \quad \Big\{u \in U(y,t) \,:\, D(y,c,u,t) < 0\Big\} \tag{3.44}$$

Because D is convex in u and usc, it follows from C1 and Corollary 2.13 that K is lsc with convex values. Also, it follows from (3.29) and (3.43) that $K(y,c,t)$ is nonempty for all $y \in \mathcal{Y}$, all $t \in I\!R$, and all $c > c_V$.

Fix $c > c_V$. The lsc property is not affected by taking closures, and it follows from Theorem 2.18 that there exists an admissible control $u(y,t)$ such that $u(y,t) \in \overline{K(y,c,t)}$ for all $(y,t) \in \mathcal{Y} \times I\!R$. Now the mapping $u \mapsto D(y,c,u,t)$ is the pointwise supremum of a family of finite convex functions, and it follows from [122, Theorems 10.1 and 9.4] that D is lsc in u. Therefore $D(y,c,u(y,t),t) \leq 0$ for all $(y,t) \in \mathcal{Y} \times I\!R$.

Let $x(t)$ denote a local solution to (3.3) for some initial condition, some admissible measurement $y(x,t)$, and some admissible disturbance. Suppose $V(x(t),t) \geq c$ at some time $t \in I\!R$ for which $x(t)$ and $\dot{x}(t)$ are defined. Then the derivative $\dot{V}(x(t),t)$ satisfies

$$\begin{aligned} \dot{V}(x(t),t) + \alpha_V(x(t),t) &\leq D(y(x(t),t),\, c,\, u(y(x(t),t),t),\, t) \\ &\leq 0 \end{aligned} \tag{3.45}$$

Thus V is strictly decreasing along trajectories whenever $V \geq c$. It follows from standard Lyapunov arguments that the solutions to Σ are RGUAS-Ω with $\Omega = \alpha_1^{-1}(c)B$, where α_1 is the class $\mathcal{K}_\infty$ function in (3.26). If $c_V = 0$, then the residual set Ω can be made arbitrarily small by an appropriate choice of $c > 0$. Finally, if Σ and V are time-invariant, then the regulation map K will be time-invariant which means the robustly stabilizing admissible control can also be chosen to be time-invariant. ∎

3.4.1 Small control property

If we wish to use the rclf to characterize the robust *asymptotic* stabilizability of a system Σ, we should assume the existence of a control which creates this equilibrium for all admissible disturbances and a measurement which detects it. Namely, we should assume that the system Σ has the following properties:

S1: $Y(0,t) = \{0\}$ for all $t \in I\!R$, and,

S2: there is a continuous function $u_0 : I\!R \to \mathcal{U}$ such that $u_0(t) \in U(0,t)$ and $f(0,u_0(t),w,t) = 0$ for all $w \in W(0,u_0(t),t)$ and all $t \in I\!R$.

These assumptions are quite restrictive because they imply that the uncertainty cannot affect the equilibrium. Assumption S2 is clearly necessary for robust asymptotic stabilizability, while assumption S1 can be relaxed at the expense of more complexity in Definition 3.12 below. We define a new regulation map $K_s : \mathcal{Y}\times I\!R_+ \times I\!R \rightsquigarrow \mathcal{U}$ as follows:

$$K_s(y,c,t) \quad := \quad \begin{cases} K(y,c,t) & \text{when } c>0 \\ \{u_0(t)\} & \text{when } c=0 \end{cases} \tag{3.46}$$

where K is the regulation map defined in (3.44).

Definition 3.12 *Let Σ satisfy* S2. *An rclf V for Σ satisfies the* **small control property** *when c_V and α_V can be chosen such that $c_V = 0$ and, for all $t \in I\!R$, the new regulation map K_s is lsc at $(0,0,t)$ and $D(0,0,u_0(t),t) \leq 0$.*

Theorem 3.13 *Let Σ satisfy* S1–S2, C1–C3, *and either* A1–A2 *or* A1′–A2′. *If there is an rclf V for Σ which satisfies the small control property, then Σ is robustly asymptotically stabilizable. Also, if Σ and V are time-invariant, then the robustly stabilizing admissible control can always be chosen to be time-invariant.*

Proof: We need only slightly modify the proof of Theorem 3.11. Because K is lsc with nonempty convex values on $\mathcal{Y}\times(0,\infty)\times I\!R$, the modified map K_s is also lsc with nonempty convex values on the set $M := [\mathcal{Y}\times(0,\infty)\times I\!R] \cup [\{0\}\times\{0\}\times I\!R]$. It follows from Theorem 2.18 that there exists a continuous function $\mu : M \to \mathcal{U}$ such that $\mu(y,c,t) \in \overline{K_s(y,c,t)}$ for all $(y,c,t) \in M$. Now it follows from C3′ and S1 that the function $\beta : \mathcal{Y}\times I\!R \to [0,1]\cup\{\infty\}$ defined by

$$\beta(y,t) \quad := \quad \inf\left\{V(x,t) \ : \ y \in Y(x,t) \ \text{ and } \ V(x,t) \leq 1\right\} \tag{3.47}$$

is zero when $y=0$ and strictly positive when $y \neq 0$. Because the set-valued map $t \rightsquigarrow \{x \in \mathcal{X} : V(x,t) \leq 1\}$ is usc with compact values, it follows from C3′, Proposition 2.5, Corollary 2.12, and Proposition 2.9 that β is lsc. Thus there is a continuous function $\rho : \mathcal{Y}\times I\!R \to [0,1]$ such that $\rho(y,t) \leq \min\{1,\beta(y,t)\}$ for all $(y,t) \in \mathcal{Y}\times I\!R$, $\rho(y,t) = 0$ when $y=0$, and $\rho(y,t) > 0$ when $y \neq 0$. Also, we have $\rho(y,t) \leq V(x,t)$ whenever $y \in Y(x,t)$. It now follows from S2 and (3.46) that the control $u(y,t) := \mu(y,\rho(y,t),t)$ is admissible and $D(y,\rho(y,t),u(y,t),t) \leq 0$ for all $(y,t) \in \mathcal{Y}\times I\!R$.

Let $x(t)$ denote a local solution to (3.3) for some initial condition, some admissible measurement $y(x,t)$, and some admissible disturbance. Let $t \in I\!R$ be such that $x(t)$ and $\dot{x}(t)$ are defined. Because $\rho(y(x(t),t),t) \leq V(x(t),t)$, the derivative $\dot{V}(x(t),t)$ satisfies

$$\begin{aligned} \dot{V}(x(t),t) + \alpha_V(x(t),t) &\leq D(y(x(t),t),\ \rho(y(x(t),t),t),\ u(y(x(t),t),t),\ t) \\ &\leq 0 \end{aligned} \tag{3.48}$$

It now follows from standard Lyapunov arguments that the solutions to Σ are RGUAS. If Σ and V are time-invariant, then the regulation maps K and K_s will be time-invariant which means the robustly stabilizing admissible control can also be time-invariant. ■

3.4.2 Output feedback

In the case of output feedback, we can only assume property C3′ rather than the stronger property C3. The only place in the proofs of Theorems 3.11 and 3.13 where we used C3 rather than C3′ was when we showed that the function D was usc. This property was then used to show that the regulation map K was lsc. We see from Proposition 2.11, however, that for K to be lsc it is sufficient that the set

$$\Big\{(y,t,u,c) \in \text{Graph}(U) \times I\!R_+ \ :\ D(y,c,u,t) < 0\Big\} \tag{3.49}$$

be open relative to $\text{Graph}(U) \times I\!R_+$. This is certainly true when D is usc, but it may also be true when D is not usc. If we can verify that the set (3.49) is open without using the compactness assumption in C3, then the conclusions of Theorems 3.11 and 3.13 will remain valid when we replace C3 by the weaker assumption C3′. Unfortunately, the openness of the set (3.49) is a property of both the system Σ and the rclf V and thus cannot be verified before an rclf is chosen. Nevertheless, this property can usually be guaranteed, possibly after a modification of parameters and constraints.

To illustrate how one can guarantee the openness of the set (3.49), let us consider the situation in which the system dynamics are affine in the unmeasured states and the rclf is quadratic in the unmeasured states. For clarity we will assume that there is no uncertainty. Suppose our state space is partitioned as $\mathcal{X} = \mathcal{X}_1 \times \mathcal{X}_2$ where $x_1 \in \mathcal{X}_1$ and $x_2 \in \mathcal{X}_2$ are the components of the state vector. We consider dynamics of the form

$$\dot{x}_1 = A_1(x_1, u)\, x_2 + B_1(x_1, u) \tag{3.50}$$

$$\dot{x}_2 = A_2(x_1, u)\, x_2 + B_2(x_1, u) \tag{3.51}$$

where A_1, A_2, B_1, and B_2 are continuous functions. Suppose also that the measurement constraint is given by $Y(x) = \{x_1\}$ which corresponds to the output equation $y = x_1$. This measurement constraint fails to satisfy C3 because the set-valued map $x_1 \rightsquigarrow \{x_1\} \times \mathcal{X}_2$ does not have compact values; note however that the weaker assumption C3′ is satisfied. Let us consider an rclf V quadratic in x_2, namely,

$$V(x) = x_2^T P x_2 + x_2^T Q(x_1) + R(x_1) \tag{3.52}$$

where P is a symmetric positive definite matrix and Q and R are C^1 functions. If $\alpha_V(x)$ can also be chosen to be quadratic in x_2, then

$$L_f V(x, u) + \alpha_V(x) = x_2^T F(x_1, u)\, x_2 + x_2^T G(x_1, u) + H(x_1, u) \tag{3.53}$$

for suitable continuous functions F, G, and H, with F having symmetric values. Now because V is an rclf, for each $x_1 \in \mathcal{X}_1$ there must exist $u \in U(x_1)$ such that $L_f V + \alpha_V < 0$ for all $x_2 \in \mathcal{X}_2 \backslash \{0\}$. It follows from (3.53) that $F(x_1, u) \leq 0$, and by replacing α_V with $\varepsilon \alpha_V$ for some $\varepsilon \in (0, 1)$ if necessary, we may further assume that $F(x_1, u) < 0$. It follows that there exists a continuous function $\delta : \mathcal{X}_1 \to (0, \infty)$ such that the new control constraint U_δ defined by

$$U_\delta(x_1) := \overline{\big\{ u \in U(x_1) \; : \; F(x_1, u) + \delta(x_1) I < 0 \big\}} \tag{3.54}$$

has nonempty values for all $x_1 \in \mathcal{X}_1$, and that V is an rclf with respect to to U_δ. It then follows from C1, A1 and Corollary 2.12 that U_δ is lsc with nonempty closed convex values and thus satisfies C1 as a control constraint. We have left to verify that the set

$$\big\{ (x_1, u, c) \in \mathrm{Graph}(U_\delta) \times I\!R_+ \; : \; D(x_1, c, u) < 0 \big\} \tag{3.55}$$

is open relative to $\mathrm{Graph}(U_\delta) \times I\!R_+$. Let $(x_1, u) \in \mathrm{Graph}(U_\delta)$ be such that the supremum of (3.53) over $\{x_2 \in \mathcal{X}_2 \, : \, V(x) \geq c\}$ is negative. Because

$L_fV + \alpha_V$ is continuous and $F(x_1, u) < 0$, this supremum will remain negative under small perturbations of x_1, u, and c, and the openness of (3.55) is thus verified. Therefore the conclusions of Theorems 3.11 and 3.13 remain valid even though assumption C3 is violated.

We have seen that we can avoid assumption C3 in the case of output feedback by tightening the control constraint, effectively "throwing out" values of the control that lie on the boundary of the set (3.49) and thereby guaranteeing the openness of this set. Such a construction will fail if this set has an empty interior relative to Graph$(U)\times I\!R_+$ for all meaningful choices of parameters and constraints. This pathological situation requires further study beyond the scope of this text.

3.4.3 Locally Lipschitz state feedback

We now give conditions under which the existence of rclf V such that L_fV is locally Lipschitz implies the existence of a locally Lipschitz robustly stabilizing admissible control. For simplicity, we will consider only the state-feedback time-invariant case. The assumptions C1 and C2 are strengthened as follows:

- L1: U is locally Lipschitz with nonempty closed convex values, and,
- L2: W is locally Lipschitz with nonempty compact values.

Theorem 3.14 *Let Σ be time-invariant, suppose $\mathcal{X} = \mathcal{Y}$, and suppose $Y(x) = \{x\}$ for all $x \in \mathcal{X}$. Let Σ satisfy* L1–L2 *and either* A1–A2 *or* A1′–A2′. *If there is a time-invariant rclf V for Σ such that $L_fV + \alpha_V$ is locally Lipschitz, then Σ is robustly stabilizable via locally Lipschitz time-invariant feedback. If furthermore $c_V = 0$, then Σ is robustly practically stabilizable via locally Lipschitz time-invariant feedback.*

Proof: The proof is similar to that of Theorem 3.11. We define the function $D : \mathcal{X}\times\mathcal{U} \to I\!R$ by

$$D(x,u) \quad := \quad \sup_{w\in W(x,u)} \left[L_fV(x,u,w) + \alpha_V(x)\right] \tag{3.56}$$

which is nothing more that (3.43) evaluated at points where $c = V(x)$. One can show as in the proof of Theorem 3.11 that D is convex in u.

Also, it follows from L2 and Proposition 2.10 that D is locally Lipschitz. We next define $K : \mathcal{X} \rightsquigarrow \mathcal{U}$ as in (3.44):

$$K(x) \quad := \quad \big\{ u \in U(x) \, : \, D(x,u) < 0 \big\} \tag{3.57}$$

From L1 and Proposition 2.17 there exists a locally Lipschitz function $r : \mathrm{Dom}(K) \to I\!R_+$ such that the set-valued map $x \rightsquigarrow \overline{K(x) \cap r(x)B}$ is locally Lipschitz with nonempty compact convex values on $\mathrm{Dom}(K)$. It then follows from Proposition 2.20 that there exists a locally Lipschitz function $\mu : \mathrm{Dom}(K) \to \mathcal{U}$ such that $\mu(x) \in \overline{K(x)}$ for all $x \in \mathrm{Dom}(K)$. Fix $c > c_V$; then from (3.30) we see that $V^{-1}[c, \infty) \subset \mathrm{Dom}(K)$. It follows from L1 and Proposition 2.20 that there exists a locally Lipschitz time-invariant admissible control $u(x)$ such that $u(x) = \mu(x)$ whenever $V(x) \geq c$. Thus $D(x, u(x)) \leq 0$ whenever $V(x) \geq c$, and the rest of the proof follows the proof of Theorem 3.11. ∎

3.5 Robust stabilizability implies rclf

Using the new converse Lyapunov theorem of [95], we now give conditions under which the robust stabilizability of a system Σ implies the existence of an rclf for Σ. The new converse theorem, quoted below as Theorem 3.15, applies to systems of the form

$$\dot{x} \;=\; F(x, d) \tag{3.58}$$

where F is locally Lipschitz and $d(t)$ is an L_∞ exogenous input taking values in a nonempty compact convex subset D of a finite-dimensional Euclidean space.

Theorem 3.15 *Suppose there exists $\beta \in \mathcal{KL}$ such that for every initial condition $(x_0, t_0) \in \mathcal{X} \times I\!R$ and every L_∞ exogenous input $d(t)$ taking values in D, the solution $x(t)$ to (3.58) exists for all $t \geq t_0$ and satisfies $|x(t)| \leq \beta(|x_0|, t - t_0)$ for all $t \geq t_0$. Then there exist time-invariant functions $V \in \mathcal{V}(\mathcal{X})$ and $\alpha_V \in \mathcal{P}(\mathcal{X})$ such that*

$$\sup_{d \in D} \Big[\nabla V(x) \cdot F(x,d) \, + \, \alpha_V(x) \Big] \;\leq\; 0 \tag{3.59}$$

for all $x \in \mathcal{X}$.

We use this converse theorem to show that, in the locally Lipschitz case, the existence of an rclf is necessary for robust stabilizability. To simultaneously illustrate the necessity of the small control property, we give the result for robust *asymptotic* stabilizability:

Theorem 3.16 *Let Σ be time-invariant, let f be locally Lipschitz, and let W and Y be locally Lipschitz with nonempty compact convex values. If Σ satisfies* S1 *and is robustly asymptotically stabilizable via locally Lipschitz time-invariant feedback, then Σ satisfies* S2 *and there exists a time-invariant rclf for Σ which satisfies the small control property.*

Proof: It follows from Proposition 2.22 that there exist locally Lipschitz functions $\phi : \mathcal{X} \times \mathcal{U} \times \mathcal{W} \to \mathcal{W}$ and $\psi : \mathcal{X} \times \mathcal{Y} \to \mathcal{Y}$ such that

$$W(x,u) = \phi(x,u,B) \tag{3.60}$$
$$Y(x) = \psi(x,B) \tag{3.61}$$

for all $(x,u) \in \mathcal{X} \times \mathcal{U}$, where B denotes the respective closed unit ball in $\mathcal{W}$ or $\mathcal{Y}$. Let $u(y)$ be a locally Lipschitz admissible control which renders the solutions to Σ RGUAS. It follows from Definition 3.1 that there exists $\beta \in \mathcal{KL}$ such that for all L_∞ signals $d_1(t)$ and $d_2(t)$ taking values in the closed unit balls of $\mathcal{Y}$ and $\mathcal{W}$ (respectively) and for all initial conditions $(x_0, t_0) \in \mathcal{X} \times I\!R$, the solution $x(t)$ to

$$\dot{x} = f\big(x,\ u(\psi(x,d_1(t))),\ \phi(x,u(\psi(x,d_1(t))),d_2(t))\big) \tag{3.62}$$

exists for all $t \geq t_0$ and satisfies $|x(t)| \leq \beta(|x_0|, t - t_0)$ for all $t \geq t_0$. From Theorem 3.15 there exist time-invariant functions $V \in \mathcal{V}(\mathcal{X})$ and $\alpha_V \in \mathcal{P}(\mathcal{X})$ such that

$$\sup_{|d_1|\leq 1}\ \sup_{|d_2|\leq 1} \Big[V_x(x) \cdot f\big(x,\ u(\psi(x,d_1)),\ \phi(x,u(\psi(x,d_1)),d_2)\big) + 2\alpha_V(x)\Big] \leq 0 \tag{3.63}$$

for all $x \in \mathcal{X}$. It follows from the parameterization (3.60) that

$$\sup_{|d_1|\leq 1}\ \sup_{w \in W(x,u(\psi(x,d_1)))} \Big[L_f V(x,\ u(\psi(x,d_1)),\ w) + 2\alpha_V(x)\Big] \leq 0 \tag{3.64}$$

for all $x \in \mathcal{X}$. Now fix $c > 0$ and $y \in \mathcal{Y}$, and suppose $x \in Q(y,c)$. Then $V(x) \geq c$, and from (3.61) there exists $d_1 \in B$ such that $y = \psi(x, d_1)$. It follows from (3.64) that

$$\sup_{w \in W(x,u(y))} \Big[L_f V(x, u(y), w) + \alpha_V(x)\Big] \;\leq\; -\delta_c \tag{3.65}$$

where $\delta_c > 0$ denotes the infimum of $\alpha_V(x)$ for $V(x) \geq c$. Taking the supremum of the left-hand side of (3.65) over all $x \in Q(y,c)$, we obtain

$$\sup_{x \in Q(y,c)} \; \sup_{w \in W(x,u(y))} \Big[L_f V(x, u(y), w) + \alpha_V(x)\Big] \;<\; 0 \tag{3.66}$$

Because $u(y)$ is admissible we have $u(y) \in U(y)$, and it follows that

$$\inf_{u \in U(y)} \; \sup_{x \in Q(y,c)} \; \sup_{w \in W(x,u)} \Big[L_f V(x, u, w) + \alpha_V(x)\Big] \;<\; 0 \tag{3.67}$$

We conclude from Definition 3.8 that V is an rclf for Σ with $c_V = 0$. It is straightforward to verify from Definition 3.12 that V satisfies the small control property. ■

3.6 Summary

The rclf introduced in this chapter provides a unified framework for robust nonlinear control. It encompasses the guaranteed stability framework developed in [89, 48, 90, 23, 11] and thus also the quadratic stability framework for linear systems [22, 78]. As shown in Section 3.2, it also encompasses the ISS framework for nonlinear disturbance attenuation and is therefore related to "nonlinear L_1 control" as well as modular adaptive nonlinear control [85].

We have shown that the existence of an rclf is equivalent to robust stabilizability, but the daunting task of constructing an rclf remains a serious obstacle to the application of this theory. Moreover, we have not yet addressed the issue of how to construct a stabilizing feedback once an rclf is known. These concerns will be the topics of the remaining chapters.

Chapter 4

Inverse Optimality

We have just established that the existence of a robust control Lyapunov function (rclf) is equivalent to robust stabilizability. This result lays the foundation for the design methods to be developed in the remainder of this book. The design tasks facing us are:

- the task of constructing an rclf for a given system, and,
- the task of constructing a stabilizing feedback once an rclf is known.

Methods for constructing rclf's will be presented in Chapter 5. In this chapter we address the second task, which is much easier than the first.

Motivated by the favorable gain and phase margins of optimal control systems, we desire that our robustly stabilizing feedback control be optimal with respect to some meaningful cost functional. At the same time, we want to avoid the unwieldy task of solving the steady-state Hamilton-Jacobi-Isaacs (HJI) partial differential equation. The apparent conflict between these two goals is resolved in our solution to an *inverse optimal robust stabilization problem* in which a known rclf for a system is used to construct an optimal control law directly and explicitly, without recourse to the HJI equation.

Inverse problems in optimal control have a long history [70, 2, 106, 110, 109, 63, 50, 64, 91, 26]. The first inverse problems to be formulated and solved were for linear time-invariant systems [70, 2, 106]. These results provide a characterization of those stabilizing control gain matrices that are also optimal with respect to some quadratic cost. Inverse problems for nonlinear systems have since been considered, but with more limited success; some solutions for open-loop stable nonlinear systems are given in [63, 64, 46], and homogeneous systems are discussed in [53, 54]. Our results in this chapter extend these existing results in two significant directions. First, we pose and solve the inverse problem in the setting of a two-person zero-sum differential game, the opposing players being the control and the disturbance. Our inverse problem thus takes system uncertainties into account as we consider *robust* stabilization (cf. Section 3.1). Second, our results are valid for all robustly stabilizable systems, including open-loop unstable systems.

This chapter is organized as follows. In Section 4.1 we give a brief review of inverse optimality and present an elementary example. In Section 4.2 we define a class of *pointwise min-norm* control laws which minimize the instantaneous control effort while maintaining some desired negativity of the worst-case Lyapunov derivative. In Sections 4.3–4.5 we show that each pointwise min-norm control law is optimal with respect to a meaningful cost functional, and that the associated rclf solves the steady-state HJI equation for the corresponding differential game. In Sections 4.3–4.4 we consider cost functionals defined over an infinite horizon, in which case we require the disturbances to vanish at the equilibrium. In Section 4.5 we allow persistent disturbances and thus consider cost functionals defined over a finite horizon determined by the time required to reach a given target set.

4.1 Optimal stabilization: obstacles and benefits

Early optimal control problems, such as the minimum fuel and minimum time problems considered in the 1950's, were motivated by guidance and trajectory optimization tasks for aerospace and space applications [4, 16]. In such problems, the time interval was finite and thus stability analysis was not a priority. A change occurred when the linear optimal regulator problem was formulated and solved by Kalman [69]. His infinite time version of what became known as the linear-quadratic (LQ) optimal control problem admits a clear and elegant solution for linear systems with quadratic cost functionals. The required computational effort is that of solving a Riccati equation, a task which today is no longer burdensome.

Numerous attempts have been made to solve, at least approximately, nonlinear analogs of the linear optimal regulator problem [1, 88, 97, 63]. The demoralizing obstacle has always been the "curse of dimensionality," so named by Bellman. For systems of dimension higher than two, there are no practical ways to solve the steady-state Hamilton-Jacobi-Bellman (HJB) partial differential equation. Worse yet, the mere existence of a solution with some desirable properties cannot be ascertained *a priori*. Most approximate methods yield only local results, that is, results valid in a region whose size must be estimated by additional computations [1, 97]. These difficulties are more pronounced in the case of the steady-state Hamilton-Jacobi-Isaacs (HJI) equation, which is the optimality condition for the robust stabilization problem.

It might appear that, for nonlinear systems, optimal stabilization is not an achievable goal in feedback design. As we shall see, however, this is not entirely the case.

4.1.1 Inverse optimality, sensitivity reduction, and stability margins

Soon after he solved the linear optimal regulator problem, Kalman asked the following question [70]: when is a stabilizing linear feedback control law optimal? The answer to this question has had as much of an impact as the solution to the optimal regulator problem itself. The discovery that not every stabilizing feedback law is necessarily optimal with respect to a meaningful cost functional brought into focus those properties of optimal control systems which other stable control systems do not possess. As a point of reference, we quote the result in [70] (see also [2, 114]):

Theorem 4.1 *For a single-input linear system of the form*

$$\dot{x} = Ax + bu \tag{4.1}$$

with (A, b) controllable, an asymptotically stabilizing linear feedback control $u = -k^{\mathrm{T}}x$ minimizes the cost functional

$$J = \int_0^\infty \left[x^{\mathrm{T}}C^{\mathrm{T}}Cx + u^2\right] dt \tag{4.2}$$

for some matrix C, with (A, C) observable, if and only if

$$\left|1 + k^{\mathrm{T}}(j\omega I - A)^{-1}b\right| \geq 1 \tag{4.3}$$

for all $\omega \in I\!R$.

The condition (4.3) is the classical sensitivity reduction condition [14, 113] which assures that the closed-loop system is less sensitive to plant and controller variations than an equivalent open-loop system. Theorem 4.1 thus states that the set of all optimal controllers for cost functionals of the form (4.2) is the same as the set of all stabilizing controllers which reduce sensitivity. This set is generally a strict subset of the set of *all* stabilizing linear feedback controls. In particular, one can conclude from (4.3) that each such optimal controller guarantees not only stability but also an infinite gain margin and a 60° phase margin, robustness properties not necessarily possessed by all stable feedback systems. Therefore the *inverse optimal stabilization problem* of constructing one of these optimal controllers is more demanding than the mere stabilization problem, and also more rewarding because its solution leads to benefits beyond stability. Our goal in this chapter is to extend such inverse results to uncertain nonlinear systems.

A key ingredient in the inverse optimal stabilization problem is the choice of the class of cost functionals. In Kalman's words, "Every control law, stable or unstable, is optimal in some sense. To avoid such a trivial conclusion, the class of performance indices must be restricted [70]." The restriction reflected in Theorem 4.1 is to disallow cross terms involving x and u in the quadratic integrand of (4.2), thereby eliminating the integrand $(k^{\mathrm{T}}x + u)^2$ which trivially renders optimal *any* linear control law $-k^{\mathrm{T}}x$. Consequently, every such restricted cost functional includes a real penalty on both the state and the control variable, and the resulting optimal feedback systems inherit benefits such as desirable gain and phase margins. We will similarly restrict the class of cost functionals in our nonlinear inverse stabilization problem, and our closed-loop systems will therefore inherit the benefits of optimality surveyed in [46]. For example, each robustly stabilizing controller we obtain as the solution of our inverse problem will possess the nonlinear analog of an infinite gain margin.

We can interpret Theorem 4.1 as providing a means of calculating an optimal control without solving a Riccati equation or even specifying a cost functional: if we can find a stabilizing control which reduces sensitivity, then this control will be optimal with respect to *some* cost functional of the form (4.2). While this observation is of questionable practical value in the linear context (solving an algebraic Riccati equation is no longer a daunting task), a similar result for nonlinear systems would allow us to obtain optimal controllers without solving steady-state HJB or HJI equations. We will provide such a result in this chapter by showing that, given an rclf for a system, we can construct a robustly stabilizing control law which is optimal with respect to *some* meaningful cost functional. In Section 4.2 we generate a class of such optimal control laws using an explicit formula which involves only the rclf, the system equations, and design parameters. The control laws given by our formula are called *pointwise min-norm* control laws because they minimize the instantaneous control effort while maintaining some desired negativity of the worst-case Lyapunov derivative. In Sections 4.3–4.5 we show that each pointwise min-norm control law is optimal with respect to a meaningful cost functional, and that the associated rclf solves the steady-state HJI equation for the corresponding differential game.

4.1.2 An introductory example

We can illustrate the main objective of this chapter by means of an elementary example. Suppose we wish to robustly stabilize the system

$$\dot{x} = -x^3 + u + wx \tag{4.4}$$

where u is an unconstrained control input (equivalently, the control constraint is $U \equiv I\!R$) and w is a disturbance input with a disturbance constraint $W \equiv [-1, 1]$. An rclf for this system is simply $V(x) = x^2$, and its derivative can be made negative with the control law

$$u = x^3 - 2x \tag{4.5}$$

This particular control law is the one suggested by feedback linearization [60, 111], and it indeed renders the solutions to the system RGUAS

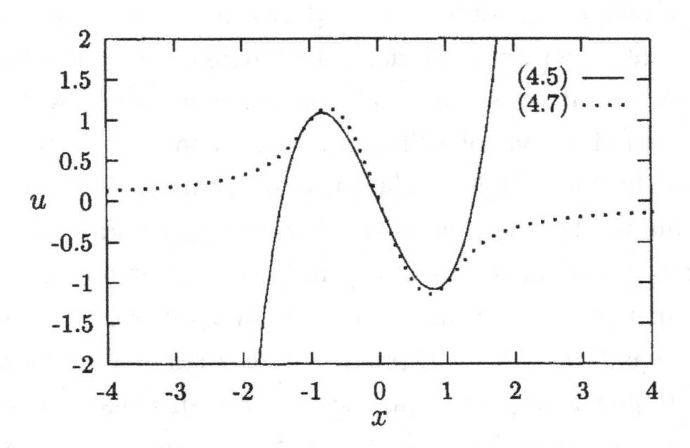

Figure 4.1: A comparison between the control laws (4.5) and (4.7)

(cf. Definition 3.1). However, it is an absurd choice because the term x^3 in (4.5) represents control effort wasted to cancel a beneficial nonlinearity. Moreover, this term is actually positive feedback which increases the risk of instability. It is easy to find a better control law for this simple system, but what we desire is a *systematic* method for choosing a reasonable control law given an rclf for a general system. One approach would be to formulate and solve an optimal robust stabilization problem with a cost functional which penalizes control effort. For the system (4.4), the quadratic cost functional

$$J \;=\; \int_0^\infty \left[x^2 + u^2\right] dt \tag{4.6}$$

results in an optimal feedback law

$$u \;=\; x^3 - x - x\sqrt{x^4 - 2x^2 + 2} \tag{4.7}$$

The control laws (4.5) and (4.7) are plotted in Figure 4.1. The optimal control law (4.7) recognizes the benefit of the nonlinearity $-x^3$ and accordingly produces little control effort for large x; moreover, this optimal control law never generates positive feedback. However, such superiority comes at the price of solving the steady-state HJI partial differential equation, a task feasible only for the simplest of nonlinear systems.

An alternative to solving the steady-state HJI equation is to calculate a pointwise min-norm control law using the explicit formula we develop in

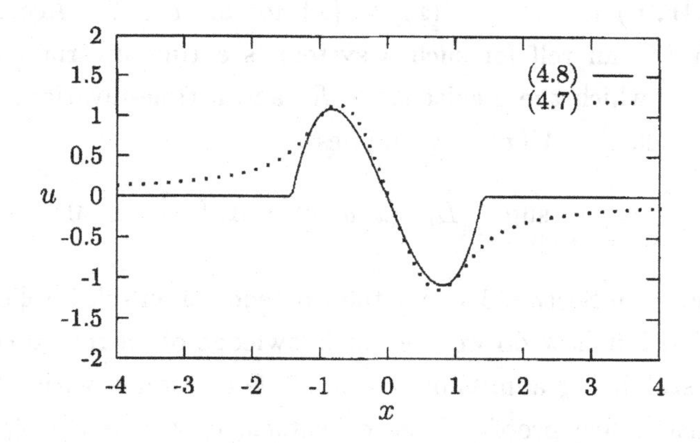

Figure 4.2: A comparison between the control laws (4.8) and (4.7)

Section 4.2. This control law will be optimal with respect to some meaningful yet *unspecified* cost functional. In Example 4.2 of Section 4.2.2, we show that one pointwise min-norm control law for the system (4.4) is

$$u = \begin{cases} x^3 - 2x & \text{when } x^2 < 2 \\ 0 & \text{when } x^2 \geq 2 \end{cases} \tag{4.8}$$

This control law is compared to the optimal control law (4.7) in Figure 4.2. We see that these two control laws, both of which are optimal with respect to a meaningful cost functional, are qualitatively the same. They both recognize the benefit of the nonlinearity $-x^3$ in (4.4) and accordingly expend little control effort for large signals; moreover, these control laws never generate positive feedback. The main difference between them lies in their synthesis: the pointwise min-norm control law (4.8) came from the simple formula provided in Section 4.2 below, while the control law (4.7) required the solution of an HJI equation. In general, the pointwise min-norm calculation is feasible but the HJI calculation is not.

4.2 Pointwise min-norm control laws

Throughout this chapter, we will consider the problem of state feedback design for time-invariant systems. Recall from Section 3.3.3 that the state feedback measurement constraint for a time-invariant system

$\Sigma = (f, U, W, Y)$ is simply $Y(x) = \{x\}$ for all $x \in \mathcal{X}$. According to Proposition 3.9, an rclf for such a system is a time-invariant function $V \in \mathcal{V}(\mathcal{X})$ for which there exist $c_V \in I\!R_+$ and a time-invariant function $\alpha_V \in \mathcal{P}(\mathcal{X})$ such that $V(x) > c_V$ implies

$$\inf_{u \in U(x)} \sup_{w \in W(x,u)} \left[L_f V(x, u, w) + \alpha_V(x) \right] < 0 \tag{4.9}$$

We have shown in Section 3.4 that the existence of an rclf implies robust stabilizability, but how do we use our knowledge of an rclf to construct a robustly stabilizing admissible control? For systems without disturbances, constructive proofs of the clf sufficiency theorem ([3], cf. Section 3.3.1), with explicit formulas for the control, are given in [132, 146, 94, 93, 92]. In these papers, different formulas are given for different *constant* control constraints (the unconstrained control case $U(x) \equiv \mathcal{U}$ is considered in [132, 146], the control constraint[1] $U(x) \equiv B$ is considered in [94], and the constraint of positive controls is considered in [93]). In this section, our goal is to provide a formula for the control which results in a *robustly* stabilizing control law for systems with disturbances, works for a general nonconstant control constraint $U(x)$, and naturally incorporates the function α_V as a design parameter. Moreover, we will show in following sections that the control laws generated by our formula, called *pointwise min-norm* control laws, are in fact optimal with respect to meaningful cost functionals.

4.2.1 General formula

As in Section 3.4, we will assume that Σ satisfies assumptions C1–C3 and A1–A2, which become as follows because Σ is time-invariant:

C1: the control constraint U is lsc with nonempty closed convex values,

C2: the disturbance constraint W is usc with nonempty compact values,

[C3: the set-valued map $y \rightsquigarrow \{x \in \mathcal{X} : y \in Y(x)\}$ is usc with compact values,]

[1] Throughout, B will denote the closed unit ball in the appropriate normed space.

A1: for each fixed $(x,w) \in \mathcal{X}\times\mathcal{W}$, the mapping $u \mapsto f(x,u,w)$ is affine, and,

A2: the disturbance constraint W is independent of the control input u, that is, $W(x,u) = W(x)$.

We are considering the state feedback case with $Y(x) = \{x\}$ for all $x \in \mathcal{X}$, so the mapping in C3 is simply $y \rightsquigarrow \{y\}$ which is trivially usc with nonempty compact values. For this reason we have put the superfluous assumption C3 in brackets. We introduce the following additional assumptions on the system Σ:

C4: the control constraint U is such that Graph(U) is closed, and,

C5: the disturbance constraint W is lsc.

Given an rclf V for Σ, we define $D : \mathcal{X}\times\mathcal{U} \to I\!R$ and $K : \mathcal{X} \rightsquigarrow \mathcal{U}$ by evaluating (3.43) and (3.44) at points where $c = V(x)$, namely,

$$D(x,u) \quad := \quad \max_{w\in W(x)} \Big[L_f V(x,u,w) + \alpha_V(x)\Big] \tag{4.10}$$

$$K(x) \quad := \quad \Big\{u \in U(x) \,:\, D(x,u) < 0\Big\} \tag{4.11}$$

It follows from C2, C5, and Proposition 2.9 that D is continuous. Moreover, we have shown in the proof of Theorem 3.11 that D is convex in u and that K is lsc with nonempty convex values on $V^{-1}(c_V,\infty)$. It follows from C1 that

$$\overline{K(x)} \quad = \quad U(x) \cap \Big\{u \in \mathcal{U} \,:\, D(x,u) \le 0\Big\} \tag{4.12}$$

for all $x \in V^{-1}(c_V,\infty)$. Each such nonempty closed convex set (4.12) has a unique element of minimum norm. We can therefore define the minimal selection of $\overline{K}$ on the set $V^{-1}(c_V,\infty)$ (cf. Section 2.4.2), namely, the function $m : V^{-1}(c_V,\infty) \to \mathcal{U}$ given by

$$m(x) \quad := \quad \arg\min\Big\{\|u\| \,:\, u \in \overline{K(x)}\Big\} \tag{4.13}$$

which is continuous by C4 and Proposition 2.19. This minimal selection can be used to generate robustly stabilizing admissible controls for Σ. Indeed, for every $c > c_V$ there is an admissible control $u(x)$ for Σ such that $u(x) = m(x)$ on $V^{-1}[c,\infty)$, and it follows from the proof of Theorem 3.11

that this admissible control renders the solutions to Σ RGUAS-Ω with a residual set $\Omega = V^{-1}[0,c]$. If furthermore V satisfies the small control property with $u_0 \equiv 0$ in S2 (cf. Section 3.4.1), then the minimal selection defines an admissible control which renders the solutions to Σ RGUAS. Such admissible controls are called *pointwise min-norm* control laws because at each point x (except possibly inside some sublevel set of V), their value is the unique element of $\mathcal{U}$ of minimum norm which satisfies the control constraint $U(x)$ and makes the worst-case Lyapunov derivative at least as negative as $-\alpha_V(x)$. These control laws naturally incorporate the function α_V as a design parameter because different choices for α_V generate different minimal selections m.

We can compute the value of a pointwise min-norm control law at any point x by solving the static minimization problem (4.13). This is a convex programming problem on the control space $\mathcal{U}$ and is completely determined by the data Σ, V, and α_V. One of the constraints in this problem depends on the function D, and the calculation of $D(x,u)$ in (4.10) for any fixed $(x,u) \in \mathcal{X} \times \mathcal{U}$ is itself a static nonlinear programming problem on the disturbance space $\mathcal{W}$. We will show in the next sections that every pointwise min-norm control law is optimal for a meaningful differential game, and therefore our formula (4.13) allows us to compute such a control law by solving a static rather than dynamic programming problem. Furthermore, as we will see in Section 4.2.2 below, this static programming problem has a simple explicit solution when Σ satisfies A1$'$, that is, when the system is jointly affine in u and w.

We end this section with a discussion on the role of the function α_V. This function represents the desired negativity of the Lyapunov derivative, and it can be adjusted to achieve a tradeoff between the control effort and the rate of convergence of the state to zero. For example, if some function α_V is a valid choice in Definition 3.8, then so is $\varepsilon\alpha_V$ for every $\varepsilon \in (0,1)$. For each such ε we then obtain a different pointwise min-norm control law for Σ. In general, smaller of values of ε will lead to smaller control magnitudes and slower convergence. Moreover, by adjusting the shape of α_V we can place more cost on some states and less on others. Thus the function α_V should be regarded as a design parameter to be adjusted according to design specifications.

4.2.2 Jointly affine systems

Suppose Σ satisfies A1$'$, that is, suppose $\dot{x} = f(x, u, w)$ can be written as

$$\dot{x} = f_0(x) + f_1(x)\,u + f_2(x)\,w \tag{4.14}$$

for continuous functions f_0, f_1, and f_2 from $\mathcal{X}$ to the appropriate space. Suppose also that the control and disturbance constraints are given by $U(x) \equiv \mathcal{U}$ and $W(x) \equiv B$, respectively. Let V be an rclf for this system; then from (4.10) we have

$$\begin{aligned} D(x,u) &= \nabla V(x) \cdot f_0(x) + \nabla V(x) \cdot f_1(x)\,u \\ &\quad + \left\| \nabla V(x) \cdot f_2(x) \right\| + \alpha_V(x) \\ &= \psi_0(x) + \psi_1^{\mathrm{T}}(x)\,u \end{aligned} \tag{4.15}$$

where

$$\psi_0(x) := \nabla V(x) \cdot f_0(x) + \left\| \nabla V(x) \cdot f_2(x) \right\| + \alpha_V(x) \tag{4.16}$$

$$\psi_1(x) := \left[\nabla V(x) \cdot f_1(x) \right]^{\mathrm{T}} \tag{4.17}$$

We substitute (4.15) into (4.11) to obtain

$$K(x) = \left\{ u \in \mathcal{U} : \psi_0(x) + \psi_1^{\mathrm{T}}(x)\,u < 0 \right\} \tag{4.18}$$

It now follows from (4.12), (4.13), and the projection theorem that

$$m(x) = \begin{cases} -\dfrac{\psi_0(x)\,\psi_1(x)}{\psi_1^{\mathrm{T}}(x)\,\psi_1(x)} & \text{when } \psi_0(x) > 0 \\ 0 & \text{when } \psi_0(x) \leq 0 \end{cases} \tag{4.19}$$

for all $x \in V^{-1}(c_V, \infty)$. This explicit formula for m depends on α_V through the function ψ_0. Note that there is never division by zero in (4.19) because the set $K(x)$ is nonempty for all $x \in V^{-1}(c_V, \infty)$. Because of the symmetry of the unit ball, this expression is also valid under control constraints of the form $U(x) = p(x)B$ for a continuous function $p : \mathcal{X} \to \mathbb{R}_+$.

Example 4.2 Let us return to the example of Section 4.1.2 and show how our formula (4.19) generates the control law

$$u = \begin{cases} x^3 - 2x & \text{when } x^2 < 2 \\ 0 & \text{when } x^2 \geq 2 \end{cases} \tag{4.20}$$

for the system

$$\dot{x} = -x^3 + u + wx \tag{4.21}$$

Here we have $f_0(x) = -x^3$, $f_1(x) = 1$, and $f_2(x) = x$. We choose $V(x) = \frac{1}{2}x^2$ so that $\nabla V(x) = x$, and we choose $\alpha_V(x) = x^2$. Then we obtain $\psi_0(x) = -x^4 + 2x^2$ and $\psi_1(x) = x$, and the formula (4.19) yields the control law (4.20).

By construction, the formula (4.19) gives the minimal selection of the set-valued map $K(x)$ in (4.18). One can obtain a different continuous selection of $K(x)$ (and thus a different robustly stabilizing admissible control) using the results of [132]. The control law defined in [132] is

$$u(x) = \begin{cases} -\dfrac{\left[\psi_0(x) + \sqrt{[\psi_0(x)]^2 + [\psi_1^{\mathrm{T}}(x)\,\psi_1(x)]^2}\right]\psi_1(x)}{\psi_1^{\mathrm{T}}(x)\,\psi_1(x)} & \text{when } \psi_1(x) \neq 0 \\ 0 & \text{when } \psi_1(x) = 0 \end{cases} \tag{4.22}$$

for $x \in V^{-1}[c, \infty)$, where $c > c_V$. This control (4.22) is clearly a selection of $K(x)$ in (4.18), and it follows from results in [132] that it is continuous (in fact, it is real-analytic in ψ_0 and ψ_1 on the set of interest). However, unlike the formula (4.19) for the minimal selection, this formula (4.22) is not easily generalized to systems which are not jointly affine in u and w or which have nontrivial control constraints.

4.2.3 Feedback linearizable systems

We now apply the formula (4.19) to the class of feedback linearizable systems. Suppose there exists a diffeomorphism $\xi = \Phi(x)$ with $\Phi(0) = 0$ which transforms our system into

$$\dot{\xi} = F\xi + G\left[\ell_0(\xi) + \ell_1(\xi)\,u + \ell_2(\xi)\,w\right] \tag{4.23}$$

where the matrix pair (F, G) is controllable and the continuous functions ℓ_0, ℓ_1, and ℓ_2 are such that $\ell_0(0) = 0$ and $\ell_1(\xi)$ is nonsingular for all $\xi \in \mathcal{X}$. Suppose also that the control and disturbance constraints are given by $U(x) \equiv \mathcal{U}$ and $W(x) \equiv B$, respectively. Choose symmetric positive definite matrices Q and R, and let P be the symmetric positive definite solution to the Riccati equation

$$F^{\mathrm{T}}P + PF - PGR^{-1}G^{\mathrm{T}}P + Q = 0 \tag{4.24}$$

As was shown in Example 3.10, the function $V(x) := \Phi(x)^{\mathrm{T}}P\Phi(x) = \xi^{\mathrm{T}}P\xi$ is an rclf for this system with $\alpha_V(x) := \varepsilon\xi^{\mathrm{T}}Q\xi$ for some $\varepsilon \in (0,1)$. The functions ψ_0 and ψ_1 in the formula (4.19) are given by

$$\begin{aligned}\psi_0(\xi) &:= \xi^{\mathrm{T}}[F^{\mathrm{T}}P + PF]\,\xi + 2\xi^{\mathrm{T}}PG\ell_0(\xi) \\ &\quad + \left\|2\xi^{\mathrm{T}}PG\ell_2(\xi)\right\| + \varepsilon\xi^{\mathrm{T}}Q\xi \qquad (4.25)\\ \psi_1(\xi) &:= 2\,\ell_1^{\mathrm{T}}(\xi)\,G^{\mathrm{T}}P\xi \qquad (4.26)\end{aligned}$$

and the pointwise min-norm control law is

$$u(x) = m(x) = \begin{cases} -\dfrac{\psi_0(\xi)\,\ell_1^{\mathrm{T}}(\xi)G^{\mathrm{T}}P\xi}{2\xi^{\mathrm{T}}PG\,\ell_1(\xi)\,\ell_1^{\mathrm{T}}(\xi)G^{\mathrm{T}}P\xi} & \text{when } \psi_0(\xi) > 0 \\ 0 & \text{when } \psi_0(\xi) \le 0 \end{cases} \tag{4.27}$$

The design parameters in the derivation of this control law are ε, Q, and R. Moreover, one can verify that the rclf V satisfies the small control property and that this control law (4.27) is well-defined and continuous for *all* $x \in \mathcal{X}$, not just those outside some sublevel set of V.

When there is no disturbance (that is, when $\ell_2 \equiv 0$), we can compare the pointwise min-norm control law (4.27) with the feedback linearizing control law given by

$$u(x) = \left[\ell_1(\xi)\right]^{-1}\left[-\,\ell_0(\xi) - R^{-1}G^{\mathrm{T}}P\xi\right] \tag{4.28}$$

Although both control laws (4.27) and (4.28) globally asymptotically stabilize the system (4.23), the pointwise min-norm control law (4.27) is optimal with respect to a meaningful cost functional (as we show in the next sections) whereas, in general, the feedback linearizing control law (4.28) is not. The potential performance advantage of (4.27) over (4.28) was illustrated in the comparison of the control laws (4.8) and (4.5) in Section 4.1.2. Also, there may be a computational advantage of the pointwise min-norm control law because it does not require the inversion of the matrix function $\ell_1(\xi)$.

4.3 Inverse optimal robust stabilization

Our goal in the remainder of this chapter is to show that every pointwise min-norm control law is optimal with respect to a meaningful cost functional. We accomplish this by showing that every rclf solves the steady-state HJI equation associated with a meaningful differential game. These results represent a solution to an inverse optimal robust stabilization problem for nonlinear systems with disturbances. As a consequence of these results, we can use the formulas (4.13) and (4.19) to compute optimal robustly stabilizing control laws *without* solving the HJI equation for the value function, provided that an rclf is known.

In this section we consider the case in which the rclf satisfies the small control property (cf. Section 3.4.1). In this case the disturbance vanishes at the point $x = 0$, and it is possible to achieve *asymptotic* convergence to this equilibrium point. We can therefore consider cost functionals defined over an infinite horizon. In Section 4.5 we remove the assumption that the rclf satisfies the small control property. In this case we allow persistent disturbances and must therefore consider cost functionals defined over a finite horizon determined by the time required to reach a given target set.

4.3.1 A preliminary result

We introduce the concept of inverse optimality by examining a system with no uncertainty and no control constraints. Consider

$$\dot{x} \;=\; f_0(x) \,+\, f_1(x)\,u \tag{4.29}$$

where f_0 and f_1 are continuous vector fields on $\mathcal{X}$ and u is a single unconstrained control input. Let V be a clf for this system, and define scalar functions $\psi_0(x) := \nabla V(x) \cdot f_0(x)$ and $\psi_1(x) := \nabla V(x) \cdot f_1(x)$. It follows from the definition of a clf that whenever $x \neq 0$ and $\psi_1(x) = 0$ we have $\psi_0(x) < 0$. Therefore the scalar functions

$$q(x) \;:=\; [\psi_1(x)]^2 + \sqrt{[\psi_0(x)]^2 + [\psi_1(x)]^4} \tag{4.30}$$

$$r(x) \;:=\; \frac{\frac{1}{4}[\psi_1(x)]^2}{[\psi_1(x)]^2 + \psi_0(x) + \sqrt{[\psi_0(x)]^2 + [\psi_1(x)]^4}} \tag{4.31}$$

are strictly positive for $x \neq 0$ (the apparent singularity in $r(x)$ at points where $\psi_1(x) = 0$ is removable [132]). Consider now the cost functional

$$J(u; x_0) \quad := \quad \int_0^\infty \left[q(x) \,+\, r(x)\, u^2\right] dt \tag{4.32}$$

The steady-state Hamilton-Jacobi-Bellman (HJB) equation associated with the system (4.29) and the cost functional (4.32) is

$$0 \equiv \min_u \left[q(x) \,+\, r(x)\, u^2 \,+\, \nabla V(x) \cdot f_0(x) \,+\, \nabla V(x) \cdot f_1(x)\, u\right] \tag{4.33}$$

One can easily verify that the clf V satisfies this HJB equation, and we conclude that V is the value function for this optimal control problem. The optimal feedback control is given by $u(x) = -\frac{1}{2}\psi_1(x)/r(x)$ which is similar to the control (4.22) proposed in [132].[2]

To summarize, we have shown that every clf for the system (4.29) is also the value function associated with the cost (4.32), and that furthermore the feedback control $u(x) = -\frac{1}{2}\psi_1(x)/r(x)$ minimizes this cost. Our goal is to extend this simple inverse optimality result to systems with uncertainties and control constraints.

4.3.2 A differential game formulation

As in Section 3.4.1, we will assume that Σ satisfies assumption S2, which becomes as follows because Σ is time-invariant and satisfies A2:

S2: there exists $u_0 \in U(0)$ such that $f(0, u_0, w) = 0$ for all $w \in W(0)$.

We assume that the small control property is satisfied with $u_0 = 0$ in S2. As in Section 4.2, we assume that the time-invariant state feedback system Σ satisfies C1–C5 and A1–A2, plus the following assumption on the control constraint:

C6: there exists a C^0 function $\pi : \mathcal{X} \to (0, \infty)$ such that $\pi(x)B \subset U(x)$ for all $x \in \mathcal{X}$.

One can show that, under assumption C1, assumption C6 is equivalent to the assumption that $0 \in \operatorname{int} U(x)$ for all $x \in \mathcal{X}$. Our cost functionals will

[2]This optimal control will not be continuous at $x = 0$ unless V satisfies the small control property.

be characterized by functions $q : \mathcal{X} \to I\!R_+$ and $r : \mathcal{X} \times \mathcal{U} \to I\!R_+$ which satisfy the following specifications:

J1: $q \in \mathcal{P}(\mathcal{X})$ (in particular, $q(x) \geq \chi(\|x\|)$ for some $\chi \in \mathcal{K}$), and,

J2: r is continuous and, for each fixed $x \in \mathcal{X}$, $r(x,u) = \gamma_x(\|u\|_x)$ for some convex class $\mathcal{K}$ function γ_x and some norm $\|\cdot\|_x$.

Using such a pair (q,r), we form a two-person zero-sum differential game $\mathcal{G}(q,r)$ by considering a cost functional J parameterized by the initial condition $x_0 \in \mathcal{X}$. In this game, the control tries to minimize J and the disturbance tries to maximize J. Given a control u, a disturbance w, and an initial condition $x_0 \in \mathcal{X}$ for Σ, we define the cost

$$J(u,w,x;x_0) \quad := \quad \int_0^\infty \Big[q(x) + r(x,u)\Big]\,dt \tag{4.34}$$

where the integration is taken along the solution $x(t)$ to Σ starting from the initial condition x_0. Because such solutions are not necessarily unique, we have included in our notation $J(u,w,x;x_0)$ the dependence on the particular state trajectory x along which we integrate. If the solution $x(t)$ cannot be extended for all $t \geq 0$, then we set $J(u,w,x;x_0) := \infty$. Also, because q is bounded from below by a class $\mathcal{K}$ function of the norm, $J < \infty$ implies $x(t) \to 0$ as $t \to \infty$. We define the *upper value function* $\bar{J} : \mathcal{X} \to I\!R_+ \cup \{\infty\}$ of the game by the equation

$$\bar{J}(x_0) \quad := \quad \inf_u \sup_w \sup_x J(u,w,x;x_0) \tag{4.35}$$

The first supremum is taken over all solutions $x(t)$ to Σ starting from x_0 (this supremum is superfluous if solutions are unique), the second supremum is taken over all *admissible* disturbances w for Σ, and the infimum is taken over *all* controls u for Σ (not just the admissible controls). A time-invariant admissible control for Σ which achieves a finite infimum in (4.35) for every $x_0 \in \mathcal{X}$ is said to be *optimal for* $\mathcal{G}(q,r)$. Such an optimal control law minimizes the worst-case cost for every initial condition. Also, because $J < \infty$ only if $x(t) \to 0$ as $t \to \infty$, every optimal control law drives the state (robustly) asymptotically to zero from any initial condition.

Let V be a time-invariant rclf for Σ which satisfies the small control property, and let u^* be a pointwise min-norm control law associated with V. Theorem 4.3 below states that there exists a pair (q, r) satisfying J1 and J2 such that u^* is optimal for $\mathcal{G}(q, r)$ with V being the corresponding upper value function. In simple terms, every pointwise min-norm control law is optimal and every rclf is an upper value function. This theorem is only of interest when such a game $\mathcal{G}(q, r)$ is meaningful, and we claim that this is indeed the case. First of all, it follows from J1 and J2 that the integrand in (4.34) is bounded below by a class $\mathcal{K}$ function of $\|x\|$; thus $J < \infty$ only if the objective of driving the state to zero is achieved. Furthermore, for each fixed $x \in \mathcal{X}$ the integrand is a convex function of u with a global minimum at the point $u = 0$ (there are no other global or local minima); thus there is always a higher penalty for values of u farther away from zero. Integrands satisfying J1 and J2 are a generalization of the familiar quadratic integrand $x^{\mathrm{T}}Qx + u^{\mathrm{T}}Ru$ for symmetric positive definite matrices Q and R. Next, the control law u^* is optimal with respect to *all* controls for Σ, not just the admissible ones which satisfy the control constraint; in other words, we do not allow the possibility of reducing the guaranteed cost by relaxing the control constraint. Finally, the disturbance w is given two advantages in the game consistent with the goal of robust stabilization: there is no direct cost on w in (4.34), and w is allowed to base its strategy on knowledge of the strategy of the control u (we consider the *upper* value of the game).

4.3.3 Main theorem

To prove optimality rather than suboptimality, we must assume that the effect of any *worst-case disturbance* (one that maximizes the Lyapunov derivative) can be approximated arbitrarily closely by admissible disturbances. A worst-case disturbance is a function $w^* : \mathcal{X} \times \mathcal{U} \to \mathcal{W}$ such that

$$L_f V(x, u, w^*(x, u)) = \max_{w \in W(x)} L_f V(x, u, w) \tag{4.36}$$

and furthermore $w^*(x, u) \in W(x)$ for all $(x, u) \in \mathcal{X} \times \mathcal{U}$. Such a function w^* may be discontinuous and thus not be an admissible disturbance for Σ. Our approximation condition is

AC: for every $x_0 \in \mathcal{X}$, every control u for Σ, and every $\Delta > 0$ there exist an admissible disturbance w_Δ for Σ and a solution $x_\Delta(t)$

to Σ starting from x_0 such that either $J(u, w_\Delta, x_\Delta; x_0) = \infty$ or for every $T \geq 0$ we have

$$\int_0^T L_f V(x_\Delta, u, w_\Delta)\, dt \;\geq\; \int_0^T \left[\max_{w \in W(x_\Delta)} L_f V(x_\Delta, u, w) \right] dt \;-\; \Delta$$

If this condition is not true for our system, then we can only prove sub-optimality. We are now ready to state the main result of this chapter:

Theorem 4.3 *Let Σ be a time-invariant state feedback system which satisfies* C1–C6 *and* A1–A2. *Let V be a time-invariant rclf for Σ which satisfies the small control property with $u_0 = 0$ in* S2, *and let u^* be a pointwise min-norm control law associated with V. Then there exists a pair (q, r) satisfying* J1–J2 *such that $J(u^*, w, x; x_0) \leq V(x_0)$ for every $x_0 \in \mathcal{X}$, every admissible disturbance w for Σ, and every solution $x(t)$ to Σ starting from x_0. If furthermore* AC *is true, then $\bar{J}(x_0) = V(x_0)$ for all $x_0 \in \mathcal{X}$ which means u^* is optimal for $\mathcal{G}(q, r)$.*

The proof of this theorem involves several steps and will be presented in the next section. The main idea is to construct the functions q and r in such a way that V satisfies the steady-state HJI equation

$$0 \;=\; \min_{u \in \mathcal{U}} \; \max_{w \in W(x)} \left[q(x) + r(x,u) + L_f V(x,u,w) \right] \qquad (4.37)$$

for all $x \in \mathcal{X}$. We accomplish this by constructing a continuous function r which satisfies J2 and which has two additional properties: first, r must be such that the function q defined by

$$q(x) \;:=\; -\, r(x, u^*(x)) \;-\; \max_{w \in W(x)} L_f V(x, u^*(x), w) \qquad (4.38)$$

satisfies J1, and second, r must be such that the equality

$$\min_{u \in \mathcal{U}} \; \max_{w \in W(x)} \left[r(x,u) + L_f V(x,u,w) \right]$$
$$=\; r(x, u^*(x)) + \max_{w \in W(x)} L_f V(x, u^*(x), w) \qquad (4.39)$$

holds for all $x \in \mathcal{X}$. Once we find such a function r, the HJI equation (4.37) will follow from (4.38) and (4.39).

4.4 Proof of the main theorem

The proof of Theorem 4.3 is long and technical. Some terminology and technical lemmas are given in Section 4.4.1, the construction of the function r is given in Section 4.4.2, the key optimality condition (4.39) is proved in Section 4.4.3, and the proof of optimality itself is given in Section 4.4.4.

4.4.1 Terminology and technical lemmas

We use the following terminology throughout this proof. For a subset C of a metric space Z, we let $\partial C = \overline{C} \backslash \operatorname{int} C$ denote the boundary of C. For a sequence $\{C_i\}$ of such subsets, we let $\operatorname{Limsup} C_i$ denote the set $\{z \in Z : \liminf d(z, C_i) = 0\}$ of cluster points of sequences $z_i \in C_i$. For a subset D of a normed space, we let $\|D\|$ denote the number $\sup\{\|x\| : x \in D\}$. If D is a convex subset of a Hilbert space and $x \in D$, we let $N_D(x) \subset X$ denote the normal cone to D at x. For a nonzero member x of a normed space, we let $\operatorname{sgn}(x)$ denote the vector $x/\|x\|$. For a convex function $h : X \to I\!R$ we let $\partial h : X \rightsquigarrow X$ denote the subdifferential of h, that is,

$$\partial h(x) := \left\{ w \in \mathcal{X} \ : \ h(\xi) \geq h(x) + \langle w, \xi - x \rangle \quad \forall \xi \in \mathcal{X} \right\} \tag{4.40}$$

We will use the following simple lemmas about subdifferentials:

Lemma 4.4 *Let Z be a metric space, let Y be a Hilbert space of finite dimension, and let $h : Z \times Y \to I\!R$ be continuous. If the mapping $y \mapsto h(z, y)$ is convex for every fixed $z \in Z$, then the partial subdifferential $\partial_y h : Z \times Y \rightsquigarrow Y$ is usc on $Z \times Y$ with nonempty, convex, compact values.*

Proof: It follows from [122, Theorem 23.4] that $\partial_y h$ has nonempty, convex, compact values. To prove upper semicontinuity, fix $(z_0, y_0) \in Z \times Y$ and let $\{(z_i, y_i)\} \in Z \times Y$ converge to (z_0, y_0). Let $\{w_i\} \in Y$ be any sequence such that $w_i \in \partial_y h(z_i, y_i)$ for all $i \geq 1$. It follows from [79, Theorem II.2.2] that we need only show that $\{w_i\}$ has a subsequence converging to some $w_0 \in \partial_y h(z_0, y_0)$. First we show that $\{w_i\}$ is bounded. From the definition of $\partial_y h$ we have $\langle w_i, v - y_i \rangle \leq h(z_i, v) - h(z_i, y_i)$ for all $v \in Y$ and all $i \geq 1$. It follows from the continuity of h and the compactness of the unit sphere in Y that there exists $M \in I\!R$ such that $\langle w_i, e\rangle \leq M$ for all unit vectors $e \in Y$ and all $i \geq 1$, and we conclude that $\{w_i\}$ is bounded. Let $\{w_{i_j}\}$ be a convergent subsequence of $\{w_i\}$ with limit $w_0 \in Y$. Fix $v \in Y$; then for all $j \geq 1$ we have $\langle w_{i_j}, v - y_{i_j}\rangle \leq h(z_{i_j}, v) - h(z_{i_j}, y_{i_j})$. It follows from the continuity of the inner product and h that $\langle w_0, v - y_0\rangle \leq h(z_0, v) - h(z_0, y_0)$. This holds for all $v \in Y$, and so $w_0 \in \partial_y h(z_0, y_0)$ as desired. ∎

Lemma 4.5 *Let Y be a finite-dimensional Hilbert space, and let h be a sublinear function on Y. For each $y \in Y$, let $E_y = \{\xi \in Y : h(\xi) \leq h(y)\}$ denote the sublevel set of h at y and let $N_E(y)$ denote the normal cone to E_y at y. Then $\partial h(y) \neq \emptyset$ for all $y \in Y$, and for every $y \in Y$ such that $h(y) \neq 0$ we have*

$$\partial h(y) \quad = \quad N_E(y) \cap \left[p + \{y\}^\perp\right] \tag{4.41}$$

where p is any member of $\partial h(y)$.

Proof: It follows from [122, Theorem 23.4] that $\partial h(y)$ is nonempty for all $y \in Y$. Let $y \in Y$ be such that $h(y) \neq 0$, and let $p \in \partial h(y)$. We first show that $\langle w, y\rangle = h(y)$ for all $w \in \partial h(y)$. Indeed, $w \in \partial h(y)$ implies $h(\xi) \geq h(y) + \langle w, \xi - y\rangle$ for all $\xi \in Y$. Taking $\xi = 0$ we obtain $0 \geq h(y) + \langle w, -y\rangle$ which means $\langle w, y\rangle \geq h(y)$, and taking $\xi = 2y$ we obtain $h(2y) = 2h(y) \geq h(y) + \langle w, y\rangle$ which means $\langle w, y\rangle \leq h(y)$. It then follows that for any $w \in \partial h(y)$ we have $\langle w - p, y\rangle = h(y) - h(y) = 0$ which means $w \in [p + \{y\}^\perp]$. Now because h is sublinear and $h(y) \neq 0$ we know that y does not minimize h, and it follows from [122, Corollary 23.7.1] that $\partial h(y) \subset N_E(y)$. We have thus shown that $\partial h(y) \subset N_E(y) \cap [p + \{y\}^\perp]$.

Next suppose $w \in N_E(y) \cap [p + \{y\}^\perp]$. Because $\langle p, y\rangle = h(y) \neq 0$ we have $0 \notin [p + \{y\}^\perp]$ which means $w \neq 0$. It then follows from [122, Corollary 23.7.1] that $\lambda w \in \partial h(y)$ for some $\lambda > 0$. From above we have $\lambda w \in [p + \{y\}^\perp]$, and it follows that $(\lambda w - w) \in \{y\}^\perp$ and so $(\lambda - 1)\langle w, y\rangle = 0$. Now $\langle w, y\rangle = \langle \lambda w, y\rangle / \lambda = h(y)/\lambda \neq 0$, and thus $\lambda = 1$. Hence $w = \lambda w \in \partial h(y)$ which means $N_E(y) \cap [p + \{y\}^\perp] \subset \partial h(y)$. ∎

4.4.2 Construction of the function r

We begin with a careful construction of the function r in (4.34) according to the requirements in J2, (4.38), and (4.39) above. This function r will be of the form

$$r(x, u) \quad := \quad \gamma(x, \sigma(x, u)) \tag{4.42}$$

where γ and σ are continuous functions, $\gamma(x, \cdot)$ is a convex class $\mathcal{K}$ function for each $x \in \mathcal{X}$, and $\sigma(x, \cdot)$ is a norm on $\mathcal{U}$ for each $x \in \mathcal{X}$.

Let us define a set-valued map $C : \mathcal{X} \rightsquigarrow \mathcal{U}$ as follows:

$$C(x) \;=\; \mathrm{co}\Big(\big[\pi(x)B\big] \cup \big\{u^*(x),\, -u^*(x)\big\}\Big) \tag{4.43}$$

The values of C are compact and convex, $0 \in \mathrm{int}\, C(x)$, and $C(x) = -C(x)$ for each $x \in \mathcal{X}$. Furthermore, it follows from [79, Theorem II.2.7 and Theorem II.2.10] that C is continuous on $\mathcal{X}$. Associated with C is the Minkowski distance functional $\sigma : \mathcal{X} \times \mathcal{U} \to I\!R_+$ defined as follows:

$$\sigma(x, u) \quad := \quad \inf\big\{\lambda \geq 0 \,:\, u \in \lambda C(x)\big\} \tag{4.44}$$

Proposition 4.6 *For each $x \in \mathcal{X}$, $\sigma(x, \cdot)$ is a norm on $\mathcal{U}$. Also, σ is continuous.*

Proof: The first statement follows from [122, Theorem 15.2], and we have left to prove that σ is continuous. Fix $(x_0, u_0) \in \mathcal{X} \times \mathcal{U}$ and let $\{(x_i, u_i)\} \in \mathcal{X} \times \mathcal{U}$ be a sequence converging to (x_0, u_0). Define $\sigma_i := \sigma(x_i, u_i)$ and $\sigma_0 := \sigma(x_0, u_0)$; we need to show that $\sigma_i \to \sigma_0$ or equivalently that every subsequence of $\{\sigma_i\}$ has in turn a subsequence which converges to σ_0. Let $\{\sigma_{i_j}\}$ be a subsequence of $\{\sigma_i\}$, and define $C_j := C(x_{i_j})$ and $C_0 := C(x_0)$. It follows from (4.44) that there exists a sequence $\{w_j\} \in \mathcal{U}$ such that $w_j \in \partial C_j$ and $u_{i_j} = \sigma_{i_j} w_j$ for all $j \geq 1$. It then follows from [79, Theorem II.2.2] that the sequence $\{w_j\}$ has a subsequence $\{w_{j_k}\}$ which converges to some $w_0 \in C_0$. We claim that $w_0 \in \partial C_0$. Indeed, because $w_{j_k} \in \partial C_{j_k}$ we know there exists a sequence of unit vectors $\{v_k\} \in \mathcal{U}$ such that $v_k \in N_{C_{j_k}}(w_{j_k})$ for all $k \geq 1$. The unit sphere in $\mathcal{U}$ is compact, and thus $\mathrm{Limsup}\, N_{C_{j_k}}(w_{j_k})$ contains a unit vector. It then follows from [7, Corollary 7.6.5] that $N_{C_0}(w_0) \neq \{0\}$ which means $w_0 \in \partial C_0$. We can now show that $\sigma_{i_{j_k}} \to \sigma_0$. First suppose $u_0 = 0$;

then $\sigma_0 = 0$ and $u_{i_{j_k}} = \sigma_{i_{j_k}} w_{j_k} \to 0$, and because $w_{j_k} \to w_0 \neq 0$ it follows that $\sigma_{i_{j_k}} \to 0 = \sigma_0$. Next suppose $u_0 \neq 0$. By the continuity of the inner product and the continuity of sgn($\cdot$) away from zero, we have $\langle \text{sgn}(w_0), \text{sgn}(u_{i_{j_k}}) \rangle = \langle \text{sgn}(w_0), \text{sgn}(w_{j_k}) \rangle \to \langle \text{sgn}(w_0), \text{sgn}(w_0) \rangle = 1$ and also $\langle \text{sgn}(w_0), \text{sgn}(u_{i_{j_k}}) \rangle \to \langle \text{sgn}(w_0), \text{sgn}(u_0) \rangle$, and it follows that $u_0 = \lambda w_0$ for some $\lambda > 0$. Because $w_0 \in \partial C_0$ we have $\lambda = \sigma_0$, and thus $u_{i_{j_k}} \to u_0$ implies $\sigma_{i_{j_k}} w_{j_k} \to \sigma_0 w_0$. Now because $w_{j_k} \to w_0 \neq 0$ we have $\sigma_{i_{j_k}} \to \sigma_0$ as desired. ■

Recall from (4.10) that we defined the function $D : \mathcal{X} \times \mathcal{U} \to I\!R$ by

$$D(x,u) \quad := \quad \max_{w \in W(x)} \Big[L_f V(x,u,w) + \alpha_V(x) \Big] \tag{4.45}$$

and showed that D is continuous in (x, u) and convex in u. We can therefore define a set-valued map $\Pi : \mathcal{X} \rightsquigarrow \mathcal{U}$ as follows:

$$\begin{aligned} \Pi(x) \quad &:= \quad \partial_u D(x, u^*(x)) &(4.46)\\ &= \quad \Big\{ w \in \mathcal{U} \,:\, D(x,u) \geq D(x,u^*(x)) + \langle w,\, u - u^*(x) \rangle \quad \forall u \in \mathcal{U} \Big\} \end{aligned}$$

Thus $\Pi(x)$ is the partial subdifferential (with respect to u) of the function D in (4.45), evaluated at $(x, u^*(x))$. It follows from Lemma 4.4 and [79, Corollary II.2.1] that Π is usc on $\mathcal{X}$ and has nonempty, convex, compact values. The next two propositions follow from the pointwise min-norm property of u^*:

Proposition 4.7 *If $x \in \mathcal{X}$ is such that $u^*(x) \neq 0$, then $D(x, u^*(x)) = 0$ and $0 \notin \Pi(x)$.*

Proof: By assumption, u^* is the minimal selection associated with V. Fix $x \in \mathcal{X}$ and suppose $u^*(x) \neq 0$. It follows from C6, (4.12), and (4.13) that $D(x, 0) > 0$. Because D is continuous and $U(x)$ is convex, the intermediate value theorem gives $D(x, u^*(x)) = 0$. Suppose $0 \in \Pi(x)$; then from (4.46) we have $0 = D(x, u^*(x)) \leq D(x, u)$ for all $u \in \mathcal{U}$. However, $u^*(x) \neq 0$ implies $x \neq 0$ (Definition 3.12), which in turn implies that $K(x)$ in (4.11) is nonempty, a contradiction. ■

Proposition 4.8 *There exist continuous functions $\mu, \nu : \mathcal{X} \to I\!R_+$ such that for all $x \in \mathcal{X}$ we have $\mu(x) \leq d(0, \Pi(x))$ and $\|\Pi(x)\| < \nu(x)$, with μ having the additional property that $\mu(x) > 0$ if and only if $u^*(x) \neq 0$.*

Proof: Because Π is usc with nonempty compact values, it follows from Proposition 2.9 that the map $x \mapsto \|\Pi(x)\|$ is usc and the map $x \mapsto d(0, \Pi(x))$ is lsc. The existence of ν then follows from [74, Problem 5X]. Let $G \subset \mathcal{X}$ denote the open set $G = \{x \in \mathcal{X} : u^*(x) \neq 0\}$; it then follows from Proposition 4.7 and [74, Problem 5X] that there exists a continuous function $\mu_0 : G \to I\!R_+$ such that $0 < \mu_0(x) \leq d(0, \Pi(x))$ for all $x \in G$. We define μ by setting $\mu(x) := 0$ when $u^*(x) = 0$ and otherwise

$$\mu(x) \quad := \quad \frac{\|u^*(x)\|}{1 + \|u^*(x)\| + \mu_0(x)}\, \mu_0(x) \tag{4.47}$$

This function μ has the desired properties. ■

These functions μ and ν should be regarded as continuous "lower" and "upper" bounds (respectively) on the set-valued map Π. We will now use them to construct the function r in the cost functional (4.34). Let $\alpha_0 \in \mathcal{P}(\mathcal{X})$ be time-invariant and such that $(\alpha_V - \alpha_0) \in \mathcal{P}(\mathcal{X})$; for example, take $\alpha_0 := \varepsilon \alpha_V$ for some $\varepsilon \in (0, 1)$. We construct continuous functions $a, b : \mathcal{X} \to I\!R_+$ as follows:

$$a(x) \quad := \quad \begin{cases} \min\left\{\mu(x)\pi(x),\ \frac{\alpha_0(x)}{\sigma(x, u^*(x))}\right\} & \text{when } u^*(x) \neq 0 \\ 0 & \text{when } u^*(x) = 0 \end{cases} \tag{4.48}$$

$$b(x) \quad := \quad \nu(x) \max\left\{\|u^*(x)\|,\ \pi(x)\right\} \tag{4.49}$$

It follows from Propositions 4.6 and 4.8 that a is continuous with $a(x) > 0$ whenever $u^*(x) \neq 0$. Also, we see from Proposition 4.8 that b is continuous with $b(x) > 0$ for all $x \in \mathcal{X}$. Furthermore, for all $x \in \mathcal{X}$ we have $0 \leq a(x) \leq \mu(x)\pi(x) < \nu(x)\pi(x) \leq b(x)$. We use a and b to define a function $\gamma : \mathcal{X} \times I\!R_+ \to I\!R_+$ as follows:

$$\gamma(x, s) \quad := \quad \begin{cases} a(x)s & \\ & \text{when } 0 \leq s \leq \sigma(x, u^*(x)) \\ b(x)s + \sigma(x, u^*(x))\left[a(x) - b(x)\right] & \\ & \text{when } \sigma(x, u^*(x)) < s \end{cases} \tag{4.50}$$

It follows from the continuity of a, b, σ, and u^* that γ is continuous. Now fix $x \in \mathcal{X}$ and consider the function $\gamma_x : I\!R_+ \to I\!R_+$ given by $\gamma_x(s) = \gamma(x, s)$. Suppose $u^*(x) = 0$; then for all $s \geq 0$ we have $\gamma_x(s) = b(x)s$ with $b(x) > 0$, and thus γ_x is a (linear) convex class $\mathcal{K}$ function. Now suppose

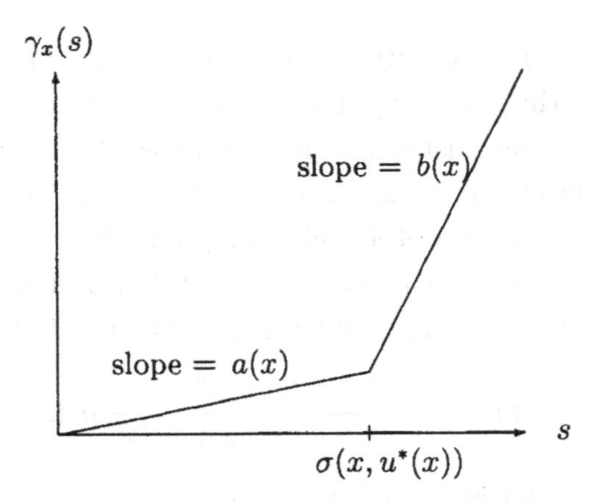

Figure 4.3: The function $\gamma_x(s)$ when $u^*(x) \neq 0$

$u^*(x) \neq 0$; then $0 < a(x) < b(x)$ and we see from (4.50) that γ_x is a (piecewise linear) convex class $\mathcal{K}$ function (see Figure 4.3). Our choice for γ is not unique; in fact, redefining γ for $s \geq \sigma(x, u^*(x)) + 1$ to be anything preserving convexity and continuity (for example replacing the linear growth with quadratic growth) will not alter our results.

We now choose the functions q and r in (4.34) as follows:

$$r(x,u) \quad := \quad \gamma(x, \sigma(x,u)) \tag{4.51}$$

$$q(x) \quad := \quad -r(x, u^*(x)) - D(x, u^*(x)) + \alpha_V(x) \tag{4.52}$$

for all $(x,u) \in \mathcal{X} \times \mathcal{U}$. Clearly q and r are continuous, and it follows from Proposition 4.6 and the properties of γ discussed above that r satisfies J2. We next show that q satisfies J1. We see from (4.50) and (4.48) that $r(x, u^*(x)) = a(x)\,\sigma(x, u^*(x)) \leq \alpha_0(x)$ for all $x \in \mathcal{X}$. Now $D(x, u^*(x)) \leq 0$ for all $x \in \mathcal{X}$, and it follows from (4.52) that $q(x) \geq \alpha_V(x) - \alpha_0(x)$ for all $x \in \mathcal{X}$. By inspection we have $q(0) = 0$, and because $(\alpha_V - \alpha_0) \in \mathcal{P}(\mathcal{X})$ it follows that $q \in \mathcal{P}(\mathcal{X})$.

4.4.3 Proof of the key proposition

We now show that the function r satisfies the key optimality condition we presented in (4.39).

Proposition 4.9 *The following is true for all $x \in \mathcal{X}$:*

$$\min_{u \in \mathcal{U}} \Big[r(x,u) + D(x,u) \Big] \quad = \quad r(x, u^*(x)) + D(x, u^*(x)) \tag{4.53}$$

Proof: We fix $x \in \mathcal{X}$. Because $r + D$ is convex in u, (4.53) is true if and only if $0 \in \partial_u(r + D)(x, u^*(x))$, where ∂_u denotes the partial subdifferential with respect to u. From [122, Theorem 23.8] and (4.46) we see that this condition is equivalent to the condition $0 \in \partial_u r(x, u^*(x)) + \Pi(x)$.

First suppose $u^*(x) = 0$. Then from (4.43) we have $C(x) = \pi(x)B$, and it follows from (4.44) that $\sigma(x, u) = \|u\|/\pi(x)$ for all $u \in \mathcal{U}$. Thus $\partial_u\sigma(x, 0) = (1/\pi(x))B$. Now we have $\sigma(x, u^*(x)) = \sigma(x, 0) = 0$ which means $\gamma(x, s) = b(x)s$ for all $s \geq 0$, and it follows from (4.51) and (4.49) that $\partial_u r(x, u^*(x)) = b(x)\,\partial_u\sigma(x, u^*(x)) = (b(x)/\pi(x))B = \nu(x)B$. Now $\|\Pi(x)\| < \nu(x)$ and so $0 \in \nu(x)B + \Pi(x) = \partial_u r(x, u^*(x)) + \Pi(x)$ as desired.

Next suppose $u^*(x) \neq 0$. We need to compute $\partial_u r(x, u^*(x))$. Now $\sigma(x, u^*(x)) > 0$ and it follows from (4.50) (see Figure 4.3) that

$$\partial_s\gamma\big(x, \sigma(x, u^*(x))\big) \;=\; [a(x),\, b(x)] \tag{4.54}$$

where ∂_s denotes the partial subdifferential with respect to s. We next compute $\partial_u\sigma(x, u^*(x))$. Let $E := \{u \in \mathcal{U} : \sigma(x, u) \leq \sigma(x, u^*(x))\}$ denote the sublevel set of σ at $u^*(x)$, and let N_E denote the normal cone to E at $u^*(x)$. From (4.43) and (4.44) we have $E \subset \|u^*(x)\|B$, and it follows that $u^*(x) \in N_E$. It then follows from [122, Corollary 23.7.1] that there exists $\lambda > 0$ such that $\lambda\,\mathrm{sgn}(u^*(x)) \in \partial_u\sigma(x, u^*(x))$. We now calculate the value of λ. From the definition of the subdifferential, it follows that $\sigma(x, \xi) \geq \sigma(x, u^*(x)) + \langle\lambda\,\mathrm{sgn}(u^*(x)),\, \xi - u^*(x)\rangle$ for all $\xi \in \mathcal{U}$. Setting $\xi = 0$ we obtain $0 \geq \sigma(x, u^*(x)) + \langle\lambda\,\mathrm{sgn}(u^*(x)),\, -u^*(x)\rangle$ which implies $\sigma(x, u^*(x)) \leq \lambda\|u^*(x)\|$. Setting $\xi = 2u^*(x)$ we obtain $2\,\sigma(x, u^*(x)) \geq \sigma(x, u^*(x)) + \langle\lambda\,\mathrm{sgn}(u^*(x)),\, u^*(x)\rangle$ which implies $\sigma(x, u^*(x)) \geq \lambda\|u^*(x)\|$. We have thus shown that $\lambda = \sigma(x, u^*(x))/\|u^*(x)\|$. Now if $\|u^*(x)\| \leq \pi(x)$ we have $\sigma(x, u^*(x)) = \|u^*(x)\|/\pi(x)$ and thus $\lambda = 1/\pi(x)$. Otherwise $\|u^*(x)\| > \pi(x)$ which means $\sigma(x, u^*(x)) = 1$ and we have $\lambda = 1/\|u^*(x)\|$. It follows from (4.49) that $\lambda = \nu(x)/b(x)$, and thus Lemma 4.5 gives

$$\partial_u\sigma(x, u^*(x)) \;=\; N_E \cap \left[\frac{\nu(x)}{b(x)}\,\mathrm{sgn}(u^*(x)) + \{u^*(x)\}^\perp\right] \tag{4.55}$$

It then follows from the projection theorem that $d(0, \partial_u\sigma(x, u^*(x))) = \nu(x)/b(x)$. We next show that $\|\partial_u\sigma(x, u^*(x))\| \leq 1/\pi(x)$. If $\|u^*(x)\| \leq \pi(x)$, then $\partial_u\sigma(x, u^*(x))$ is a singleton which means $\|\partial_u\sigma(x, u^*(x))\| = d(0, \partial_u\sigma(x, u^*(x))) = \nu(x)/b(x) = 1/\pi(x)$. Next suppose $\|u^*(x)\| >$

$\pi(x)$ and let $w \in \partial_u\sigma(x, u^*(x))$; then from (4.55) we have $w \in N_E\backslash\{0\}$ which means $H := u^*(x) + \{w\}^\perp$ is a supporting hyperplane of E. Let $v = d(0, H)\,\mathrm{sgn}(w)$; then from the projection theorem we have $v \in H$ which means $H = v + \{w\}^\perp$. Because $\pi(x)B \subset E$ we have $\|v\| = d(0, H) \geq \pi(x) > 0$, which means $H = v + \{v\}^\perp$. Now $\nu(x)/b(x) = 1/\|u^*(x)\|$ for $\|u^*(x)\| > \pi(x)$, and it follows from (4.55) that $w = \mathrm{sgn}(u^*(x))/\|u^*(x)\| + w_1$ for some $w_1 \in \{u^*(x)\}^\perp$. Thus $\langle w, u^*(x)\rangle = \langle \mathrm{sgn}(u^*(x))/\|u^*(x)\|, u^*(x)\rangle = 1$. Also, because $u^*(x) \in H$ we have $u^*(x) = v + v_1$ for some $v_1 \in \{v\}^\perp$, and so $\langle v, u^*(x)\rangle = \langle v, v\rangle = \|v\|^2$. Now because $\mathrm{sgn}(w) = \mathrm{sgn}(v)$ we have $\langle w, u^*(x)\rangle / \|w\| = \langle v, u^*(x)\rangle / \|v\|$, and substituting from above we obtain $1/\|w\| = \|v\|$. Because $\|v\| \geq \pi(x)$ we have $\|w\| \leq 1/\pi(x)$, and because w was arbitrary we conclude that $\|\partial_u\sigma(x, u^*(x))\| \leq 1/\pi(x)$. We summarize these results as follows:

$$\frac{\nu(x)}{b(x)} \;=\; d(0, \partial_u\sigma(x, u^*(x))) \;\leq\; \|\partial_u\sigma(x, u^*(x))\| \;\leq\; \frac{1}{\pi(x)} \qquad (4.56)$$

It follows from (4.51), (4.54), and the chain rule [20, Theorem 2.3.9] that

$$\begin{aligned}
\partial_u r(x, u^*(x)) &= \overline{\mathrm{co}}\Big\{\eta\zeta \,:\, \eta \in \partial_s\gamma(x, \sigma(x, u^*(x))),\ \zeta \in \partial_u\sigma(x, u^*(x))\Big\} \\
&= \overline{\mathrm{co}}\Big\{\eta\zeta \,:\, \eta \in [a(x), b(x)]\,,\ \zeta \in \partial_u\sigma(x, u^*(x))\Big\} \qquad (4.57)
\end{aligned}$$

We now use (4.56) and (4.57) to show that

$$N_E \cap \Big\{u \in \mathcal{U} \,:\, \mu(x) \leq \|u\| \leq \nu(x)\Big\} \;\subset\; \partial_u r(x, u^*(x)) \qquad (4.58)$$

Let $w \in N_E$ be such that $\mu(x) \leq \|w\| \leq \nu(x)$. Because $u^*(x) \neq 0$ we have $\mu(x) > 0$ which means $w \neq 0$. It then follows from [122, Corollary 23.7.1] that there exists $\delta > 0$ such that $\delta w \in \partial_u\sigma(x, u^*(x))$. From (4.56) we have $\delta\mu(x) \leq \delta\|w\| \leq 1/\pi(x)$, and it follows from (4.48) that $(1/\delta) \geq \mu(x)\pi(x) \geq a(x)$. Also from (4.56) we have $\nu(x)/b(x) \leq \delta\|w\| \leq \delta\nu(x)$ which means $(1/\delta) \leq b(x)$. Thus $(1/\delta) \in [a(x), b(x)]$ and it follows from (4.57) that $w = (1/\delta)\delta w \in \partial_u r(x, u^*(x))$. The choice for w was arbitrary and thus (4.58) is true.

Now $u^*(x) \neq 0$, and from Proposition 4.7 we have $D(x, u^*(x)) = 0$. We define the sublevel set $L(x) := \{u \in \mathcal{U} \,:\, D(x, u) \leq D(x, u^*(x))\}$ of D at $u^*(x)$ and note from (4.12) that $\overline{K(x)} = U(x) \cap L(x)$. Let N_L denote the normal cone to $L(x)$ at $u^*(x)$, let N_U denote the normal cone to $U(x)$

at $u^*(x)$, and let N_K denote the normal cone to $\overline{K(x)}$ at $u^*(x)$. Now because $K(x) \neq \varnothing$ we have $0 \in \text{int}\,[L(x) - U(x)]$, and so from [7, Table 4.3] we have $N_K = N_U + N_L$. Recall that $u^*(x)$ is the element of $\overline{K(x)}$ of minimum norm; this together with the projection theorem implies $-u^*(x) \in N_K$. Also, from C6 and (4.43) we have $E \subset C(x) \subset U(x)$, and therefore $N_U \subset N_E$. It then follows that $-u^*(x) \in N_E + N_L$. Thus there exist $v_E \in N_E$ and $v_L \in N_L$ such that $-u^*(x) = v_E + v_L$. Our goal is to show that $-v_L \in N_E \backslash \{0\}$. Now $-u^*(x) \notin N_E$ which means $v_L \neq 0$. Suppose $v_E = 0$; then $-v_L = u^*(x) \in N_E$. Otherwise we use the linearity of the inner product to obtain $\langle v_E, \text{sgn}(u^*(x))\rangle + \langle v_L, \text{sgn}(u^*(x))\rangle = \langle -u^*(x), \text{sgn}(u^*(x))\rangle = -\|u^*(x)\|$, and therefore

$$\begin{aligned} \langle \text{sgn}(-v_L), \text{sgn}(u^*(x))\rangle &= \frac{\|v_E\|\,\langle \text{sgn}(v_E), \text{sgn}(u^*(x))\rangle + \|u^*(x)\|}{\|v_L\|} \\ &\geq \frac{\|v_E\|\,\langle \text{sgn}(v_E), \text{sgn}(u^*(x))\rangle + \|u^*(x)\|}{\|v_E\| + \|u^*(x)\|} \\ &\geq \langle \text{sgn}(v_E), \text{sgn}(u^*(x))\rangle \end{aligned}$$

Thus the angle between $-v_L$ and $u^*(x)$ is smaller than the angle between v_E and $u^*(x)$. Now $v_E \in N_E$, and it follows from the convexity of N_E and the symmetry of N_E around $u^*(x)$ that $-v_L \in N_E$. We have thus shown that $-v_L \in N_E \backslash \{0\}$.

Now $v_L \in N_L \backslash \{0\}$, and it follows from [122, Corollary 23.7.1] that there exists $\delta > 0$ such that $\delta v_L \in \Pi(x)$. From Proposition 4.8 we have $\mu(x) \leq \|\delta v_L\| \leq \nu(x)$, and because $-\delta v_L \in N_E$ we have from (4.58) that $-\delta v_L \in \partial_u r(x, u^*(x))$. Therefore $0 = -\delta v_L + \delta v_L \in \partial_u r(x, u^*(x)) + \Pi(x)$ and the proof is complete. ∎

4.4.4 Proof of optimality

An immediate consequence of (4.52) and Proposition 4.9 are the equalities

$$\begin{aligned} 0 &= q(x) + r(x, u^*(x)) + D(x, u^*(x)) - \alpha_V(x) && (4.59) \\ &= q(x) - \alpha_V(x) + \min_{u \in \mathcal{U}} \left[r(x,u) + D(x,u) \right] && (4.60) \\ &= \min_{u \in \mathcal{U}} \max_{w \in W(x)} \left[q(x) + r(x,u) + L_f V(x,u,w) \right] && (4.61) \end{aligned}$$

which hold for all $x \in \mathcal{X}$. Therefore the rclf V satisfies the steady-state HJI equation associated with the cost functional J.

Let $J_T(u, w, x; x_0)$ denote the cost J in (4.34) truncated at time T, again setting $J_T := \infty$ when the solution does not exist over $[0, T]$:

$$J_T(u, w, x; x_0) \quad := \quad \int_0^T \Big[q(x) \; + \; r(x, u)\Big] \, dt \tag{4.62}$$

Fix $x_0 \in \mathcal{X}$ and an admissible disturbance w for Σ, and let $x(t)$ be a solution to Σ starting from x_0 with the control u^*. Because u^* renders the solutions to Σ RGUAS, the solution $x(t)$ exists for all $t \geq 0$ and furthermore $x(t) \to 0$ as $t \to \infty$. Thus we can integrate $L_f V$ along $x(t)$ and use (4.45) and (4.59) to obtain the following for all $T \geq 0$:

$$\begin{aligned} V(x_0) \quad &= \quad V(x(T)) \; - \; \int_0^T L_f V(x, u^*(x), w) \, dt \\ &\geq \quad V(x(T)) \; - \; \int_0^T \Big[D(x, u^*(x)) \; - \; \alpha_V(x)\Big] \, dt \\ &\geq \quad V(x(T)) \; + \; \int_0^T \Big[q(x) \; + \; r(x, u^*(x))\Big] \, dt \end{aligned} \tag{4.63}$$

Thus $V(x(T)) + J_T(u^*, w, x; x_0) \leq V(x_0)$ for all $T \geq 0$, and because $V(x(T)) \to 0$ as $T \to \infty$, in the limit we obtain $J(u^*, w, x; x_0) \leq V(x_0)$.

Next fix $x_0 \in \mathcal{X}$, let u be a control for Σ, let $\Delta > 0$, and suppose AC is true. Then there exist an admissible disturbance w_Δ for Σ and a solution $x_\Delta(t)$ to Σ starting from x_0 such that either $J(u, w_\Delta, x_\Delta; x_0) = \infty$ or for every $T \geq 0$ we have

$$\int_0^T L_f V(x_\Delta, u, w_\Delta) \, dt \;\; \geq \;\; \int_0^T \Big[D(x_\Delta, u) - \alpha_V(x_\Delta)\Big] \, dt \; - \; \Delta \tag{4.64}$$

If $J(u, w_\Delta, x_\Delta; x_0) = \infty$, then trivially $J(u, w_\Delta, x_\Delta; x_0) \geq V(x_0) - \Delta$. Otherwise $x_\Delta(t) \to 0$ as $t \to \infty$ and we can integrate $L_f V$ along the solution x_Δ as above and use (4.64) and (4.60) to obtain the following for all $T \geq 0$:

$$\begin{aligned} V(x_0) \quad &= \quad V(x_\Delta(T)) \; - \; \int_0^T L_f V(x_\Delta, u, w_\Delta) \, dt \\ &\leq \quad V(x_\Delta(T)) \; + \; \Delta \; - \; \int_0^T \Big[D(x_\Delta, u) \; - \; \alpha_V(x_\Delta)\Big] \, dt \\ &\leq \quad V(x_\Delta(T)) \; + \; \Delta \; + \; \int_0^T \Big[q(x_\Delta) + r(x_\Delta, u)\Big] \, dt \end{aligned} \tag{4.65}$$

Therefore $V(x_\Delta(T)) + J_T(u, w_\Delta, x_\Delta; x_0) \geq V(x_0) - \Delta$ for all $T \geq 0$, and because $V(x_\Delta(T)) \to 0$ as $T \to \infty$, we obtain $J(u, w_\Delta, x_\Delta; x_0) \geq V(x_0) - \Delta$ in the limit. Because Δ was arbitrary, it follows from (4.35) that $\bar{J}(x_0) \geq V(x_0)$. Recall from above that u^* guarantees $J \leq V(x_0)$; it follows that $\bar{J}(x_0) = V(x_0)$. The initial condition x_0 was arbitrary, and we conclude that u^* is optimal for $\mathcal{G}(q, r)$.

4.5 Extension to finite horizon games

We next extend our inverse optimality results to the case where the small control property is *not* satisfied. In this case, persistent disturbances may make convergence to the point $x = 0$ impossible; indeed, this point may no longer be an equilibrium point. For example, the trivial system

$$\dot{x} = u + w \tag{4.66}$$

with control constraint $U \equiv [-2, 2]$ and disturbance constraint $W \equiv [-1, 1]$ admits an rclf $V(x) = x^2$ with $\alpha_V(x) = |x|$ and $c_V = 0$:

$$\begin{aligned} \inf_{u \in U} \sup_{w \in W} \Big[L_f V(x, u, w) + \alpha_V(x) \Big] &= \inf_{u \in [-2,2]} \sup_{w \in [-1,1]} \Big[2xu + 2xw + |x| \Big] \\ &= -|x| \end{aligned} \tag{4.67}$$

We conclude from Theorem 3.11 that this system is robustly practically stabilizable, that is, that the residual set Ω can be made arbitrarily small by our choice of an admissible control. However, because we require admissible controls to be continuous functions of the state, no such control can create a robustly asymptotically stable equilibrium at $x = 0$. Consequently, no admissible control can achieve a finite value of the infinite horizon cost functional (4.34) for all admissible disturbances, and the associated inverse optimal robust stabilization problem has no solution.

We conclude that to obtain inverse optimality results for systems such as (4.66) which do not satisfy the small control property, we must modify the cost functional (4.34) either by relaxing conditions J1 and J2 on the functions q and r or by introducing a *terminal time* for the differential game. We explore the latter option in this section.

4.5.1 A finite horizon differential game

A common method of defining the termination of a differential game of variable duration is to introduce a *target set* and end the game when the state trajectory reaches this set [9]. We will adopt a similar definition here, but instead of ending the game when the state trajectory *first* reaches the target set, we will end the game when the state trajectory enters the target set *without ever again leaving.* This definition of the terminal time is consistent with our goal of robust stabilization.

Given a nonempty bounded *target set* $\Lambda \subset \mathcal{X}$ and a solution $x(t)$ to Σ, we define the terminal time T_Λ as follows:

$$T_\Lambda \quad := \quad \inf\left\{T \geq 0 \,:\, x(t) \in \Lambda \text{ for all } t \geq T\right\} \tag{4.68}$$

Thus T_Λ represents the first time the solution enters the target set Λ without ever again leaving. If the solution $x(t)$ does not exist for all $t \geq 0$ or is not eventually contained in Λ, then we set $T_\Lambda := \infty$. In addition to a terminal time for our game, we introduce a terminal cost $q_f : \mathcal{X} \to I\!R_+$ according to the specification

J3: $q_f \in \mathcal{P}(\mathcal{X})$

Using a triple (q, r, q_f) satisfying J1–J3, we will define a game $\mathcal{G}_\Lambda(q, r, q_f)$ for each nonempty bounded set $\Lambda \subset \mathcal{X}$ by considering a cost functional J_Λ parameterized by the initial condition $x_0 \in \mathcal{X}$. Given a control u, a disturbance w, and an initial condition $x_0 \in \mathcal{X}$ for Σ, we define the cost

$$J_\Lambda(u, w, x; x_0) \quad := \quad q_f(x(T_\Lambda)) + \int_0^{T_\Lambda} \Big[q(x) + r(x, u)\Big]\, dt \tag{4.69}$$

where the integration is taken along the (possibly nonunique) solution $x(t)$ to Σ starting from the initial condition x_0. If $T_\Lambda = \infty$, then we set $J_\Lambda(u, w, x; x_0) := \infty$. We define $\bar{J}_\Lambda : \mathcal{X} \to I\!R_+ \cup \{\infty\}$, the upper value function of the game, by the equation

$$\bar{J}_\Lambda(x_0) \quad := \quad \inf_u \sup_w \sup_x J_\Lambda(u, w, x; x_0) \tag{4.70}$$

The first supremum is taken over all solutions $x(t)$ to Σ starting from x_0 (this supremum is superfluous if solutions are unique), the second supremum is taken over all admissible disturbances w for Σ, and the infimum is taken over all (not necessarily admissible) controls u for Σ. A time-invariant admissible control for Σ which achieves a finite infimum in (4.70) for every $x_0 \in \mathcal{X}$ is said to be *optimal for* $\mathcal{G}_\Lambda(q, r, q_f)$. Note that every optimal control law is robustly stabilizing in the sense that every closed-loop trajectory is eventually contained in Λ.

Let V be a time-invariant rclf for Σ. We say a nonempty bounded set $\Lambda \subset \mathcal{X}$ is an *admissible target set for* V when there exists $c > c_V$ such that $V^{-1}[0, c] \subset \Lambda$. If $c_V = 0$, then any bounded set Λ containing a neighborhood of the origin is an admissible target set for V; this is consistent with practical stabilizability. For each admissible target set Λ^* for V, we define a nonempty set $\Xi(V, \Lambda^*)$ of pointwise min-norm control laws as follows: we say a control u for Σ belongs to $\Xi(V, \Lambda^*)$ when it is pointwise min-norm on $V^{-1}[c, \infty)$ for some $c > c_V$ satisfying $V^{-1}[0, c] \subset \Lambda^*$. Theorem 4.10 below states that for every control law $u^* \in \Xi(V, \Lambda^*)$ there exists a triple (q, r, q_f) satisfying J1–J3 such that u^* is optimal for $\mathcal{G}_\Lambda(q, r, q_f)$ for every $\Lambda \supset \Lambda^*$, with V being the corresponding upper value function. Each game $\mathcal{G}_\Lambda(q, r, q_f)$ is meaningful for the same reasons the game of Section 4.3 was meaningful; the difference here is the terminal time T_Λ and the terminal cost q_f. Our choice for the terminal time in (4.68) is consistent with the stabilization objective: the game ends when we enter the target set Λ permanently; if we enter and then leave again, the game continues. We must allow for the possibility of the trajectory leaving Λ because there does not necessarily exist a control which renders Λ positively invariant.

4.5.2 Main theorem: finite horizon

Our approximation condition on the admissible disturbances, which states that the effect of a worst-case disturbance can be approximated arbitrarily closely by admissible disturbances, now depends on the target set Λ:

AC$_\Lambda$: for every $x_0 \in \mathcal{X}$, every control u for Σ, and every $\Delta > 0$ there exist an admissible disturbance w_Δ for Σ and a solution $x_\Delta(t)$ to Σ starting from x_0 such that either $J_\Lambda(u, w_\Delta, x_\Delta; x_0) = \infty$ or

$$\int_0^{T_\Lambda} L_f V(x_\Delta, u, w_\Delta)\, dt \geq \int_0^{T_\Lambda} \left[\max_{w \in W(x_\Delta)} L_f V(x_\Delta, u, w) \right] dt - \Delta$$

As in the infinite horizon problem, if AC_Λ is not true then we may achieve suboptimality rather than optimality. We now state the main result of this section:

Theorem 4.10 *Let Σ be a time-invariant state feedback system which satisfies* C1–C6 *and* A1–A2. *Let V be a time-invariant rclf for Σ, let Λ^* be an admissible target set for V, and let $u^* \in \Xi(V, \Lambda^*)$. Then there exists a triple (q, r, q_f) satisfying* J1–J3 *such that $J_\Lambda(u^*, w, x; x_0) \leq V(x_0)$ for every $\Lambda \supset \Lambda^*$, every $x_0 \in \mathcal{X}$, every admissible disturbance w for Σ, and every solution $x(t)$ to Σ starting from x_0. If furthermore* AC_Λ *is true, then $\bar{J}_\Lambda(x_0) = V(x_0)$ for all $x_0 \in \mathcal{X}$ which means u^* is optimal for $\mathcal{G}_\Lambda(q, r, q_f)$.*

4.5.3 Proof of the main theorem

The proof of Theorem 4.10 is similar to the proof of Theorem 4.3. From the definition of $\Xi(V, \Lambda^*)$ above, the control law u^* is pointwise min-norm on $V^{-1}[c, \infty)$ for some $c > c_V$ satisfying $V^{-1}[0, c] \subset \Lambda^*$. We choose $q_f(x) := V(x)$ for all $x \in \mathcal{X}$; this choice satisfies J3. We define $C(x)$, $\sigma(x, u)$, and $\Pi(x)$ as in (4.43)–(4.46). Proposition 4.6 remains true, but Propositions 4.7 and 4.8 become weaker as follows:

Proposition 4.11 *If $x \in V^{-1}(c, \infty)$ and $u^*(x) \neq 0$, then $D(x, u^*(x)) = 0$ and $0 \notin \Pi(x)$.*

Proposition 4.12 *There exist continuous functions $\mu, \nu : \mathcal{X} \to I\!R_+$ such that for all $x \in V^{-1}(c, \infty)$ we have $\mu(x) \leq d(0, \Pi(x))$ and for all $x \in \mathcal{X}$ we have $\|\Pi(x)\| < \nu(x)$, with μ having the additional property that $\mu(x) > 0$ if and only if $u^*(x) \neq 0$.*

Because $\alpha_V(x)$ is bounded away from zero for large x, there exist $\alpha_0 \in \mathcal{P}(\mathcal{X})$ and $\hat{\alpha} > 0$ such that $\alpha_V(x) - \alpha_0(x) \geq \hat{\alpha}$ for all $x \in V^{-1}[c, \infty)$. We then define $a(x)$, $b(x)$, $\gamma(x, s)$, and $r(x, u)$ as in (4.48)–(4.51). It follows that r satisfies J2. Our next task is to construct the function q. We first show that there exist continuous functions $d_1, d_2 : \mathcal{X} \to I\!R$ such that $d_1(x) > 0$ for all $x \in \mathcal{X}$ and

$$r(x, u) + D(x, u) \;\; \geq \;\; d_1(x)\, \|u\| + d_2(x) \tag{4.71}$$

for all $(x,u) \in \mathcal{X} \times \mathcal{U}$. First, it follows from (4.44) and (4.49) that $b(x)\,\sigma(x,u) \geq \nu(x)\,\|u\|$ for all $(x,u) \in \mathcal{X} \times \mathcal{U}$. It then follows from (4.50) and (4.51) that

$$\begin{aligned} r(x,u) &\geq b(x)\,\sigma(x,u) + \sigma(x,u^*(x))\,[a(x)-b(x)] \\ &\geq \nu(x)\,\|u\| + \sigma(x,u^*(x))\,[a(x)-b(x)] \end{aligned} \tag{4.72}$$

for all $(x,u) \in \mathcal{X} \times \mathcal{U}$. For each $x \in \mathcal{X}$ choose $w_x \in \Pi(x)$; then it follows from (4.46) and Proposition 4.12 that

$$\begin{aligned} D(x,u) &\geq D(x,u^*(x)) + \langle w_x,\, u-u^*(x)\rangle \\ &\geq D(x,u^*(x)) - \|w_x\|\cdot\|u\| - \|w_x\|\cdot\|u^*(x)\| \\ &\geq D(x,u^*(x)) - \|\Pi(x)\|\cdot\|u\| - \nu(x)\,\|u^*(x)\| \end{aligned} \tag{4.73}$$

for all $(x,u) \in \mathcal{X} \times \mathcal{U}$. We add the inequalities (4.72) and (4.73) to obtain

$$\begin{aligned} r(x,u) + D(x,u) \geq\; & \big[\nu(x) - \|\Pi(x)\|\big]\,\|u\| + \sigma(x,u^*(x))\,[a(x)-b(x)] \\ & + D(x,u^*(x)) - \nu(x)\,\|u^*(x)\| \end{aligned} \tag{4.74}$$

for all $(x,u) \in \mathcal{X} \times \mathcal{U}$. It follows from Proposition 4.12, [74, Problem 5X], and the lower semicontinuity of the mapping $x \mapsto \nu(x) - \|\Pi(x)\|$ that there exists a continuous function $d_1 : \mathcal{X} \to I\!R$ such that $0 < d_1(x) \leq \nu(x) - \|\Pi(x)\|$ for all $x \in \mathcal{X}$. Thus if we define $d_2(x) := \sigma(x,u^*(x))\,[a(x)-b(x)] + D(x,u^*(x)) - \nu(x)\,\|u^*(x)\|$ for all $x \in \mathcal{X}$, then d_2 is continuous and (4.74) implies (4.71).

We next define a function $\omega : \mathcal{X} \to I\!R_+$ as follows for all $x \in \mathcal{X}$:

$$\omega(x) \;:=\; \frac{r(x,u^*(x)) + D(x,u^*(x)) - d_2(x)}{d_1(x)} \tag{4.75}$$

Thus ω is continuous and it follows from (4.71) that $\|u^*(x)\| \leq \omega(x)$ for all $x \in \mathcal{X}$. Also, from (4.71) we see that for all $(x,u) \in \mathcal{X} \times \mathcal{U}$ such that $\|u\| > \omega(x)$, we have $r(x,u) + D(x,u) \geq d_1(x)\,\|u\| + d_2(x) > r(x,u^*(x)) + D(x,u^*(x))$. Therefore

$$\begin{aligned} \inf_{u\in\mathcal{U}} \big[r(x,u) + D(x,u)\big] &= \inf_{u\in\omega(x)B} \big[r(x,u) + D(x,u)\big] \\ &= \min_{u\in\omega(x)B} \big[r(x,u) + D(x,u)\big] \end{aligned} \tag{4.76}$$

for all $x \in \mathcal{X}$, where the second line follows from the continuity of $r + D$ and the compactness of $\omega(x)B$. Now Proposition 2.9 implies that the right-hand side of (4.76) is continuous on $\mathcal{X}$, and so the mapping

$$x \ \mapsto \ \min_{u \in \mathcal{U}} \left[r(x,u) + D(x,u) \right] \tag{4.77}$$

is well-defined and continuous on $\mathcal{X}$. We choose $\alpha_1 \in \mathcal{P}(\mathcal{X})$ such that $\alpha_1(x) \leq \hat{\alpha}$ for all $x \in \mathcal{X}$, and we define q as follows for all $x \in \mathcal{X}$:

$$q(x) \ := \ \max \Big\{ \alpha_1(x), \ \alpha_V(x) - \min_{u \in \mathcal{U}} \left[r(x,u) + D(x,u) \right] \Big\} \tag{4.78}$$

Thus q is continuous with $q(0) = 0$ and $q \geq \alpha_1$, and it follows that $q \in \mathcal{P}(\mathcal{X})$ as required in J1. With this choice for q, the following inequality is true for all $x \in \mathcal{X}$:

$$q(x) \ - \ \alpha_V(x) \ + \ \min_{u \in \mathcal{U}} \left[r(x,u) + D(x,u) \right] \ \geq \ 0 \tag{4.79}$$

One can use Propositions 4.11 and 4.12 in the proof of Proposition 4.9 to show that (4.53) is true for $x \in V^{-1}(c, \infty)$, and so we have

$$q(x) \ = \ \max \left\{ \alpha_1(x), \ \alpha_V(x) - r(x, u^*(x)) - D(x, u^*(x)) \right\} \tag{4.80}$$

for all such x. Now $\alpha_V(x) - r(x,u^*(x)) - D(x,u^*(x)) \geq \alpha_V(x) - \alpha_0(x) \geq \hat{\alpha} \geq \alpha_1(x)$ for all such x, which means

$$q(x) \ = \ \alpha_V(x) \ - \ r(x, u^*(x)) \ - \ D(x, u^*(x)) \tag{4.81}$$

for all such x. Therefore the following is true for all $x \in V^{-1}(c, \infty)$:

$$\begin{aligned} 0 \ &= \ q(x) \ - \ \alpha_V(x) \ + \ r(x, u^*(x)) \ + \ D(x, u^*(x)) && (4.82) \\ &= \ q(x) \ - \ \alpha_V(x) \ + \ \min_{u \in \mathcal{U}} \left[r(x,u) + D(x,u) \right] && (4.83) \end{aligned}$$

Thus the HJI *inequality* (4.79) is true for all $x \in \mathcal{X}$, whereas the HJI *equality* (4.83) is true whenever $x \in V^{-1}(c, \infty)$. We are now ready to prove optimality.

Let $\Lambda \supset \Lambda^*$, let $x_0 \in \mathcal{X}$, let w be an admissible disturbance for Σ, and let $x(t)$ be a solution to Σ starting from x_0 with the control u^*. Because u^* renders the solutions to Σ RGUAS-Ω for $\Omega = V^{-1}[0, c] \subset \Lambda$, it follows from (4.68) that $T_\Lambda < \infty$. Thus we can integrate L_fV along $x(t)$ to obtain

$$0 = V(x_0) - V(x(T_\Lambda)) + \int_0^{T_\Lambda} L_fV(x, u^*(x), w)\, dt \tag{4.84}$$

Because $q_f = V$, we can add this zero quantity (4.84) to (4.69) and use (4.45) to obtain

$$\begin{aligned} J_\Lambda(u^*, w, x; x_0) &= V(x_0) + \int_0^{T_\Lambda} \Big[q(x) + r(x, u^*(x)) \\ &\qquad + L_fV(x, u^*(x), w)\Big]\, dt \\ &\leq V(x_0) + \int_0^{T_\Lambda} \Big[q(x) - \alpha_V(x) \\ &\qquad + r(x, u^*(x)) + D(x, u^*(x))\Big]\, dt \end{aligned} \tag{4.85}$$

Now u^* renders the set $V^{-1}[0, c]$ robustly positively invariant, and so from (4.68) we have $x(t) \in V^{-1}(c, \infty)$ for all $t \in [0, T_\Lambda)$. It then follows from (4.82) that the integrand in (4.85) is zero for all $t \in [0, T_\Lambda)$, and thus we have $J_\Lambda(u^*, w, x; x_0) \leq V(x_0)$.

Next fix $x_0 \in \mathcal{X}$, let u be a control for Σ, let $\Delta > 0$, and suppose AC_Λ is true. Then there exist an admissible disturbance w_Δ for Σ and a solution $x_\Delta(t)$ to Σ starting from x_0 such that either $J_\Lambda(u, w_\Delta, x_\Delta; x_0) = \infty$ or

$$\int_0^{T_\Lambda} L_fV(x_\Delta, u, w_\Delta)\, dt \geq \int_0^{T_\Lambda} \Big[D(x_\Delta, u) - \alpha_V(x_\Delta)\Big]\, dt - \Delta \tag{4.86}$$

If $J_\Lambda(u, w_\Delta, x_\Delta; x_0) = \infty$, then trivially we have $J_\Lambda(u, w_\Delta, x_\Delta; x_0) \geq V(x_0) - \Delta$. Otherwise $T_\Lambda < \infty$ and we can integrate L_fV along the solution x_Δ as above and use (4.86) and (4.79) to obtain the following:

$$\begin{aligned} J_\Lambda(u, w_\Delta, x_\Delta; x_0) &= V(x_0) + \int_0^{T_\Lambda} \Big[q(x_\Delta) + r(x_\Delta, u) \\ &\qquad + L_fV(x_\Delta, u, w_\Delta)\Big]\, dt \\ &\geq V(x_0) - \Delta + \int_0^{T_\Lambda} \Big[q(x_\Delta) - \alpha_V(x_\Delta) \\ &\qquad + r(x_\Delta, u) + D(x_\Delta, u)\Big]\, dt \\ &\geq V(x_0) - \Delta \end{aligned} \tag{4.87}$$

Because Δ was arbitrary, it follows from (4.70) that $\bar{J}_\Lambda(x_0) \geq V(x_0)$. Recall from above that u^* guarantees $J_\Lambda \leq V(x_0)$; it follows that $\bar{J}_\Lambda(x_0) = V(x_0)$. The initial condition x_0 was arbitrary, and we conclude that u^* is optimal for $\mathcal{G}_\Lambda(q, r, q_f)$.

4.6 Summary

We have formulated and solved an inverse optimal robust stabilization problem by showing that every rclf is an upper value function for a meaningful game and that every pointwise min-norm control law is optimal for such a game. Our formulas (4.13) and (4.19) can be used to generate control laws which have the desirable properties of optimality and yet do not require the solution of an HJB or HJI equation. In the following chapters, we will address the important issue of how to construct an rclf for a given system.

Chapter 5

Robust Backstepping

Thus far our path has brought us to a base camp at which the design of a robustly stabilizing control law appears deceptively simple. All we need to do is find a robust control Lyapunov function (rclf); the remaining task of selecting a control law to make the Lyapunov derivative negative is straightforward. As we demonstrated in Chapter 4, explicit formulas are available for control laws which are optimal with respect to meaningful cost functionals.

Unfortunately, in reality a robust control Lyapunov function is often not known. Our path now climbs to a higher lookout point with a clearer view on the construction of rclf's. Although we will not reach the highest peak, our climb will nevertheless be rewarding. In this chapter we will develop *robust backstepping*—a systematic procedure with which we can construct families of rclf's (and hence also stabilizing controls laws) for a significant class of uncertain nonlinear systems.

We present the basic robust backstepping procedure in Section 5.2. A major practical improvement which results in "softer" control laws is presented in Section 5.3, and then critical smoothness assumptions are relaxed in Section 5.4.

Before we begin our climb, we dedicate Section 5.1 to the inspection of the climbing gear available at the base camp. This includes the now classical *Lyapunov redesign* [89, 48, 90, 23, 11, 139, 75] in which a known clf for the nominal system (that is, the system without uncertainties) is used as the rclf for the uncertain system. We will see that this design is essentially limited to the case in which the uncertainties enter through the same channels as the control variables. This restrictive condition, known as the *matching condition*, had been a stumbling block for many years. Several attempts to weaken the matching condition [144, 12, 19, 154, 120, 18] met with limited success. Early results for nonlinear systems [12, 19] required that the unmatched uncertainties be sufficiently small. For the special case of quadratic stability for uncertain linear systems, surveyed in [22], a generalized matching condition was proposed in [144] and extended in [154].

Although helpful, none of these tools would assure a safe climb. What encouraged us to attempt this climb, which we initiated in [28, 31, 32], was a breakthrough in adaptive nonlinear control in [72] that employed the technique of "adding an integrator." This technique was introduced in [136, 17, 146], and its recursive application in [81, 124] eventually led to the adaptive control results of [72] currently known as *integrator backstepping* or simply *backstepping* [73, 80, 85]. For the nonlinear robust control problem of Chapter 3, backstepping led to the discovery of a structural *strict feedback condition*, much weaker than the matching condition, under which the systematic construction of an rclf is always possible. These robust backstepping results first appeared in [31]. They were also obtained independently in [101, 129, 121].

The robust backstepping procedure presented in Section 5.2 leads to the construction of an rclf which is quadratic in a set of transformed coordinates. It was discovered in [34] that, as a rule, such quadratic rclf's can result in control laws with undesirable high-gain properties. In Section 5.3 we alleviate this difficulty by introducing a non-quadratic, *flattened* rclf which dramatically reduces the control gains required to make the Lyapunov derivative negative. A major practical achievement is that these "softer" control laws lead to improved performance with less control effort.

Finally, in Section 5.4 we show how to use the robust backstepping tools of Sections 5.2 and 5.3 to weaken the critical smoothness assumptions in backstepping designs. In particular, we introduce *nonsmooth* rclf's and illustrate how they lead to the construction of nonsmooth robust control laws.

5.1 Lyapunov redesign

A standard method for finding an rclf for an uncertain system, developed in [89, 48, 90, 23, 11] and known as *Lyapunov redesign* or *min-max design*, has been incorporated in several textbooks [75, Section 5.5], [139, Chapter 10]. The key idea of this method is to employ a clf for the nominal system (the system without uncertainties) as an rclf for the uncertain system. This re-use of the same Lyapunov function is alluded to by the term "redesign" which goes back to an early paper in adaptive control [112].

5.1.1 Matched uncertainty

To illustrate Lyapunov redesign, we consider the system

$$\dot{x} \;=\; F(x) + G(x)\,u + H(x)\,w \tag{5.1}$$

where F, G, and H are (known) continuous functions, u is the control input, and w is the disturbance input (cf. Section 3.1.1). We assume that the nominal system $\dot{x} = F(x) + G(x)\,u$ is stabilizable, and for simplicity we assume that the state is available for feedback ($Y(x) = \{x\}$), that there is no constraint on the control ($U \equiv \mathcal{U}$), and that the disturbance constraint is the closed unit ball ($W \equiv B$). Our crucial assumption is that a clf is known for the nominal system. Namely, we assume knowledge of a C^1, positive definite, proper function $V : \mathcal{X} \to I\!R_+$ such that

$$\inf_{u \in U} \nabla V(x) \cdot \big[F(x) + G(x)\,u\big] \;<\; -\alpha_V(x) \tag{5.2}$$

for all $x \neq 0$ and for some function $\alpha_V \in \mathcal{P}(\mathcal{X})$. Another way to write the clf condition (5.2) is

$$\nabla V(x) \cdot G(x) = 0 \quad\Longrightarrow\quad \nabla V(x) \cdot F(x) \;<\; -\alpha_V(x) \tag{5.3}$$

for all $x \neq 0$. A stabilizing control for the nominal system can be designed using this clf. This "nominal" control law is to be "redesigned" to account for the uncertainty w in the actual system. The key idea of the redesign is to use the clf V as an rclf for the uncertain system (5.1). We stress that this rclf is chosen independently of the uncertainty and thus disregards any knowledge about the structure of the disturbance input matrix $H(x)$.

We know that to be an rclf, the function V must satisfy

$$\inf_{u\in U}\sup_{w\in B} \nabla V(x)\cdot\Big[F(x) + G(x)\,u + H(x)\,w\Big] \;<\; -\alpha_V(x) \tag{5.4}$$

for all $x \neq 0$, or equivalently

$$\nabla V(x)\cdot G(x) = 0 \;\Longrightarrow\; \nabla V(x)\cdot F(x) + \Big\|\nabla V(x)\cdot H(x)\Big\| < -\alpha_V(x) \tag{5.5}$$

for all $x \neq 0$. It is clear that Lyapunov redesign is feasible for this choice V if and only if, given (5.3), the implication (5.5) also holds. In light of (5.3), we can view the rclf condition (5.5) as a restriction on the effective size $H(x)$ of the system uncertainty. Unfortunately, this condition depends on our choice for V and cannot be checked *a priori* on the system (5.1). To overcome this obstacle, let us suppose that $H(x)$ satisfies the structural condition

$$H(x) \;=\; G(x)\,E(x) \tag{5.6}$$

for some continuous function $E(x)$. Then (5.5) is a *consequence* of (5.3), which means the Lyapunov redesign method is guaranteed to work. This structural condition (5.6) is the *matching condition* mentioned earlier. Under this condition, the system (5.1) can be written in the form

$$\dot{x} \;=\; F(x) + G(x)\Big[u + E(x)\,w\Big] \tag{5.7}$$

Here the uncertainty is said to be *matched* with the control input u because the disturbance enters through the same input channel as the control. For example, of the two systems

$$\begin{aligned} \dot{x}_1 &= x_2 \\ \dot{x}_2 &= u + x_1^3\,w \end{aligned} \qquad\qquad \begin{aligned} \dot{x}_1 &= x_2 + x_1^3\,w \\ \dot{x}_2 &= u \end{aligned} \tag{5.8}$$

the one to the left satisfies the matching condition while the one to the right does not.

The matching condition, which guarantees the success of Lyapunov redesign, has also been its most severe restriction. To significantly weaken it we must abandon the key idea of Lyapunov redesign by *taking the uncertainty into account during the construction of the Lyapunov function.*

5.1.2 Beyond the matching condition

We just showed that an rclf V for the system (5.1) must be such that

$$\nabla V(x) \cdot G(x) = 0 \quad \Longrightarrow \quad \nabla V(x) \cdot F(x) + \left\| \nabla V(x) \cdot H(x) \right\| \leq 0 \tag{5.9}$$

In the Lyapunov redesign method, this was viewed as a constraint on the uncertainty size H. Now let us instead view this as a constraint on the rclf V. This new look at (5.9) will lead us beyond Lyapunov redesign: our construction of V will be based on (5.9) rather than on the nominal system. In other words, we will take the uncertainty into account during the construction of V itself.

To illustrate our departure from Lyapunov redesign, let us consider the second-order single-input uncertain system

$$\dot{x}_1 = x_2 + x_1 \phi(x_1)\, w\,, \qquad \phi(x_1) \geq 0 \tag{5.10}$$

$$\dot{x}_2 = u \tag{5.11}$$

Again, for simplicity we assume a disturbance constraint $w \in W \equiv B$. The function ϕ, which determines the size of the uncertainty, is assumed to be smooth and nonnegative. The scalar disturbance input w represents unmatched uncertainty because in this case $G(x) = [0 \;\; 1]^{\mathrm{T}}$, $H(x) = [x_1\phi(x_1) \;\; 0]^{\mathrm{T}}$, and the matching condition (5.6) cannot be satisfied for any E.

Our goal is to find an rclf for this system. The Lyapunov redesign method tells us to find a clf for the nominal system ($w = 0$) and then to check if it is an rclf for the uncertain system. Because the nominal system in this case is linear, we can construct a quadratic clf $V(x) = x^{\mathrm{T}}Px$ by solving a Riccati equation. The necessary condition (5.9) becomes

$$x^{\mathrm{T}}P \begin{bmatrix} 0 \\ 1 \end{bmatrix} = 0 \quad \Longrightarrow \quad x^{\mathrm{T}}P \begin{bmatrix} x_2 \\ 0 \end{bmatrix} + \left| x^{\mathrm{T}}P \begin{bmatrix} x_1\phi(x_1) \\ 0 \end{bmatrix} \right| \leq 0 \tag{5.12}$$

By the construction of P, the linear feedback $-[0 \;\; 1]Px$ will stabilize the linear nominal system. The coefficients multiplying x_1 and x_2 in this linear feedback must both be negative to insure stability, and we conclude that $-[0 \;\; 1]Px = -\bar{c}(x_2 + cx_1)$ for some constants $\bar{c} > 0$ and $c > 0$. Therefore $x^{\mathrm{T}}P\,[0 \;\; 1]^{\mathrm{T}} = 0$ if and only if $x_2 = -cx_1$. We evaluate (5.12) at points where $x_2 = -cx_1$, and, after some algebra, this inequality becomes

$$\phi(x_1) \leq c \tag{5.13}$$

for all $x_1 \in I\!R$. The key observation is that the nonnegative function ϕ must be bounded by the constant c. In other words, if the size of the uncertainty is such that (5.13) is violated, then this particular Lyapunov redesign fails. This is a consequence of the fact that the uncertainty in the system (5.10)–(5.11) does not satisfy the matching condition (5.6).

The above Lyapunov redesign failed because it was based on the linear nominal system which suggested a quadratic clf V. Let us now ignore the nominal system and base our search for V directly on the inequality (5.9). With a smooth function $s(x_1)$ to be determined later, we consider

$$V(x) \;=\; x_1^2 + \left[x_2 - x_1\, s(x_1)\right]^2 \tag{5.14}$$

Being smooth, positive definite, and radially unbounded, this function V qualifies as a candidate Lyapunov function for our system (5.10)–(5.11). We will motivate this structure for V in the next section; here we illustrate how the freedom to select $s(x_1)$ can be used to remove the restrictive condition (5.13) on the size of the uncertainty. For V in (5.14) we have $\nabla V(x)\cdot G(x) = 0$ if and only if $x_2 = x_1\, s(x_1)$, which means the necessary condition (5.9) now becomes

$$x_1^2\, s(x_1) + x_1^2 \phi(x_1) \;\leq\; 0 \tag{5.15}$$

for all $x_1 \in I\!R$. Because we have left the choice for $s(x_1)$ open, this inequality can be viewed as a guide for the construction of V (through s) rather than as a constraint on the function ϕ. It is clear that we can always satisfy (5.15): a simple choice for s is

$$s(x_1) \;=\; -1 - \phi(x_1) \tag{5.16}$$

In our attempt to go beyond the matching condition and remove restrictions on $\phi(x_1)$, this is a breakthrough: the design with $s(x_1)$ succeeds (that is, V is an rclf) for any function ϕ, regardless of its growth. The restriction (5.13) on the size of ϕ which appeared in the above Lyapunov redesign is gone; this new design allows arbitrary growth of the function ϕ. Note that the rclf (5.14) depends on the uncertainty size ϕ through $s(x_1)$, but it does not depend on the uncertainty w itself.

We have thus shown how the limitations of Lyapunov redesign can be overcome through a re-interpretation of the necessary condition (5.9) as a guide for the construction of the rclf V rather than as a constraint on the size H of the uncertainty. The structure of V in (5.14) comes from the recursive Lyapunov design we now present.

5.2 Recursive Lyapunov design

In the preceding section we have shown that the function

$$V(x) = x_1^2 + \left[x_2 - x_1\, s(x_1)\right]^2 \tag{5.17}$$

with

$$s(x_1) = -1 - \phi(x_1) \tag{5.18}$$

is an rclf for the system

$$\dot{x}_1 = x_2 + x_1\, \phi(x_1)\, w \tag{5.19}$$

$$\dot{x}_2 = u \tag{5.20}$$

for $w \in [-1, 1]$. We now briefly explain the reasoning which led to the above structure of V. Let us examine the first system equation (5.19) and pretend that x_2 is its control input. Were we to choose for this control input the conceptual control law $x_2 = x_1 s_1(x_1)$, then the conceptual closed-loop system would be

$$\dot{x}_1 = -x_1 - x_1\, \phi(x_1)\, [1 - w] \tag{5.21}$$

This scalar system is clearly robustly asymptotically stable with a conceptual Lyapunov function $V_c(x_1) = x_1^2$. For each of the three ingredients of this design—the control law, the closed-loop system, and the Lyapunov function—we have used the term "conceptual." We have done this to stress that the control law $x_2 = x_1 s_1(x_1)$ cannot be implemented because x_2 is not a control variable. However, this conceptual design has helped us to recognize the benefit of the state x_2 being close to $x_1 s_1(x_1)$. We therefore add to our conceptual Lyapunov function V_c a term penalizing the difference between x_2 and $x_1 s_1(x_1)$, the result being the rclf (5.17). This simple construction, when applied repeatedly, leads to the recursive Lyapunov design known as *robust backstepping* that will be the topic of the remainder of this chapter.

5.2.1 Class of systems: strict feedback form

We begin by defining the class of uncertain nonlinear systems to which our backstepping design will apply. We consider an nth-order system

$$\dot{x} \;=\; F(x,w) + G(x,w)\,u \tag{5.22}$$

where F and G are continuous functions. We assume that we have state feedback ($Y(x) = \{x\}$), that there is a single unconstrained control input ($U(x) \equiv \mathcal{U} = I\!R$), and that the disturbance constraint $W(x)$ is independent of the control u. If we assume that W is continuous with nonempty compact convex values, then it follows from Proposition 2.22 that, without loss of generality, we may assume a constant disturbance constraint of the form $W(x) \equiv B$. We will construct an rclf for this system (5.22) under the following structural assumptions on F and G. First, we assume that F and G can be written in the form

$$F(x,w) = \begin{bmatrix} \phi_{11}(x,w) & \phi_{12}(x,w) & 0 & \cdots & 0 \\ \phi_{21}(x,w) & \phi_{22}(x,w) & \phi_{23}(x,w) & \cdots & 0 \\ \vdots & \vdots & \vdots & \ddots & \vdots \\ \phi_{n-1,1}(x,w) & \phi_{n-1,2}(x,w) & \phi_{n-1,3}(x,w) & \cdots & \phi_{n-1,n}(x,w) \\ \phi_{n1}(x,w) & \phi_{n2}(x,w) & \phi_{n3}(x,w) & \cdots & \phi_{nn}(x,w) \end{bmatrix} x \;+\; F(0,w)$$

$$G(x,w) = \begin{bmatrix} 0 \\ \vdots \\ 0 \\ \phi_{n,n+1}(x,w) \end{bmatrix} \tag{5.23}$$

for continuous scalar functions ϕ_{ij}. Note that the decomposition of F in (5.23) need not be unique. Next, we assume that each function ϕ_{ij} depends only on w and the state components x_1 through x_i, namely,

$$\phi_{ij}(x,w) \;=\; \phi_{ij}(x_1,\ldots,x_i,w) \tag{5.24}$$

for $1 \le i \le n$ and $1 \le j \le i+1$. Finally, we assume that

$$\phi_{i,i+1}(x_1,\ldots,x_i,w) \;\neq\; 0 \tag{5.25}$$

for all $x_1,\ldots,x_i \in I\!R$, for all $w \in B$, and for $1 \le i \le n$. This last condition (5.25) insures that the system (5.22) is controllable for each fixed $w \in B$. A system satisfying these structural conditions (5.23)–(5.25) is said to be in *strict feedback form*, also known as *lower triangular form*. Note that the system (5.10)–(5.11) analyzed in Section 5.1.2 is in strict feedback form. As will be demonstrated below, every system which can be transformed into strict feedback form admits an rclf.

Systems in strict feedback form generally do not satisfy the matching condition of the Lyapunov redesign method in Section 5.1. Indeed, the matching condition would require that the functions ϕ_{ij} be independent of the disturbance w for $1 \leq i \leq n-1$, a tremendous restriction. Also, unlike conditions in quadratic stability theory [22] which require nonlinearities to be (linearly) sector bounded, the strict feedback condition imposes no growth restrictions on the nonlinearities. Finally, if the functions ϕ_{ij} were all independent of the state variable x, then strict feedback form would reduce to one of the standard forms identified in the linear theory [144, 154].

Geometric necessary and sufficient conditions for the existence of a coordinate transformation which brings a system of the type (5.22) into strict feedback form are given in [101]. However, tests based on these conditions are not always conclusive because it may take more than a coordinate transformation to bring a system into strict feedback form. For example, the system

$$\dot{x}_1 = x_2 + \frac{u|u|x_1^3}{1+u^2} \tag{5.26}$$

$$\dot{x}_2 = x_3$$

$$\dot{x}_3 = u \tag{5.27}$$

is not in strict feedback form because of the presence of u in the first equation (5.26). It is not even in the general control-affine form (5.22) and will therefore fail the geometric tests in [101]. Nevertheless, this system represents one of many systems characterized by

$$\dot{x}_1 = x_2 + x_1^3 w \tag{5.28}$$

$$\dot{x}_2 = x_3$$

$$\dot{x}_3 = u \tag{5.29}$$

where w is a disturbance input with constraint $w \in [-1, 1]$. Indeed, $w = u|u|(1+u^2)^{-1}$ is an admissible disturbance according to our definition in Section 3.1.1, which means a robustly asymptotically stabilizing control law for the uncertain system (5.28)–(5.29) will be an asymptotically stabilizing control law for the system (5.26)–(5.27). Because the new system is in strict feedback form, we can construct such a control law using the backstepping technique described below. Of course, this control law may be conservative because it stabilizes all systems of the form (5.28)–(5.29), not just the original system.

5.2.2 Construction of an rclf

Let us now construct an rclf for a system in strict feedback form. We will build a diffeomorphism on the state space $\mathcal{X}$ using smooth scalar functions $s_1(x_1)$, $s_2(x_1, x_2)$, ..., $s_{n-1}(x_1, \ldots, x_{n-1})$ which are to be determined in the recursive manner outlined below. Each function s_i will depend only on the state components x_1 through x_i. Once these functions are chosen, we define a transformed state vector z as follows:

$$
\begin{aligned}
z_1 &:= x_1 && (5.30)\\
z_2 &:= x_2 - z_1 s_1(x_1) \\
z_3 &:= x_3 - z_2 s_2(x_1, x_2) \\
&\vdots \\
z_n &:= x_n - z_{n-1} s_{n-1}(x_1, \ldots, x_{n-1}) && (5.31)
\end{aligned}
$$

In matrix form, this diffeomorphism and its inverse are given by

$$
\begin{aligned}
z &:= S(x)\, x \\
&= \begin{bmatrix}
1 & 0 & 0 & 0 & \ldots & 0 \\
-s_1 & 1 & 0 & 0 & \ldots & 0 \\
s_1 s_2 & -s_2 & 1 & 0 & \ldots & 0 \\
-s_1 s_2 s_3 & s_2 s_3 & -s_3 & 1 & \ldots & 0 \\
\vdots & \vdots & \vdots & \ddots & \ddots & \vdots \\
\pm s_1 \cdots s_{n-1} & \mp s_2 \cdots s_{n-1} & \pm s_3 \cdots s_{n-1} & \ldots & -s_{n-1} & 1
\end{bmatrix} x \qquad (5.32)
\end{aligned}
$$

$$x = S^{-1}(x)\,z = \begin{bmatrix} 1 & 0 & 0 & 0 & \dots & 0 \\ s_1 & 1 & 0 & 0 & \dots & 0 \\ 0 & s_2 & 1 & 0 & \dots & 0 \\ 0 & 0 & s_3 & 1 & \dots & 0 \\ \vdots & \vdots & \vdots & \ddots & \ddots & \vdots \\ 0 & 0 & 0 & \dots & s_{n-1} & 1 \end{bmatrix} z \tag{5.33}$$

where the signs in the last row in (5.32) depend on whether the dimension n of the state vector is even or odd.

In the proof of the following theorem, we will show how to construct the functions s_i so that $V(x) := z^{\mathrm{T}}z$ is an rclf for the system (5.22). We will also provide a robustly stabilizing state feedback control law. This rclf V will satisfy the small control property when $F(0,w) \equiv 0$, that is, when the disturbance does not affect the equilibrium at $x = 0$; in this case the control law will guarantee convergence to $x = 0$ (rather than convergence to a residual set).

Theorem 5.1 *If the system* (5.22) *is in strict feedback form, then there exist suitable functions s_i such that $V(x) := z^{\mathrm{T}}z$ is an rclf for the system. If furthermore $F(0,w) \equiv 0$, then V will satisfy the small control property.*

Proof: Our first task is to calculate $\dot{z}$ from (5.32) and (5.22). Taking the derivative of $z = S(x)\,x$ in (5.32), we obtain

$$\begin{aligned} \dot{z} &= \begin{bmatrix} \frac{\partial S}{\partial x_1}x & \frac{\partial S}{\partial x_2}x & \dots & \frac{\partial S}{\partial x_n}x \end{bmatrix} \dot{x} + S(x)\,\dot{x} \\ &:= T(x)\,\dot{x} \end{aligned} \tag{5.34}$$

where $T(x)$ can easily be calculated from (5.32). If we let $\circledast_i$ denote any function depending only on the states x_1 through x_i and the functions s_1 through s_i and their partial derivatives, then we have

$$\dot{z} = T(x)\,\dot{x} = \begin{bmatrix} 1 & 0 & 0 & 0 & \dots & 0 \\ \circledast_1 & 1 & 0 & 0 & \dots & 0 \\ \circledast_2 & \circledast_2 & 1 & 0 & \dots & 0 \\ \circledast_3 & \circledast_3 & \circledast_3 & 1 & \dots & 0 \\ \vdots & \vdots & \vdots & \ddots & \ddots & \vdots \\ \circledast_{n-1} & \circledast_{n-1} & \circledast_{n-1} & \dots & \circledast_{n-1} & 1 \end{bmatrix} \dot{x} \tag{5.35}$$

Our robustly stabilizing state feedback control law will be of the form $u(x) = z_n s_n(x)$ where $s_n(x)$ is another smooth function yet to be determined. With this choice for u, we can use (5.33) to rewrite (5.22) as

$$\dot{x} = \begin{bmatrix} \phi_{11} & \phi_{12} & 0 & \dots & 0 & 0 \\ \phi_{21} & \phi_{22} & \phi_{23} & \dots & 0 & 0 \\ \vdots & \vdots & \vdots & \ddots & \vdots & \vdots \\ \phi_{n-1,1} & \phi_{n-1,2} & \phi_{n-1,3} & \dots & \phi_{n-1,n} & 0 \\ \phi_{n1} & \phi_{n2} & \phi_{n3} & \dots & \phi_{nn} & \phi_{n,n+1} \end{bmatrix} \cdot \begin{bmatrix} 1 & 0 & 0 & 0 & \dots & 0 \\ s_1 & 1 & 0 & 0 & \dots & 0 \\ 0 & s_2 & 1 & 0 & \dots & 0 \\ 0 & 0 & s_3 & 1 & \dots & 0 \\ \vdots & \vdots & \vdots & \ddots & \ddots & \vdots \\ 0 & 0 & 0 & \dots & s_{n-1} & 1 \\ 0 & 0 & 0 & \dots & 0 & s_n \end{bmatrix} z + F(0,w) \tag{5.36}$$

Substituting (5.36) for $\dot{x}$ in (5.35) we obtain

$$\begin{aligned} \dot{z} &= \begin{bmatrix} \phi_{11} + \phi_{12}s_1 & \phi_{12} & 0 & \dots & 0 \\ \bigstar_1 & \bigstar_1 + \phi_{23}s_2 & \phi_{23} & \dots & 0 \\ \vdots & \vdots & \ddots & \ddots & \vdots \\ \bigstar_{n-2} & \bigstar_{n-2} & \dots & \ddots & \phi_{n-1,n} \\ \bigstar_{n-1} & \bigstar_{n-1} & \bigstar_{n-1} & \dots & \bigstar_{n-1} + \phi_{n,n+1}s_n \end{bmatrix} z \\ &\qquad + T(x)\,F(0,w) \\ &= \begin{bmatrix} \phi_{11} & \phi_{12} & 0 & 0 & \dots & 0 \\ \bigstar_1 & \bigstar_1 & \phi_{23} & 0 & \dots & 0 \\ \bigstar_2 & \bigstar_2 & \bigstar_2 & \phi_{34} & \dots & 0 \\ \vdots & \vdots & \vdots & \ddots & \ddots & \vdots \\ \bigstar_{n-2} & \bigstar_{n-2} & \bigstar_{n-2} & \dots & \bigstar_{n-2} & \phi_{n-1,n} \\ \bigstar_{n-1} & \bigstar_{n-1} & \bigstar_{n-1} & \bigstar_{n-1} & \dots & \bigstar_{n-1} \end{bmatrix} z \\ &\quad + \begin{bmatrix} \phi_{12}s_1 & 0 & 0 & \dots & 0 \\ 0 & \phi_{23}s_2 & 0 & \dots & 0 \\ 0 & 0 & \phi_{34}s_3 & \dots & 0 \\ \vdots & \vdots & \vdots & \ddots & \vdots \\ 0 & 0 & 0 & \dots & \phi_{n,n+1}s_n \end{bmatrix} z + T(x)\,F(0,w) \end{aligned} \tag{5.37}$$

where $\bigstar_i$ denotes any function depending only on the disturbance w, the states x_1 through x_{i+1}, and the functions s_1 through s_i and their partial derivatives. Note that a function of the type $\circledast_i$ is also of the type $\bigstar_i$, but the converse is not necessarily true because a $\bigstar_i$ function is allowed to depend on w and x_{i+1}. If we let $A(x,w)$ and $D(x,w)$ denote the first and second matrices in (5.37), respectively, then (5.37) becomes

$$\dot{z} \;=\; \big[A(x,w) + D(x,w)\big]z + T(x)\,F(0,w) \tag{5.38}$$

We now calculate the derivative of $V(x) := z^{\mathrm{T}}z$ as follows:

$$\dot{V} \;=\; z^{\mathrm{T}}\big[A(x,w) + A^{\mathrm{T}}(x,w) + 2D(x,w)\big]z \;+\; 2F^{\mathrm{T}}(0,w)\,T^{\mathrm{T}}(x)\,z \tag{5.39}$$

Using Young's inequality ($2ab \le a^2 + b^2$) on the last term in (5.39),

$$\dot{V} \le z^{\mathrm{T}}\big[A(x,w) + A^{\mathrm{T}}(x,w) + 2D(x,w) + T(x)T^{\mathrm{T}}(x)\big]z + \big\|F(0,w)\big\|^2 \tag{5.40}$$

and from (5.35) we have

$$T(x)\,T^{\mathrm{T}}(x) \;=\; I_{n\times n} + \begin{bmatrix} 0 & \circledast_1 & \circledast_2 & \cdots & \circledast_{n-1} \\ \circledast_1 & \circledast_1 & \circledast_2 & \cdots & \circledast_{n-1} \\ \circledast_2 & \circledast_2 & \circledast_2 & \cdots & \circledast_{n-1} \\ \vdots & \vdots & \vdots & \ddots & \vdots \\ \circledast_{n-1} & \circledast_{n-1} & \circledast_{n-1} & \cdots & \circledast_{n-1} \end{bmatrix} \tag{5.41}$$

We combine (5.40) and (5.41) and use the definition of $A(x,w)$ from (5.37) and (5.38) to obtain

$$\begin{aligned} \dot{V} \;&\le\; z^{\mathrm{T}}z + \big\|F(0,w)\big\|^2 + 2z^{\mathrm{T}}D(x,w)\,z \\ &\quad + z^{\mathrm{T}} \begin{bmatrix} 2\phi_{11} & \bigstar_1 & \bigstar_2 & \cdots & \bigstar_{n-1} \\ \bigstar_1 & \bigstar_1 & \bigstar_2 & \cdots & \bigstar_{n-1} \\ \bigstar_2 & \bigstar_2 & \bigstar_2 & \cdots & \bigstar_{n-1} \\ \vdots & \vdots & \vdots & \ddots & \vdots \\ \bigstar_{n-1} & \bigstar_{n-1} & \bigstar_{n-1} & \cdots & \bigstar_{n-1} \end{bmatrix} z \end{aligned} \tag{5.42}$$

We now show that for any $c > 1$ there exist choices for the s_i such that

$$\max_{w \in B} \dot{V} \;\le\; -(c-1)\,z^{\mathrm{T}}z \;+\; \max_{w \in B} \big\|F(0,w)\big\|^2 \tag{5.43}$$

Using the definition of $D(x,w)$ from (5.37), it follows from (5.42) that

$$\dot{V} \;\leq\; -(c-1)\,z^{\mathrm{T}}z \;-\; z^{\mathrm{T}}M(x,w)\,z \;+\; \big\|F(0,w)\big\|^2 \qquad (5.44)$$

where the symmetric matrix $M(x,w)$ is given by

$$M=\begin{bmatrix} -c-2\phi_{11}-2\phi_{12}s_1 & \bigstar_1 & \bigstar_2 & \cdots & \bigstar_{n-1} \\ \bigstar_1 & \bigstar_1-2\phi_{23}s_2 & \bigstar_2 & \cdots & \bigstar_{n-1} \\ \bigstar_2 & \bigstar_2 & \bigstar_2-2\phi_{34}s_3 & \cdots & \bigstar_{n-1} \\ \vdots & \vdots & \vdots & \ddots & \vdots \\ \bigstar_{n-1} & \bigstar_{n-1} & \bigstar_{n-1} & \cdots & \bigstar_{n-1}-2\phi_{n,n+1}s_n \end{bmatrix} \qquad (5.45)$$

If we can choose the functions s_i such that this matrix $M(x,w)$ is positive definite for all $x \in \mathcal{X}$ and all $w \in B$, then (5.43) will follow from (5.44). It suffices to show that each leading minor of M can be made positive for all $x \in \mathcal{X}$ and all $w \in B$. This leads to the construction of the functions s_i as follows. The first leading minor of M is simply

$$M_1(x_1,w) \;\;:=\;\; -c \;-\; 2\phi_{11}(x_1,w) \;-\; 2\phi_{12}(x_1,w)\,s_1(x_1) \qquad (5.46)$$

Because the function ϕ_{12} is never zero by assumption (5.25), it is straightforward to construct a smooth function s_1 such that $M_1(x_1,w)$ is strictly positive for all $x_1 \in I\!R$ and all $w \in B$. Once s_1 has been determined, we can calculate explicitly the functions $\bigstar_1$ in (5.45). The second leading minor of M is given by the determinant of the matrix

$$M_2(x_1,x_2,w) \;\;:=\;\; \begin{bmatrix} M_1(x_1,w) & \bigstar_1 \\ \bigstar_1 & \bigstar_1 - 2\phi_{23}(x_1,x_2,w)\,s_2(x_1,x_2) \end{bmatrix}$$

Because both functions M_1 and ϕ_{23} are never zero, there exists a smooth function s_2 such that $\det M_2(x_1,x_2,w)$ is strictly positive for all $x_1, x_2 \in I\!R$ and all $w \in B$. In the same manner, we continue to construct each function s_i in turn so that the corresponding leading minor $\det M_i$ is always positive. Once all the functions s_i have been chosen, the feedback control $u(x) = z_n s_n(x)$ is known explicitly, and it achieves the inequality (5.43). We conclude that V is an rclf for the system, and that V satisfies the small control property when $F(0,w) \equiv 0$. ■

5.2.3 Backstepping design procedure

The design procedure outlined in the proof of Theorem 5.1 can be summarized as follows. We first choose the design parameter $c > 1$ and a smooth function $s_1(x_1)$ so that M_1 defined in (5.46) is always positive. As discussed in the beginning of this section, one may think of this step as constructing a conceptual control law $x_2 = x_1 s_1(x_1)$ for the first equation in (5.22), namely

$$\dot{x}_1 = \phi_{11}(x_1, w)\, x_1 + \phi_{12}(x_1, w)\, x_2 + F_1(0, w) \tag{5.47}$$

pretending that x_2 is the control variable. Next we use (5.32) and (5.34) to calculate the $\circledast_1$ entry in (5.35) (this involves computing the derivative of s_1 with respect to x_1). Algebraic manipulations then lead to an explicit construction of the $\bigstar_1$ entries in (5.45). We next choose the function s_2 such that $\det M_2$ is always positive; one can think of this step as constructing a conceptual control law $x_3 = z_2 s_2(x_1, x_2)$ for the first two equations in (5.22), namely

$$\begin{aligned} \dot{x}_1 &= \phi_{11}(x_1, w)\, x_1 + \phi_{12}(x_1, w)\, x_2 + F_1(0, w) && (5.48) \\ \dot{x}_2 &= \phi_{21}(x_1, x_2, w)\, x_1 + \phi_{22}(x_1, x_2, w)\, x_2 \\ &\quad + \phi_{23}(x_1, x_2, w)\, x_3 + F_2(0, w) && (5.49) \end{aligned}$$

pretending that x_3 is the control variable. Once s_2 is determined, we go back to (5.32) and (5.34) to calculate the $\circledast_2$ entries in (5.35) (this involves computing the partial derivatives of s_2). Algebraic manipulations then lead to an explicit construction of the $\bigstar_2$ entries in (5.45), and the construction of the function s_3 follows. So the design proceeds until the function s_n is calculated. At this point, the new state vector z, the rclf $V(x) = z^{\mathrm{T}} z$, and the control law $u(x) = z_n s_n(x)$ are all explicitly known.

Once the function s_1 is chosen, the procedure for constructing the remaining functions s_i for $i \in \{2, \ldots, n\}$ can be stated in the following recursive form:

Step i_1: use (5.32) and (5.34) to calculate the $\circledast_{i-1}$ entries in (5.35) (this involves computing the partial derivatives of s_{i-1})

Step i_2: use (5.36), (5.37), and (5.40) to calculate the $\bigstar_{i-1}$ entries in (5.45)

Step i_3: choose a smooth function $s_i(x_1, \ldots, x_i)$ so that the determinant of the matrix $M_i(x_1, \ldots, x_i, w)$ given by

$$M_i \;:=\; \begin{bmatrix} & & & \bigstar_{i-1} \\ & M_{i-1} & & \vdots \\ & & & \bigstar_{i-1} \\ \bigstar_{i-1} & \cdots & \bigstar_{i-1} & \bigstar_{i-1} - 2\phi_{i,i+1}\, s_i \end{bmatrix}$$

is strictly positive for all $x_1, \ldots, x_i \in I\!R$ and all $w \in B$

Thus the construction of each function s_i involves computing the partial derivatives of s_{i-1} (step i_1) and performing algebraic manipulations (steps i_1 through i_3).

The size of the residual set Ω in the above design can be calculated from (5.43) as follows:

$$\Omega \;=\; \left\{ x \in \mathcal{X} \;:\; z^{\mathrm{T}} z \;\leq\; \frac{1}{c-1} \max_{w \in B} \left\| F(0,w) \right\|^2 \right\} \tag{5.50}$$

Thus the residual set is $\Omega = \{0\}$ when $F(0,w) \equiv 0$, in which case we achieve convergence to the equilibrium. Otherwise the size of the residual set depends on the choice of the design parameter c. We can make Ω arbitrarily small in the z-coordinates by choosing c large, but this does *not* necessarily make Ω small in the original x-coordinates. Because $z_1 = x_1$, however, we can always make Ω arbitrarily small in the x_1-direction, possibly at the expense of making it large in other directions.

There are other design flexibilities in this backstepping procedure besides the choices for the functions s_i. For example, instead of the rclf $V(x) = z^{\mathrm{T}} z$ one could choose $V(x) = z^{\mathrm{T}} P z$ for a diagonal positive definite matrix P. This would introduce $n-1$ additional independent design parameters. Also, one could introduce weighting functions in the application of Young's inequality in (5.39) as follows:

$$\begin{aligned} 2F^{\mathrm{T}}(0,w)\, T^{\mathrm{T}}(x)\, z \;\leq\;& z^{\mathrm{T}}\, T(x)\, [\Pi(x)]^{-1}\, T^{\mathrm{T}}(x)\, z \\ &+\; F^{\mathrm{T}}(0,w)\, \Pi(x)\, F(0,w) \end{aligned} \tag{5.51}$$

where $\Pi(x)$ is a diagonal weighting matrix with positive entries. One would require that the ith entry in $\Pi(x)$ depend only on x_1 through x_i so that the structure of the matrix M in (5.45) is preserved, and that M be rendered large enough for $z^{\mathrm{T}}M(x,w)\,z$ to dominate $F^{\mathrm{T}}(0,w)\,\Pi(x)F(0,w)$ outside a compact set. This weighting function $\Pi(x)$ can be used to reduce the excessive growth of the nonlinearities caused by the quadratic term $T(x)\,T^{\mathrm{T}}(x)$ in (5.40), and it also determines a trade-off between control effort (size of the functions s_i) and disturbance attenuation (size of the residual set Ω). For the special case of vanishing disturbances in which $F(0,w) \equiv 0$, we can choose $\Pi(x) = \infty I_{n\times n}$ and thus drop the term $z^{\mathrm{T}}T(x)\,T^{\mathrm{T}}(x)z$ from the inequality (5.40). In this case we obtain

$$\dot{V} \;\leq\; -cz^{\mathrm{T}}z \;-\; z^{\mathrm{T}}M(x,w)\,z \tag{5.52}$$

where $M(x,w) = -cI_{n\times n} - A(x,w) - A^{\mathrm{T}}(x,w) - 2D(x,w)$.

5.2.4 A benchmark example

Let us now apply the robust backstepping procedure to the system (5.28)–(5.29), namely, the third-order system

$$\dot{x}_1 \;=\; x_2 + x_1^3\,w \tag{5.53}$$

$$\dot{x}_2 \;=\; x_3$$

$$\dot{x}_3 \;=\; u \tag{5.54}$$

where the disturbance w takes values in the interval $[-1,1]$. We note that this is a severe disturbance because $x_1^3\,w$ with $w > 0$ is destabilizing and can force x_1 to escape to infinity in finite time.

The functions $F(x,w)$ and $G(x,w)$ in (5.23) are given by

$$F(x,w) \;=\; \begin{bmatrix} x_1^2 w & 1 & 0 \\ 0 & 0 & 1 \\ 0 & 0 & 0 \end{bmatrix} x \qquad G(x,w) \;=\; \begin{bmatrix} 0 \\ 0 \\ 1 \end{bmatrix} \tag{5.55}$$

which means $\phi_{11} = x_1^2 w$, $\phi_{12} = \phi_{23} = \phi_{34} = 1$, and all other functions ϕ_{ij} are zero. We begin by choosing $c = 2$ and $s_1(x_1) = -2 - x_1^2$ so that M_1 defined in (5.46) is strictly positive for all $x_1 \in I\!R$ and all $w \in [-1,1]$:

$$\begin{aligned} M_1(x_1,w) \;&=\; 2 + 2x_1^2(1-w) \\ &\geq\; 2 \end{aligned} \tag{5.56}$$

Using (5.32) and (5.34) to calculate the $\circledast_1$ entry in (5.35), we obtain

$$T(x) = \begin{bmatrix} 1 & 0 & 0 \\ 2+3x_1^2 & 1 & 0 \\ \circledast_2 & \circledast_2 & 1 \end{bmatrix} \tag{5.57}$$

We next compute $\dot{z} = T(x)\,\dot{x}$ from (5.36) and (5.57):

$$\dot{z} = \begin{bmatrix} x_1^2 w + s_1 & 1 & 0 \\ (2+3x_1^2)(x_1^2 w + s_1) & 2+3x_1^2+s_2 & 1 \\ \star_2 & \star_2 & \star_2 + s_3 \end{bmatrix} z \tag{5.58}$$

Using the expression for $M(x,w)$ following (5.52), we obtain

$$M = \begin{bmatrix} M_1(x_1,w) & -1-(2+3x_1^2)(x_1^2 w + s_1) & \star_2 \\ -1-(2+3x_1^2)(x_1^2 w + s_1) & -6-6x_1^2-2s_2 & \star_2 \\ \star_2 & \star_2 & \star_2 - 2s_3 \end{bmatrix}$$

We must now choose the function $s_2(x_1,x_2)$ so that the determinant of

$$M_2 = \begin{bmatrix} M_1(x_1,w) & -1-(2+3x_1^2)(x_1^2 w + s_1) \\ -1-(2+3x_1^2)(x_1^2 w + s_1) & -6-6x_1^2-2s_2 \end{bmatrix}$$

is strictly positive for all $x_1, x_2 \in I\!R$ and all $w \in [-1,1]$. For this particular example, s_2 can be a function of x_1 alone; this is due to the fact that our system (5.53)–(5.54) is linear in x_2. We thus choose $s_2(x_1)$ to be any smooth function satisfying

$$s_2(x_1) \leq -3 - 3x_1^2 - \max_{w\in[-1,1]} \frac{1+\left[1+(2+3x_1^2)(x_1^2 w + s_1)\right]^2}{4+4x_1^2(1-w)} \tag{5.59}$$

This choice yields $M_2(x_1,w) \geq 1$ for all $x_1 \in I\!R$ and all $w \in [-1,1]$.

We continue the design by using (5.32) and (5.34) to complete the matrix in (5.57):

$$T(x) = \begin{bmatrix} 1 & 0 & 0 \\ 2+3x_1^2 & 1 & 0 \\ s_1 s_2 + \frac{d(s_1 s_2)}{dx_1}x_1 - \frac{ds_2}{dx_1}x_2 & -s_2 & 1 \end{bmatrix} \tag{5.60}$$

With (5.36) and (5.60) we complete the matrix in (5.58):

$$\dot{z} = \begin{bmatrix} x_1^2 w + s_1 & 1 & 0 \\ (2+3x_1^2)(x_1^2 w + s_1) & 2+3x_1^2+s_2 & 1 \\ t_{31}\cdot(x_1^2 w + s_1) & t_{31} - s_2^2 & -s_2 + s_3 \end{bmatrix} z \tag{5.61}$$

where

$$t_{31}(x_1,x_2) \quad := \quad s_1 s_2 + \frac{d(s_1 s_2)}{dx_1} x_1 - \frac{ds_2}{dx_1} x_2 \tag{5.62}$$

denotes the bottom left entry in the matrix (5.60). Finally, we use the expression for $M(x,w)$ following (5.52) to obtain

$$M(x,w) = \begin{bmatrix} \multicolumn{2}{c}{M_2(x_1,w)} & \begin{matrix} -t_{31}\cdot(x_1^2 w + s_1) \\ -1 - t_{31} + s_2^2 \end{matrix} \\ -t_{31}\cdot(x_1^2 w + s_1) & -1 - t_{31} + s_2^2 & -2 + 2s_2 - 2s_3 \end{bmatrix}$$

We have left to choose a smooth function s_3 so that the determinant of $M(x,w)$ is strictly positive for all $x \in I\!R^3$ and all $w \in [-1,1]$. For this particular example, s_3 can be a function of x_1 and x_2 alone. Once we have determined such a function s_3, a robustly stabilizing control law for the system (5.53)–(5.54) is simply $u = z_3 s_3(x_1,x_2)$.

5.3 Flattened rclf's for softer control laws

The design procedure outlined in Sections 5.2.2–5.2.4 is based on an rclf which is quadratic in a set of (nonlinearly) transformed z-coordinates. This quadratic-like rclf was chosen for expository convenience only; as we shall soon discover, such an rclf can lead to control laws which exhibit unnecessarily poor large-signal behavior.

5.3.1 Hardening of control laws

We now identify an undesirable property of the quadratic-like rclf used in the backstepping design of Sections 5.2.2–5.2.4. We show that control laws generated by this rclf are unnecessarily "hard," that is, they exhibit unnecessarily high local gains in some regions of the state space. These high gains can cause excessive control effort such as high-magnitude chattering in the control signal. The hardening property is best illustrated through the control design for the second-order system

$$\dot{x}_1 = x_2 + |x_1|^{1+r} w \tag{5.63}$$

$$\dot{x}_2 = u \tag{5.64}$$

where w is a scalar disturbance taking values in the interval $B = [-1, 1]$. The growth parameter $r > 0$, which appears in the exponent of the nonlinearity multiplying the disturbance, will be significant in revealing the hardening property.

The system (5.63)–(5.64) is clearly in strict feedback form, so the design of the previous section applies. We consider the rclf

$$V(x) \;=\; x_1^2 + p\big[x_2 - x_1\, s_1(x_1)\big]^2 \tag{5.65}$$

where $p > 0$ is a design parameter and $s_1(x_1)$ is smooth. This rclf is of the quadratic type $V(x) = z^{\mathrm{T}}Pz$ where $z_1 := x_1$ and $z_2 := x_2 - x_1\, s_1(x_1)$; see also (5.14). The worst-case derivative of V is given by

$$\begin{aligned} \max_{w\in B}\, \dot{V} \;&=\; \Big|2x_1 - 2pz_2[s_1 + x_1 s_1']\Big|\, |x_1|^{1+r} \\ &\quad + \Big[2x_1 - 2pz_2[s_1 + x_1 s_1']\Big] x_2 \;+\; 2pz_2 u \end{aligned} \tag{5.66}$$

where s_1' denotes the derivative of $s_1(x_1)$ with respect to x_1.

We will show below that every smooth control law $u(x_1, x_2)$ which renders (5.66) nonpositive also satisfies

$$\left|\frac{\partial u}{\partial x_2}\right| \;\geq\; \frac{p}{4}\, |x_1|^{3r} \tag{5.67}$$

on some set $M \subset \mathcal{X}$ which is unbounded in the x_1-direction. In other words, the local gain of the control law in the x_2-direction grows like $|x_1|^{3r}$ in the set M as $|x_1| \to \infty$. The exponent $3r$ which quantifies this hardening property is not affected by the choices of the function s_1 or the parameter p in (5.65). Moreover, it characterizes *every* control law which makes the Lyapunov derivative negative.

What is encouraging is that the growth of the local gain in (5.67) is not necessary for robust stabilization; it is merely an artifact of the quadratic-like rclf. Indeed, in Section 5.3.2 we will construct a different type of rclf which results in much slower growth (on the order of $|x_1|^r$).

Evaluating (5.66) at points where $x_2 = x_1 s_1(x_1)$ (that is, where $z_2 = 0$), we obtain the following:

$$\max_{w\in B}\, \dot{V}\Big|_{z_2=0} \;=\; 2|x_1|^{2+r} \;+\; 2x_1^2 s_1(x_1) \tag{5.68}$$

If V is an rclf, then (5.68) must be nonpositive for all $x_1 \in \mathbb{R}$, and it follows that $s_1(x_1) \leq -|x_1|^r$ for all $x_1 \in \mathbb{R}$. We claim that the set

$$\mathcal{I} \quad := \quad \left\{x_1 \in \mathbb{R} \,:\, x_1 s_1'(x_1) \,<\, 0\right\} \tag{5.69}$$

is unbounded. Indeed, if $\mathcal{I}$ were bounded, then there would exist $a > 0$ such that $x_1 s_1'(x_1) \geq 0$ for all $x_1 \geq a$, which would mean s_1 is nondecreasing on the interval $[a, \infty)$; however, this contradicts the fact that $s_1(x_1) \leq -|x_1|^r$ for all $x_1 \in \mathbb{R}$. Furthermore, for all $x_1 \in \mathcal{I}$ we have

$$s_1(x_1) + x_1 s_1'(x_1) \quad \leq \quad s_1(x_1) \quad \leq \quad -|x_1|^r \tag{5.70}$$

Let $u(x_1, x_2)$ be any smooth feedback law which renders (5.66) everywhere nonpositive. We will now derive an estimate of the partial derivative $\partial u/\partial x_2$. For $x_1 \neq 0$ we define

$$\sigma(x_1) \quad := \quad \frac{2}{p}\,|x_1|^{1-r} \tag{5.71}$$

Evaluating (5.66) at points where where $x_1 \neq 0$ and $z_2 = \sigma(x_1)$ and setting the result less than or equal to zero, we obtain

$$\begin{aligned} 2p\sigma u(x_1,\, x_1 s_1 + \sigma) \;\leq\; & -\Big|2x_1 \,-\, 2p\sigma[s_1 + x_1 s_1']\Big|\,|x_1|^{1+r} \\ & -\Big[2x_1 \,-\, 2p\sigma[s_1 + x_1 s_1']\Big]\,[x_1 s_1 + \sigma] \end{aligned} \tag{5.72}$$

Similarly, at points where $x_1 \neq 0$ and $z_2 = -\sigma(x_1)$ we obtain

$$\begin{aligned} -2p\sigma u(x_1,\, x_1 s_1 - \sigma) \;\leq\; & -\Big|2x_1 \,+\, 2p\sigma[s_1 + x_1 s_1']\Big|\,|x_1|^{1+r} \\ & -\Big[2x_1 \,+\, 2p\sigma[s_1 + x_1 s_1']\Big]\,[x_1 s_1 - \sigma] \end{aligned} \tag{5.73}$$

We now add the inequalities (5.72) and (5.73):

$$\begin{aligned} 2p\sigma\Big[u(x_1,\, & x_1 s_1 + \sigma) \\ - u(x_1,\, x_1 s_1 - \sigma)\Big] \;\leq\; & -\Big|2x_1 \,-\, 2p\sigma[s_1 + x_1 s_1']\Big|\,|x_1|^{1+r} \\ & -\Big|2x_1 \,+\, 2p\sigma[s_1 + x_1 s_1']\Big|\,|x_1|^{1+r} \\ & -4x_1^2 s_1 \,+\, 4p\sigma^2[s_1 + x_1 s_1'] \end{aligned} \tag{5.74}$$

for all $x_1 \neq 0$. From the inequality $2|b| \leq |a - b| + |a + b|$, we obtain

$$\begin{aligned} 2p\sigma\Big[u(x_1,\, & x_1 s_1 + \sigma) \\ - u(x_1,\, x_1 s_1 - \sigma)\Big] \;\leq\; & -4p\sigma|x_1|^{1+r}|s_1 + x_1 s_1'| \\ & -4x_1^2 s_1 \,+\, 4p\sigma^2[s_1 + x_1 s_1'] \end{aligned} \tag{5.75}$$

for all $x_1 \neq 0$. It follows from (5.70) that the last term on the right-hand side of (5.75) is negative for $x_1 \in \mathcal{I}$ and can therefore be dropped; it then follows from (5.70) that

$$\begin{aligned} 2p\sigma\Big[u(x_1,\, x_1 s_1 + \sigma) & \\ -\; u(x_1,\, x_1 s_1 - \sigma)\Big] \;&\leq\; -4|s_1 + x_1 s_1'|\Big[p\sigma|x_1|^{1+r} - x_1^2\Big] \end{aligned} \tag{5.76}$$

for all $x_1 \in \mathcal{I}$. We substitute for σ from (5.71) in the right-hand side of (5.76) to obtain

$$2p\sigma\Big[u(x_1,\, x_1 s_1 + \sigma) \;-\; u(x_1,\, x_1 s_1 - \sigma)\Big] \;\leq\; -4x_1^2|s_1 + x_1 s_1'| \tag{5.77}$$

for all $x_1 \in \mathcal{I}$. Dividing both sides by $4p\sigma^2$ we get

$$\frac{u(x_1,\, x_1 s_1 + \sigma) \;-\; u(x_1,\, x_1 s_1 - \sigma)}{2\sigma} \;\leq\; -\frac{x_1^2|s_1 + x_1 s_1'|}{p\sigma^2} \tag{5.78}$$

for all $x_1 \in \mathcal{I}$. Finally, from (5.71) and (5.70) we see that

$$\frac{u(x_1,\, x_1 s_1 + \sigma) \;-\; u(x_1,\, x_1 s_1 - \sigma)}{2\sigma} \;\leq\; -\frac{p}{4}\,|x_1|^{3r} \tag{5.79}$$

for all $x_1 \in \mathcal{I}$. It now follows from the mean value theorem that for each $x_1 \in \mathcal{I}$ there exists $\xi \in \mathbb{R}$ with $|\xi| \leq \sigma(x_1)$ such that

$$\frac{\partial u}{\partial x_2}(x_1,\, x_1 s_1 + \xi) \;\leq\; -\frac{p}{4}\,|x_1|^{3r} \tag{5.80}$$

Therefore these (x_1, ξ) pairs form a set $M \subset \mathcal{X}$ which is unbounded in x_1 (because $\mathcal{I}$ is unbounded) and is such that

$$\left|\frac{\partial u}{\partial x_2}\right| \;\geq\; \frac{p}{4}\,|x_1|^{3r} \tag{5.81}$$

on M. In other words, the local gain of the control law in the x_2-direction grows like $|x_1|^{3r}$ in the set M as $|x_1| \to \infty$. The exponent $3r$ in this growth factor is not affected by the choices of the function s_1 or the parameter p in (5.65), and it applies to *every* control law which makes the Lyapunov derivative negative.

This hardening of the control law, that is, the growth of its local gain, can be seen in Figure 5.1 on page 128. This figure shows the surface of a robustly stabilizing control law for the system (5.63)–(5.64) with $r = 2$, designed using the rclf V in (5.65). In this case, the slope of the control surface in the x_2-direction grows like $|x_1|^6$.

Hardening can become an even greater problem in higher-dimensional systems. Recall that the function s_2 in the second step of the general design in Section 5.2 is chosen so that the conceptual control law $x_3 = z_2 s_2(x_1, x_2)$ robustly stabilizes the system (5.48)–(5.49) with x_3 as the conceptual control variable. As we have just shown above, this function s_2 might be excessively large in some regions of the state space regardless of design choices. Because the $\bigstar_2$ entries in (5.45) depend on (and are amplified by) the function s_2 and its partial derivatives, these entries may also be excessively large in some areas. As a result, the function s_3, which must dominate these $\bigstar_2$ entries, will inherit the hardening of s_2. Thus hardening is propagated and amplified through each step of the recursive design, in general resulting in extremely large local gains of the control law $u = z_n s_n(x)$. The next section outlines a modification of the backstepping design which reduces hardening and results in lower gains and less control effort.

5.3.2 Flattened rclf's

We have just seen that the quadratic-like rclf (5.65) for the example system (5.63)–(5.64) leads to control laws having local gains which grow like $|x_1|^{3r}$. In this section, we will show how to modify the rclf so that the resulting control laws have local gains which grow only like $|x_1|^r$. This reduction in hardening will greatly reduce the control effort required for robust stabilization, with no sacrifice in performance.

Recall that the quadratic-like rclf for the system (5.63)–(5.64) was

$$V(x) \;=\; x_1^2 + p\big[x_2 \;-\; x_1\, s_1(x_1)\big]^2 \tag{5.82}$$

The second term in (5.82) penalizes the distance (in the x_2-direction) to the manifold defined by $x_2 = x_1 s_1$. We obtain a new rclf by "flattening" this penalty term, that is, by penalizing the distance to a region around the manifold rather than the distance to the manifold itself:

$$V(x) = x_1^2 + p\begin{cases} \big[x_2 - x_1 s_1(x_1) - \varrho_1(x_1)\big]^2 & \text{when } \; z_2 \geq \varrho_1(x_1) \\ 0 & \text{when } \; |z_2| \leq \varrho_1(x_1) \\ \big[x_2 - x_1 s_1(x_1) + \varrho_1(x_1)\big]^2 & \text{when } \; z_2 \leq -\varrho_1(x_1) \end{cases} \tag{5.83}$$

where $z_2 := x_2 - x_1 s_1$ as above, and ϱ_1 is a smooth nonnegative scalar function to be determined along with the function s_1 in the design. By requiring $\varrho_1(0) = 0$, we ensure that this choice for V is C^1, positive definite, and radially unbounded. This new rclf (5.83) reduces to the old one (5.82) when $\varrho_1(x_1) \equiv 0$, and so (5.83) represents a generalization of the quadratic-like rclf of the previous section. Our goal is to show that the introduction of ϱ_1 as a new design flexibility allows us to dramatically reduce the undesirable hardening property (5.81) associated with the quadratic-like rclf (5.82).

We first evaluate the derivative of (5.83) along solutions to (5.63)–(5.64) at points where $|z_2| \leq \varrho_1(x_1)$ as follows:

$$\begin{aligned}\dot{V}\Big|_{|z_2|\leq\varrho_1} &= 2x_1\left[z_2 + x_1 s_1(x_1) + |x_1|^{1+r}w\right] \\ &\leq 2|x_1|\varrho_1(x_1) + 2x_1^2 s_1(x_1) + 2|x_1|^{2+r} \end{aligned} \tag{5.84}$$

Note that $\dot{V}$ is independent of the control u in the region where $|z_2| \leq \varrho_1$. For $\dot{V}$ to be negative in this region, it is clearly sufficient that

$$s_1(x_1) \;\leq\; -1 - |x_1|^r - \frac{\varrho_1(x_1)}{|x_1|} \tag{5.85}$$

for all $x_1 \in I\!R$. We can satisfy (5.85) by choosing, for example,

$$\varrho_1(x_1) = |x_1|^{1+r} \tag{5.86}$$

$$s_1(x_1) = -1 - 2\left(1 + x_1^2\right)^{\frac{r}{2}} \tag{5.87}$$

which yields

$$\dot{V}\Big|_{|z_2|\leq\varrho_1} \;\leq\; -2x_1^2 \tag{5.88}$$

We next evaluate the derivative of (5.83) at points where $z_2 \geq \varrho_1$:

$$\begin{aligned}\dot{V}\Big|_{z_2\geq\varrho_1} &= 2x_1\left[z_2 + x_1 s_1 + |x_1|^{1+r}w\right] \\ &\quad + 2p\left[z_2 - \varrho_1\right]\left[u - [s_1 + x_1 s_1' + \varrho_1'][x_2 + |x_1|^{1+r}w]\right] \\ &\leq -2x_1^2 + 2p\left[z_2 - \varrho_1\right]\left[\frac{1}{p}x_1 + u \right. \\ &\qquad \left. - [s_1 + x_1 s_1' + \varrho_1']\left[x_2 + |x_1|^{1+r}w\right]\right]\end{aligned} \tag{5.89}$$

where the second line follows from (5.88). If we let $u^+(x_1, x_2)$ denote the control law in the region where $z_2 \geq \varrho_1$, then for (5.89) to be negative it is sufficient that

$$\begin{aligned} u^+(x_1, x_2) \;\leq\; & -\frac{1}{p}x_1 + [s_1 + x_1 s_1' + \varrho_1']\,x_2 \\ & - |x_1|^{1+r}|s_1 + x_1 s_1' + \varrho_1'| \end{aligned} \tag{5.90}$$

Similarly, we evaluate the derivative of (5.83) at points where $z_2 \leq -\varrho_1$:

$$\begin{aligned} \dot{V}\Big|_{z_2 \leq -\varrho_1} \;\leq\; & -2x_1^2 + 2p\big[z_2 + \varrho_1\big]\bigg[\frac{1}{p}x_1 + u \\ & - [s_1 + x_1 s_1' - \varrho_1']\big[x_2 + |x_1|^{1+r}w\big]\bigg] \end{aligned} \tag{5.91}$$

If we let $u^-(x_1, x_2)$ denote the control law in the region where $z_2 \leq -\varrho_1$, then for (5.91) to be negative it is sufficient that

$$\begin{aligned} u^-(x_1, x_2) \;\geq\; & -\frac{1}{p}x_1 + [s_1 + x_1 s_1' - \varrho_1']\,x_2 \\ & + |x_1|^{1+r}|s_1 + x_1 s_1' - \varrho_1'| \end{aligned} \tag{5.92}$$

Recall from (5.88) that $\dot{V}$ is negative in the region where $|z_2| \leq \varrho_1$, regardless of the control law. We let $u^\circ(x_1, x_2)$ denote the control law in this region, and choose it to be

$$\begin{aligned} u^\circ(x_1, x_2) \;=\; & \frac{u^+(x_1,\, x_1 s_1 + \varrho_1) \;-\; u^-(x_1,\, x_1 s_1 - \varrho_1)}{2\varrho_1}\, z_2 \\ & + \frac{u^+(x_1,\, x_1 s_1 + \varrho_1) \;+\; u^-(x_1,\, x_1 s_1 - \varrho_1)}{2} \end{aligned} \tag{5.93}$$

so that the control law $u(x_1, x_2)$ obtained by patching together the control laws u^+, u^-, and u° from the different regions is continuous and piecewise smooth. Let us now derive a bound on the partial derivative $\partial u/\partial x_2$. From (5.90), (5.92), and (5.93) we have

$$\begin{aligned} \frac{\partial u^\circ}{\partial x_2} \;=\; & \frac{u^+(x_1,\, x_1 s_1 + \varrho_1) \;-\; u^-(x_1,\, x_1 s_1 - \varrho_1)}{2\varrho_1} \\ \;\leq\; & \frac{\varrho_1[s_1 + x_1 s_1'] + x_1 s_1 \varrho_1' - |x_1|^{1+r}|\varrho_1'|}{\varrho_1} \end{aligned} \tag{5.94}$$

If u^+ and u^- are chosen such that the inequalities (5.90) and (5.92) are sufficiently tight (that is, such that the differences between the respective left- and right-hand sides are bounded), then it follows from (5.94), (5.86), and (5.87) that

$$\left|\frac{\partial u^{\circ}}{\partial x_2}\right| \leq \gamma|x_1|^r + \kappa \tag{5.95}$$

for some constants γ and κ. One can derive similar estimates for $\partial u^+/\partial x_2$ and $\partial u^-/\partial x_2$ from (5.90) and (5.92), and we conclude that

$$\left|\frac{\partial u}{\partial x_2}\right| \leq \gamma|x_1|^r + \kappa \tag{5.96}$$

for some (possibly new) constants γ and κ. Let us compare this bound on the local gain of the control law u with the bound

$$\left|\frac{\partial u}{\partial x_2}\right| \geq \frac{p}{4}|x_1|^{3r} \tag{5.97}$$

we obtained in (5.81) using the quadratic-like rclf of the previous section. Recall that the new bound (5.96) is valid over the entire state space, whereas the old bound (5.97) is valid on the set M which is unbounded in the x_1-direction. Both designs exhibit hardening (that is, the growth of the local gain), but the design using the flattened rclf (5.83) is much softer in the sense that the growth of its local gain is much slower ($|x_1|^r$ compared to $|x_1|^{3r}$).

5.3.3 Design example: elimination of chattering

The flattened rclf is used to design a robustly stabilizing control law for the system (5.63)–(5.64) with $r = 2$, and Figure 5.2 shows the control surface resulting from (5.90), (5.92), and (5.93). The flattened rclf itself is pictured in Figure 5.3. Figure 5.1 shows the surface of a control law $u = z_2 s_2(x_1, x_2)$ designed for the same system using the quadratic-like rclf (5.82).

Although both control laws are smooth, the high local gain of the one pictured in Figure 5.1 makes it look discontinuous compared to the one pictured in Figure 5.2. This comparison carries over to simulations: Figure 5.4 shows the control signals of the two closed-loop systems from the same initial condition. Only the control law of Figure 5.1 results in high-magnitude chattering. The state trajectories created by the two control laws are nearly identical for this initial condition, so there is no difference in (state-space) performance.

Simulations from many other initial conditions allow us to conclude that, when compared to the control law designed using the quadratic-like rclf, the control law designed using the flattened rclf achieves the same (and sometimes better) performance in the state space with much less control effort.

We will show in the next chapter how flattening can be performed at each step of the general backstepping procedure of Section 5.2. As explained at the end of Section 5.3.1, the resulting reduction in hardening for higher-dimensional systems may result in performance improvements even more significant than those observed for this second-order example (5.63)–(5.64).

5.4 Nonsmooth backstepping

In Chapter 3, we defined an rclf V to be a *continuously differentiable* function on the state space, and the rclf's we have constructed thus far in this chapter have indeed all been C^1. For some systems, however, it may be easier to find a control law by relaxing this differentiability requirement rather than searching for a C^1 rclf. Indeed, suppose we wish to construct a globally asymptotically stabilizing control law for the system

$$\dot{x}_1 = x_1^2 + x_1 + 2|x_1| - |x_2| \tag{5.98}$$

$$\dot{x}_2 = x_3$$

$$\dot{x}_3 = u \tag{5.99}$$

We might try the recursive backstepping design of Section 5.2 and propose a clf V of the form

$$\begin{aligned} V(x_1, x_2, x_3) &= z^{\mathrm{T}} z \\ &= x_1^2 + \big[x_2 - x_1 s_1(x_1)\big]^2 \\ &\quad + \big[x_3 - [x_2 - x_1 s_1(x_1)] s_2(x_1, x_2)\big]^2 \end{aligned} \tag{5.100}$$

where the functions s_1 and s_2 are yet to be determined. We know from Section 5.2 that (5.100) will be a clf for our system only if the conceptual

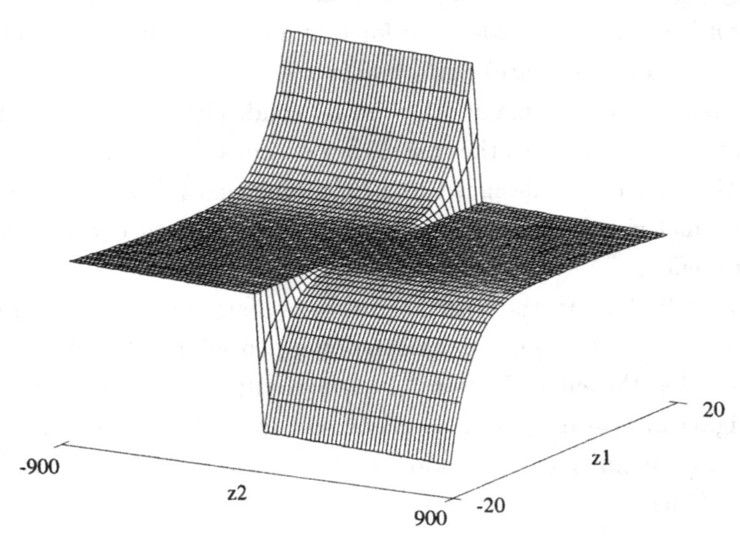

Figure 5.1: Control law designed using quadratic-like rclf $V(x)$ in (5.82). *Source:* Freeman, R. A. and Kokotović, P. V. 1993. Design of 'softer' robust nonlinear control laws. *Automatica* **29**(6), 1425–1437. With permission.

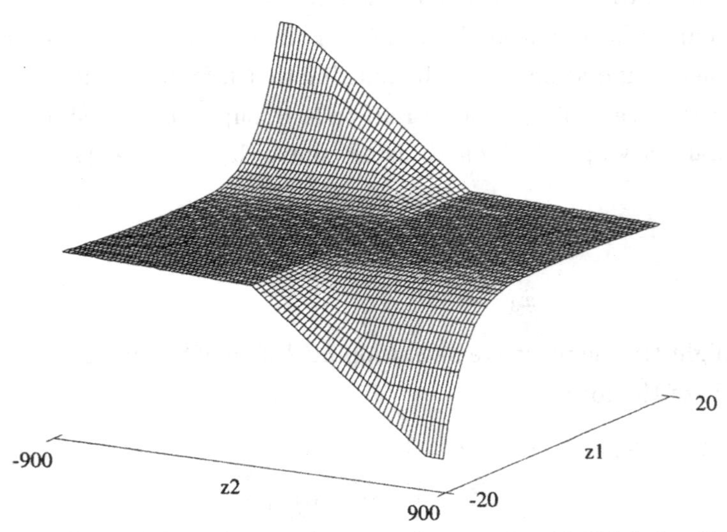

Figure 5.2: Control law designed using flattened rclf $V(x)$ in (5.83). *Source:* Freeman, R. A. and Kokotović, P. V. 1993. Design of 'softer' robust nonlinear control laws. *Automatica* **29**(6), 1425–1437. With permission.

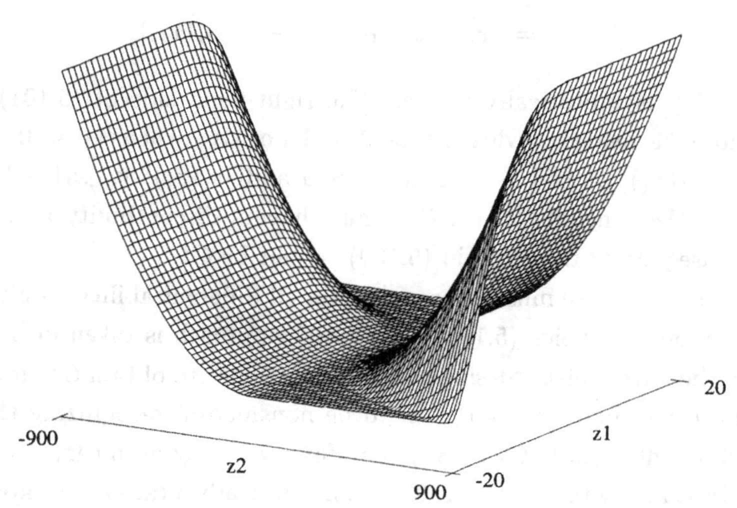

Figure 5.3: Flattened rclf $V(x)$ in (5.83). *Source:* Freeman, R. A. and Kokotović, P. V. 1993. Design of 'softer' robust nonlinear control laws. *Automatica* **29**(6), 1425–1437. With permission.

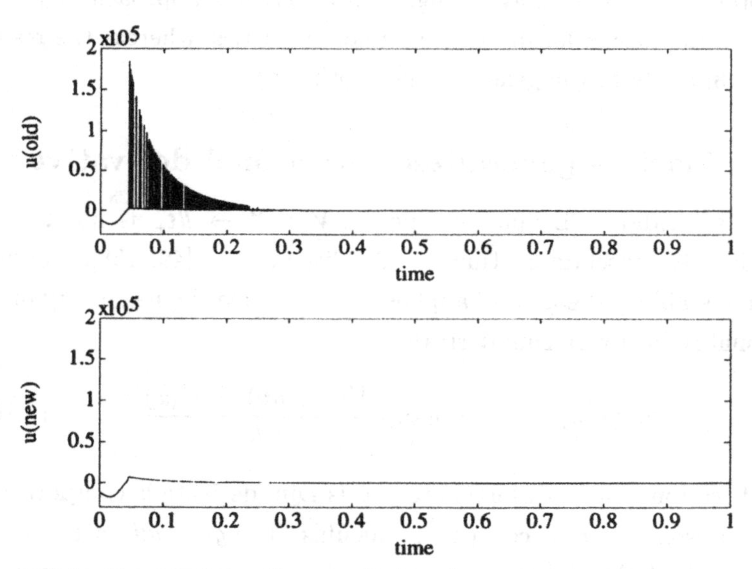

Figure 5.4: Comparison of control signals. *Source:* Freeman, R. A. and Kokotović, P. V. 1993. Design of 'softer' robust nonlinear control laws. *Automatica* **29**(6), 1425–1437. With permission.

closed-loop system

$$\dot{x}_1 \;=\; x_1^2 + x_1 + 2|x_1| - |x_1 s_1(x_1)| \qquad (5.101)$$

is globally asymptotically stable. The right-hand side of (5.101) must therefore be negative when $x_1 > 0$ and positive when $x_1 < 0$, which means $|s_1(x_1)| > |x_1| + 3$ when $x_1 > 0$ and $|s_1(x_1)| < |x_1| + 1$ when $x_1 < 0$. We conclude that $s_1(x_1)$ must have a discontinuity at $x_1 = 0$, and consequently the clf V in (5.100) cannot be C^1.

At this point we must either relax our clf differentiability requirement or abandon the choice (5.100). The latter approach is taken in [24, 118] where the authors use "desingularizing functions" to obtain C^1 clf's. Our approach will be to allow the clf to be nonsmooth, recognizing the fact that differentiability of the Lyapunov function is by no means a requirement in stability theory [49, 158, 5]. The main advantage of our approach is that it allows an explicit construction of the control law following the same steps outlined in Section 5.2. In contrast, the "desingularizing functions" of [24, 118] can be difficult to obtain explicitly and are not easily incorporated into a recursive design. However, our approach is yet restricted to the case of locally Lipschitz nonlinearities, whereas the results in [24, 118] apply to the general nonsmooth case.

5.4.1 Clarke's generalized directional derivative

Suppose a candidate Lyapunov function $V : \mathcal{X} \to I\!R_+$ is not C^1 but merely locally Lipschitz continuous. As discussed in [49, 158], standard Lyapunov stability theory still applies if we examine the negativity of the directional right upper Dini derivate

$$V^+(x; v) \;:=\; \limsup_{h \to 0^+} \frac{V(x + hv) - V(x)}{h} \qquad (5.102)$$

in the direction $v = \dot{x}$. Unfortunately, this Dini derivate is computationally cumbersome with no convenient calculus. The *generalized directional derivative* of Clarke [20] is more suitable for our purposes; it is given by

$$V^\circ(x; v) \;:=\; \limsup_{y \to x,\, h \to 0^+} \frac{V(y + hv) - V(y)}{h} \qquad (5.103)$$

In Clarke's derivative (5.103), the base point y is not fixed at x as it is in the Dini derivate (5.102), and as a result we have $V^+(x;v) \leq V^\circ(x;v)$ for all $x, v \in \mathcal{X}$. While this inequality would seem to make Clarke's derivative less attractive for testing stability ($V^\circ(x;\dot{x}) < 0$ is more restrictive than $V^+(x;\dot{x}) < 0$), the computational advantages of (5.103) compensate for such conservativeness.

We can compute Clarke's generalized directional derivative of a locally Lipschitz function $V : \mathcal{X} \to I\!R_+$ by first constructing his *generalized gradient* [20]. The generalized gradient of V is the set-valued map $\partial V : \mathcal{X} \rightsquigarrow \mathcal{X}^*$ defined by

$$\partial V(x) \;:=\; \Big\{\xi \in \mathcal{X}^* \;:\; \xi \cdot v \leq V^\circ(x;v) \quad \forall v \in \mathcal{X}\Big\} \tag{5.104}$$

This set-valued map ∂V is usc with nonempty compact convex values, and it can be computed directly from V via the formula

$$\partial V(x) \;=\; \mathrm{co}\Big\{\xi \in \mathcal{X}^* \;:\; \exists\{x_i\} \in \mathcal{D}_V,\;\; x_i \to x,\;\; \nabla V(x_i) \to \xi\Big\} \tag{5.105}$$

where $\mathcal{D}_V \subset \mathcal{X}$ denotes the set of points at which the gradient ∇V exists.[1] Roughly speaking, the set $\partial V(x)$ is the convex hull of all gradients of V near x. If V is differentiable at x then $\nabla V(x) \in \partial V(x)$, and if V is *continuously* differentiable at x then $\partial V(x) = \{\nabla V(x)\}$. The computation of the generalized gradient ∂V is facilitated by its basic calculus which includes sum and product formulas, chain rules, and mean-value theorems [20, Section 2.3]. Once ∂V is known, the generalized directional derivative of V is given by the "inverse" of (5.104):

$$\begin{aligned} V^\circ(x;v) \;&=\; \max \partial V(x) \cdot v \\ &=\; \max\Big\{\xi \cdot v \;:\; \xi \in \partial V(x)\Big\} \end{aligned} \tag{5.106}$$

Using Clarke's machinery, we can extend most of the material of Chapter 3 to the case where V is not C^1 but only locally Lipschitz continuous.

5.4.2 Nonsmooth rclf's

Recall that in the C^1 case we defined the Lyapunov derivative of V to be $L_f V(x,u,w) := \nabla V(x) \cdot f(x,u,w)$, where the continuous function f

[1] The set $\mathcal{D}_V$ is dense in $\mathcal{X}$ by Rademacher's theorem, which states that a locally Lipschitz function on $\mathcal{X}$ is differentiable almost everywhere.

described the system dynamics, that is, $\dot{x} = f(x, u, w)$. If V is not C^1, we simply modify this definition according to (5.106) as follows:

$$\begin{aligned} L_f V(x, u, w) &:= V^{\circ}(x; f(x, u, w)) \\ &= \max \partial V(x) \cdot f(x, u, w) \end{aligned} \tag{5.107}$$

This modification preserves the crucial properties of $L_f V$, namely, upper semicontinuity in (x, u, w) (which follows from Proposition 2.9) and convexity in u (assuming f is affine in u). Therefore the main results in Chapter 3, such as Theorem 3.11, are also valid in this nonsmooth case.

The worst-case value of the Lyapunov derivative (5.107) for all values of the uncertainty $w \in W(x, u)$ can be written as

$$\sup_{w \in W(x,u)} L_f V(x, u, w) = \sup_{w \in W(x,u)} \max_{\xi \in \partial V(x)} \xi \cdot f(x, u, w) \tag{5.108}$$

A major advantage of this approach is that we may regard nonsmoothness in V as simply another source of system uncertainty: the variable ξ in (5.108) plays precisely the same role as the disturbance w. We should therefore expect the recursive backstepping design to proceed as before, even though the rclf is not C^1.

5.4.3 Backstepping with nonsmooth nonlinearities

Rather than give the general recursive design for the nonsmooth case, we will illustrate the main ideas by constructing a control law for the system (5.98)–(5.99) using the clf (5.100), written in a more convenient form as

$$V(x_1, x_2, x_3) = x_1^2 + \left[x_2 - \mu(x_1)\right]^2 + \left[x_3 - \nu(x_1, x_2)\right]^2 \tag{5.109}$$

where the functions μ and ν are yet to be determined. The interested reader may refer to [39] for more details on the general design.

We can approximate $\partial V(x)$ from (5.109) using the sum and chain rules from [20, Section 2.3]:

$$\begin{aligned} \partial V(x) \subset\ & 2x_1 \cdot [1\ \ 0\ \ 0] + 2\left[x_2 - \mu(x_1)\right] [-\partial \mu(x_1)\ \ 1\ \ 0] \\ & + 2\left[x_3 - \nu(x_1, x_2)\right] [-\partial \nu(x_1, x_2)\ \ 1] \end{aligned} \tag{5.110}$$

Note that $\partial\mu(x_1)$ will be a set of scalars and $\partial\nu(x_1, x_2)$ will be a set of row vectors of length two. The first step in the design is to choose the function $\mu(x_1)$ such that the conceptual system

$$\dot{x}_1 = x_1^2 + x_1 + 2|x_1| - |\mu(x_1)| \tag{5.111}$$

is globally asymptotically stable. This can be accomplished with the locally Lipschitz choice

$$\mu(x_1) = \begin{cases} x_1^2 + 4x_1 & \text{when } x_1 > 0 \\ 0 & \text{when } x_1 \leq 0 \end{cases} \tag{5.112}$$

Indeed, the function $\phi(x_1) := x_1 \cdot (x_1^2 + x_1 + 2|x_1| - |\mu(x_1)|)$, which should be negative definite for stability, is given by

$$\phi(x_1) = -x_1^2 + \min\{0, x_1^3\} \tag{5.113}$$

Using (5.105), we see that the generalized gradient of μ is

$$\partial\mu(x_1) = \begin{cases} \{2x_1 + 4\} & \text{when } x_1 > 0 \\ [0, 4] & \text{when } x_1 = 0 \\ \{0\} & \text{when } x_1 < 0 \end{cases} \tag{5.114}$$

To facilitate the construction of the function ν, we will rewrite the first system equation (5.98) as

$$\begin{aligned} \dot{x}_1 &= x_1^2 + x_1 + 2|x_1| - |x_2| \\ &= x_1^2 + x_1 + 2|x_1| - |\mu(x_1)| - \psi(x_2, \mu(x_1)) \big[x_2 - \mu(x_1)\big] \end{aligned} \tag{5.115}$$

where the discontinuous function ψ is given by

$$\psi(a, b) := \begin{cases} \dfrac{|a| - |b|}{a - b} & \text{when } a \neq b \\ \operatorname{sgn}(a) & \text{when } a = b \end{cases} \tag{5.116}$$

Note that $|\psi(a, b)| \leq 1$ for all $a, b \in I\!R$. We let $f(x, u)$ denote the dynamics of our system, that is, the right-hand side of (5.98)–(5.99). Our next task is to choose the function $\nu(x_1, x_2)$ so that the Lyapunov derivative $L_f V$, evaluated at points where $x_3 = \nu(x_1, x_2)$, is negative definite. We begin from (5.107) and calculate from (5.110) as follows:

$$\begin{aligned} \partial V(x) \cdot f(x,u)\Big|_{x_3=\nu} \subset\; & 2\Big[x_1 - \big[x_2 - \mu(x_1)\big]\,\partial\mu(x_1)\Big]\cdot\Big[x_1^2 + x_1 + 2|x_1| \\ & - |\mu(x_1)| - \psi(x_2, \mu(x_1))\,\big[x_2 - \mu(x_1)\big]\Big] \\ & + 2\big[x_2 - \mu(x_1)\big]\,\nu(x_1, x_2) \end{aligned} \tag{5.117}$$

Rearranging the right-hand side of (5.117) and substituting for the function ϕ, we obtain

$$\begin{aligned} \partial V(x) \cdot f(x,u)\Big|_{x_3=\nu} \subset\; & 2\phi(x_1) - 2\big[x_2 - \mu(x_1)\big]\Big[x_1\psi(x_2, \mu(x_1)) \\ & + \partial\mu(x_1)\big[x_1^2 + x_1 + 2|x_1| - |\mu(x_1)|\big]\Big] \\ & + 2\big[x_2 - \mu(x_1)\big]^2\,\partial\mu(x_1)\,\psi(x_2, \mu(x_1)) \\ & + 2\big[x_2 - \mu(x_1)\big]\,\nu(x_1, x_2) \end{aligned} \tag{5.118}$$

If we now define $z_1 := x_1$ and $z_2 := x_2 - \mu(x_1)$ and choose the function ν to be of the form $\nu(x_1, x_2) = z_2 s_2(x_1, x_2)$, then we see that (5.118) can be written in the form

$$\partial V \cdot f\Big|_{x_3=\nu} \subset -\begin{bmatrix} z_1 & z_2 \end{bmatrix}\begin{bmatrix} 2 - 2\min\{0, x_1\} & \bigstar_1 \\ \bigstar_1 & \bigstar_1 - 2s_2 \end{bmatrix}\begin{bmatrix} z_1 \\ z_2 \end{bmatrix} \tag{5.119}$$

where the $\bigstar_1$ entries depend on x_1, x_2, ψ, μ, and $\partial\mu$. Because the set-valued map $\partial\mu$ has compact values, we can use (5.114) to choose a smooth function $s_2(x_1, x_2)$ such that the maximum of the right-hand side of (5.119) (which is nothing more than the Lyapunov derivative $L_f V$) is negative definite, perhaps bounded from above by $-z_1^2 - z_2^2$:

$$L_f V(x,u)\Big|_{x_3=\nu} = \max\, \partial V(x) \cdot f(x,u)\Big|_{x_3=\nu} \leq -z_1^2 - z_2^2 \tag{5.120}$$

Now that we have determined the function ν, our clf V in (5.109) is complete. If we define $z_3 := x_3 - \nu(x_1, x_2)$, then from (5.109) we have $V(x) = z^{\mathrm{T}} z$ just as in Section 5.2; the difference here is that the z-variables are related to the x-variables through a *nonsmooth* transformation.

We have left to construct the control law from the clf V. To form the Lyapunov derivative $L_f V$ in (5.107), we must first compute $\partial V(x) \cdot f(x,u)$

from (5.110) as follows:

$$
\begin{aligned}
\partial V \cdot f \;&=\; \partial V \cdot f\Big|_{x_3=\nu} + \partial V\Big|_{x_3=\nu} \cdot \begin{bmatrix} 0 \\ z_3 \\ 0 \end{bmatrix} + 2z_3\Big[-\partial\nu \;\; 1\Big] \cdot f \\
&=\; \partial V \cdot f\Big|_{x_3=\nu} + 2z_2z_3 + 2z_3u \\
&\quad -\; 2z_3\,\partial\nu \cdot \begin{bmatrix} x_1^2 + x_1 + 2|x_1| - |\mu(x_1)| - z_2\psi(x_2,\mu) \\ z_3 + z_2 s_2(x_1,x_2) \end{bmatrix}
\end{aligned}
\tag{5.121}
$$

We choose a control law of the form $u = z_3 s_3(x_1, x_2, x_3)$ where the function s_3 is yet to be determined. With this choice, we can use (5.120) and (5.121) to write the Lyapunov derivative $L_f V$ in the form

$$
\begin{aligned}
L_f V(x,u) \;&=\; \max\, \partial V(x) \cdot f(x,u) \\
&\le\; -z^{\mathrm{T}} \begin{bmatrix} 1 & 0 & \bigstar_2 \\ 0 & 1 & \bigstar_2 \\ \bigstar_2 & \bigstar_2 & \bigstar_2 - 2s_3 \end{bmatrix} z
\end{aligned}
\tag{5.122}
$$

where the $\bigstar_2$ entries depend on x_1, x_2, ψ, μ, s_2, and $\partial\nu$. Because the set-valued map $\partial\nu$ has compact values, there exists a smooth function s_3 such that the right-hand side of (5.122) is negative definite.

To construct the function s_3, we must first compute (or approximate) the generalized gradient $\partial\nu$. Using the product formula [20, Proposition 2.3.13], we obtain

$$
\partial\nu(x_1,x_2) \subset \left[z_2\frac{\partial s_2}{\partial x_1} - \partial\mu(x_1)\, s_2(x_1,x_2) \qquad z_2\frac{\partial s_2}{\partial x_2} + s_2(x_1,x_2) \right] \tag{5.123}
$$

From this inclusion we can obtain bounds on the $\bigstar_2$ entries in (5.122). We then choose $s_3(x_1, x_2, x_3)$ to be smooth and sufficiently large and negative at each point (x_1, x_2, x_3) so that the right-hand side of (5.122) is negative definite. Once the function s_3 is chosen, the final form of the stabilizing control law in the original x-variables is

$$
u \;=\; \Big[x_3 - \big[x_2 - \mu(x_1)\big]\, s_2(x_1,x_2)\Big]\, s_3(x_1,x_2,x_3) \tag{5.124}
$$

This control law is not C^1 because of the nonsmooth function μ, but it is locally Lipschitz continuous.

We have seen that the nonsmooth, locally Lipschitz backstepping design follows the same steps outlined in Section 5.2 for the C^1 case, the only real difference being the presence of the generalized gradients in the computation of the Lyapunov derivative. These generalized gradients do not need to be calculated precisely; approximations such as (5.123) are sufficient for control design.

5.5 Summary

In this chapter we presented various methods for constructing rclf's for systems having specific structures. We showed that the Lyapunov redesign method is essentially limited to systems whose uncertainties satisfy a restrictive matching condition; this motivated us to abandon Lyapunov redesign and take the uncertainty into account during the construction of the Lyapunov function itself. We then identified a class of strict feedback systems for which the systematic construction of rclf's is possible through a recursive backstepping procedure.

We demonstrated that the quadratic-like rclf's used in Section 5.2 can potentially yield controllers with undesirable high-gain properties. We therefore introduced flattened rclf's which substantially improve the behavior of the resulting control laws; these flattened rclf's will be examined in further detail in the next chapter.

Finally, we illustrated how the recursive backstepping design can be easily modified to allow for nonsmooth nonlinearities in the construction of the rclf. At the time of this writing, this particular nonsmooth extension of backstepping is limited to the locally Lipschitz case.

Chapter 6

Measurement Disturbances

Our robust stabilization results thus far were obtained under the assumption of perfect state feedback. In particular, in Chapter 5 we gave a constructive proof of the fact that every nonlinear system in strict feedback form admits a robust control Lyapunov function (rclf) and is therefore robustly stabilizable with perfect state measurements. In this chapter, we show that such systems remain robustly stabilizable when the state measurement is corrupted by disturbances (such as sensor noise). To be precise, we show that strict feedback systems can be made (globally) input-to-state stable (cf. Definition 3.3) with respect to additive state measurement disturbances.

We have seen that bounded disturbances in nonlinear systems can cause severe forms of instability (such as finite escape times). Thus far we have considered only those disturbances which cannot influence our knowledge of the system state. In contrast, the problem of counteracting state measurement disturbances is fundamentally more difficult. This will become evident in Section 6.1 when we show that such disturbances, even small ones which decay to zero in finite time, can actually destroy the (global) stabilizability of an otherwise stabilizable system [29]. We are therefore motivated to search for classes of systems for which stabilizability is preserved in the presence of state measurement disturbances. Strict feedback systems comprise one such class; we established this fact in [35] and will give its constructive proof in Section 6.2. Larger classes of such nonlinear systems have yet to be identified.

6.1 Effects of measurement disturbances

We provide two examples in this section which illustrate the destabilizing effects of additive state measurement disturbances in nonlinear control systems. The first example, given in Section 6.1.1, demonstrates the well-known fact that such disturbances can destroy global stability. The second example, given in Section 6.1.2, establishes the recently proved result that such disturbances can even destroy global *stabilizability*.

6.1.1 Loss of global stability

In general, a control law which guarantees global stability under perfect state feedback will *not* provide global robustness to state measurement disturbances. To illustrate this fact, we consider the scalar system

$$\dot{x} \;=\; x\exp(x^2) + u \tag{6.1}$$

A globally stabilizing control law under perfect state feedback is

$$u(x) \;=\; -kx\exp(x^2) \tag{6.2}$$

for some design parameter $k > 1$. Now suppose the measurement of x is corrupted by an additive disturbance d so that the actual measurement y seen by the controller is

$$y \;=\; x + d \tag{6.3}$$

The resulting closed-loop system with $u = u(y)$ is

$$\begin{aligned} \dot{x} &= x\exp(x^2) - k(x+d)\exp((x+d)^2) \\ &= \exp(x^2)\left[x - k(x+d)\exp(2xd+d^2)\right] \end{aligned} \tag{6.4}$$

Given an arbitrarily small constant disturbance $d \neq 0$, it should be clear that this system exhibits finite escape times from initial conditions x_0 which are sufficiently large in magnitude and of the opposite sign of d. In other words, every nonzero (constant) measurement disturbance d will destroy global stability. For this particular system, however, the disturbance d does not destroy *stabilizability*: as will be seen in Section 6.2, there is a different control law for this system which provides global input-to-state stability (ISS) with respect to the measurement disturbance d.

6.1.2 Loss of global stabilizability

The above discussion leads to the following open question: if a nonlinear system is robustly stabilizable under perfect state feedback, is it also robustly stabilizable (perhaps with a different controller) in the presence of state measurement disturbances? If one considers only memoryless, time-invariant feedback control laws, the answer is *negative*; a counterexample was provided in [29] and will be discussed below. The question remains open for the broader class of time-varying and/or dynamic controllers. However, we will show in Section 6.2 that every system in strict feedback form is indeed robustly stabilizable in the presense of additive state measurement disturbances.

We now construct a second-order single-input system which is globally exponentially stabilizable via time-invariant perfect state feedback, but for which no time-invariant admissible control can prevent finite escape times when small measurement disturbances are present.

Let $\Theta : I\!R \to I\!R^{2\times 2}$ be the smooth matrix function

$$\Theta(\theta) \quad := \quad \begin{bmatrix} \cos\theta & -\sin\theta \\ \sin\theta & \cos\theta \end{bmatrix} \tag{6.5}$$

so that for any $x \in I\!R^2$, the vector $\Theta(\theta)\,x$ is the result of rotating x by the angle θ. The rotation function Θ satisfies $\Theta(0) = I$ along with the following identities:

$$\Theta(\theta_1)\,\Theta(\theta_2) \equiv \Theta(\theta_1+\theta_2), \qquad \frac{d\Theta}{d\theta}(\theta) \equiv \Theta(\theta+\tfrac{\pi}{2}), \qquad x^{\mathrm{T}}\Theta(\tfrac{\pi}{2})\,x \equiv 0$$

This rotation function plays an important role in the description of our second-order system:

$$\dot{x} = \Big[I_{2\times 2} + 2\Theta(\tfrac{\pi}{2})\,xx^{\mathrm{T}}\Big]\Theta(x^{\mathrm{T}}x)\left(\begin{bmatrix} -1 & 0 \\ 0 & x^{\mathrm{T}}x \end{bmatrix}\Theta(-x^{\mathrm{T}}x)\,x + \begin{bmatrix} 0 \\ 1 \end{bmatrix} u\right) \tag{6.6}$$

$$y = x + d \tag{6.7}$$

There is a memoryless, time-invariant feedback law for this system that provides global exponential stability when $d \equiv 0$ (that is, when $y = x$), but there is *no* such feedback law that can prevent finite escape times in the presence of a particular class of small disturbances d which vanish in finite time. To be precise, we have the following:

Theorem 6.1 *There exists a smooth perfect state feedback control law $u = K_0(x)$ such that the resulting closed-loop system has a globally exponentially stable equilibrium at $x = 0$. However, for any time-invariant admissible control which, when $d(t) \equiv 0$, renders the solutions* RGUAS-Ω *for some compact residual set $\Omega \subset \mathbb{R}^2$, there exists for any $\varepsilon > 0$ an instant $t_f \in (0, \infty)$, an initial condition $x_0 \in \mathbb{R}^2$, and a continuous measurement disturbance $d(t)$ with $\|d\|_\infty < \varepsilon$ such that one solution $x(t)$ from x_0 exhibits a finite escape time at $t = t_f$. Moreover, d can be chosen such that $d(t) \to 0$ as $t \to t_f$.*

Proof: The mapping $x \mapsto \Theta(-x^{\mathrm{T}}x)\,x$ from $\mathbb{R}^2$ onto $\mathbb{R}^2$ is a diffeomorphism, and so we may define new coordinates $z := \Theta(-x^{\mathrm{T}}x)\,x$ which satisfy $z^{\mathrm{T}}z = x^{\mathrm{T}}x$ and $x = \Theta(z^{\mathrm{T}}z)\,z$. We compute $\dot{z}$ from (6.6) as follows:

$$\begin{aligned}
\dot{z} &= \Theta(-x^{\mathrm{T}}x)\,\dot{x} + \Theta(\tfrac{\pi}{2} - x^{\mathrm{T}}x)\,(-2x^{\mathrm{T}}\dot{x})\,x \\
&= \left[\Theta(-x^{\mathrm{T}}x) - 2\Theta(\tfrac{\pi}{2} - x^{\mathrm{T}}x)\,xx^{\mathrm{T}}\right]\dot{x} \\
&= \begin{bmatrix} -1 & 0 \\ 0 & z^{\mathrm{T}}z \end{bmatrix} z + \begin{bmatrix} 0 \\ 1 \end{bmatrix} u \qquad (6.8)
\end{aligned}$$

Thus in the new coordinates $z := [z_1\ z_2]^{\mathrm{T}}$, the system (6.6)–(6.7) becomes

$$\begin{aligned}
\dot{z}_1 &= -z_1 &(6.9)\\
\dot{z}_2 &= (z_1^2 + z_2^2)\,z_2 + u &(6.10)\\
y &= \Theta(z^{\mathrm{T}}z)\,z + d &(6.11)
\end{aligned}$$

We now define the smooth function K_0 by

$$K_0(x) \;:=\; \Lambda(\Theta(-x^{\mathrm{T}}x)\,x) \qquad (6.12)$$

where the function $\Lambda : \mathbb{R}^2 \to \mathbb{R}$ is given by

$$\Lambda(z) \;:=\; -(1 + z_1^2 + z_2^2)\,z_2 \qquad (6.13)$$

It is then clear from (6.10) and (6.13) that the system (6.6) with control $u = K_0(x)$ is globally exponentially stable. Indeed, the resulting closed-loop system is simply $\dot{z} = -z$, and because $|x| = |z|$ we have $|x(t)| = |x_0|\,e^{-t}$ for any closed-loop solution $x(t)$ starting from an initial condition $x_0 \in \mathbb{R}^2$.

Now let $K(y)$ be any time-invariant admissible control which, when $d(t) \equiv 0$ (that is, when $y = x$), renders the solutions to (6.6) RGUAS-Ω for some compact residual set $\Omega \subset I\!R^2$. For simplicity we will assume that $\Omega = \{0\}$; the proof for a nontrivial residual set requires a slight modification of the arguments given below and will be omitted. We wish to examine the resulting closed-loop system for nonzero disturbances d. We see from (6.9) that the set $S := \{x \in I\!R^2 : z_1 = 0\}$ represents an invariant set for the system (6.9)–(6.11) for any inputs u and d. In Figure 6.1 we have plotted this set S in the x-plane; the solid and dotted lines represent positive and negative values of z_2, respectively. The dynamics on this invariant set S under the feedback $u = K(y)$ are determined by setting $z_1 = 0$ in (6.10)–(6.11) as follows:

$$\dot{z}_2 = z_2^3 + K(y) \tag{6.14}$$

$$y = \begin{bmatrix} -z_2 \sin(z_2^2) \\ z_2 \cos(z_2^2) \end{bmatrix} + d \tag{6.15}$$

From now on we consider only initial conditions on the invariant set S and restrict our analysis to these dynamics (6.14)–(6.15). Because the controller K provides global asymptotic stability when $d(t) \equiv 0$, the scalar system

$$\dot{z}_2 = z_2^3 + K(x) \tag{6.16}$$

must be globally asymptotically stable when x is restricted to lie on S. It follows that for $x \in S$ we must have $K(x) < 0$ when $z_2 > 0$ and $K(x) > 0$ when $z_2 < 0$. In other words, the function K must have negative values on the solid line and positive values on the dotted line in Figure 6.1. Because the solid and dotted arms of the spiral S get arbitrarily close to each other for large values of x (as seen in Figure 6.1), a small error d in the measurement of x can lead to an error in the sign of the control signal. This means that $K(y)$ can be positive when $K(x)$ is negative, even for an arbitrarily small error d between y and x. Such an error in the sign of the control signal, if persistent, will produce a finite escape time for this system. Our strategy, therefore, is to construct the measurement disturbance $d(t)$ so that the measurement y lies on the dotted line when the actual state x lies on the nearby solid line. The details are given next.

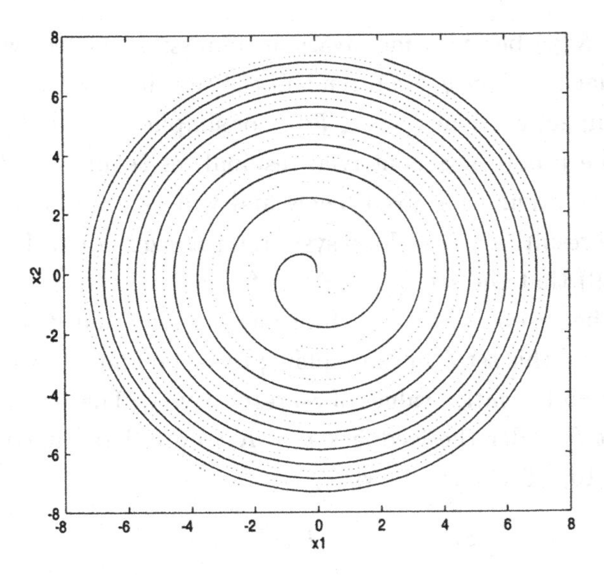

Figure 6.1: The invariant set S where $z_1 = 0$: solid for $z_2 > 0$ and dotted for $z_2 < 0$. *Source:* Freeman, R. A. 1995. Global internal stabilizability does not imply global external stabilizability for small sensor disturbances. *IEEE Transactions on Automatic Control* **40**(12), 2119–2122. ©1995 IEEE.

We consider the following auxiliary system defined for $r > 0$:

$$\dot{r} = r^3 + K(\bar{y}(r)) \tag{6.17}$$

$$\bar{y}(r) := \begin{bmatrix} \sqrt{r^2+\pi}\,\sin(r^2+\pi) \\ -\sqrt{r^2+\pi}\,\cos(r^2+\pi) \end{bmatrix} \tag{6.18}$$

One can check that for every $r > 0$, the point $\bar{y}(r)$ lies on the dotted line in Figure 6.1 (when plotted in the x-plane). Because K has positive values on the dotted line, we see that $K(\bar{y}(r)) > 0$ for every $r > 0$. Now all nonzero solutions of $\dot{r} = r^3$ escape to infinity in finite time, and because $K(\bar{y}(r)) > 0$, positive solutions of (6.17) can only escape sooner. Thus every solution $r(t)$ of (6.17)–(6.18) escapes to infinity in finite time. We use this fact to construct a measurement disturbance $d(t)$ as follows. Let the initial condition $x_0 \neq 0$ lie on the solid line in Figure 6.1, and let $r_0 > 0$ denote the z_2-coordinate of x_0. Let $r(t)$ denote the solution of (6.17)–(6.18) starting from r_0; then there exists $t_f \in (0, \infty)$ such that $r(t) \to \infty$ as $t \to t_f$. We now use this solution $r(t)$ to construct the measurement disturbance $d(t)$ on $[0, t_f)$ as follows:

$$d(t) \quad := \quad \bar{y}(r(t)) - \begin{bmatrix} -r(t)\sin(r^2(t)) \\ r(t)\cos(r^2(t)) \end{bmatrix} \tag{6.19}$$

It is clear that $d(t)$ is continuous on $[0, t_f)$ and that one solution $z_2(t)$ of (6.14)–(6.15) starting from r_0 satisfies $z_2(t) \equiv r(t)$ and thus escapes to infinity at $t = t_f$. Because $|x| = |z|$, the corresponding solution $x(t)$ of (6.6)–(6.7) escapes to infinity at $t = t_f$. We have left to show that $d(t) \to 0$ as $t \to t_f$ (we can then satisfy $\|d\|_\infty < \varepsilon$ by choosing the initial condition $x_0 \in S$ sufficiently large):

$$\begin{aligned} |d(t)|^2 &= \left[\sqrt{r^2(t)+\pi}\,\sin(r^2(t)+\pi) \;+\; r(t)\sin(r^2(t))\right]^2 \\ &\quad + \left[-\sqrt{r^2(t)+\pi}\,\cos(r^2(t)+\pi) \;-\; r(t)\cos(r^2(t))\right]^2 \\ &= 2r^2(t) \;+\; \pi \;+\; 2r(t)\sqrt{r^2(t)+\pi}\,\cdot \\ &\quad \left[\sin(r^2(t)+\pi)\sin(r^2(t)) + \cos(r^2(t)+\pi)\cos(r^2(t))\right] \\ &= 2r^2(t) \;+\; \pi \;-\; 2r(t)\sqrt{r^2(t)+\pi} \\ &= \left(r(t) \;-\; \sqrt{r^2(t)+\pi}\right)^2 \end{aligned} \tag{6.20}$$

Because $r(t) \to \infty$ as $t \to t_f$, we see that $d(t) \to 0$ as $t \to t_f$. ∎

6.2 Design for strict feedback systems

We have just shown that, in general, state measurement disturbances can destroy global stabilizability. However, this is not the case for systems in strict feedback form: we show in this section that such systems can be made (globally) input-to-state stable (ISS) with respect to additive state measurement disturbances.

We begin in Section 6.2.1 by formulating our problem in terms of the robust control Lyapunov function (rclf) defined in Chapter 3. We then state our main results in Section 6.2.2 and give constructive proofs of these results in Sections 6.2.3 and 6.2.4.

6.2.1 Measurement constraint for ISS

In Chapter 3, we defined a system $\Sigma = (f, U, W, Y)$ to be a function f describing the dynamics $\dot{x} = f(x, u, w)$ together with three set-valued constraints U, W, and Y characterizing the admissible controls, disturbances, and measurements (respectively). In this chapter we will assume that f is in strict feedback form (reviewed below), $U(x) \equiv \mathcal{U} = \mathbb{R}$ (unconstrained control input), and $W(x) \equiv B$ (where B denotes the closed unit ball in $\mathcal{W}$). We have left to choose an appropriate measurement constraint $Y(x)$. For perfect state feedback, the measurement constraint is $Y(x) = \{x\}$ for all $x \in \mathcal{X}$, which means the only admissible measurement is the one identical to the state trajectory. However, we instead wish to allow the state measurement to be corrupted by an additive disturbance $d(t)$, that is, we want the measured output y to be

$$y \;=\; x + d \tag{6.21}$$

If the disturbance d were allowed to take values in some set $D \subset \mathcal{X}$, then the measurement constraint would be simply $Y(x) = x + D$. For example, the choice $D = pB$ for some $p > 0$ (where B denotes the unit ball in $\mathcal{X}$) would correspond to the set $\{d \in L_\infty : \|d\|_\infty \leq p\}$ of admissible measurement disturbances $d(t)$. In this case, the distance between any admissible measurement y and the true state x at any point in time would satisfy $|y - x| \leq p$.

We are not yet satisfied with the choice $Y(x) = x + pB$, however, because we wish to consider *all* bounded disturbances $d \in L_\infty$, not just those satisfying a particular bound p. We therefore let p vary with the magnitude of the state x and choose the measurement constraint

$$Y(x) \;=\; x + \rho(|x|)B \tag{6.22}$$

where ρ is a class $\mathcal{K}_\infty$ function. With this new measurement constraint, the allowable distance between the measurement y and the true state x depends on the size of x. According to the results of Sections 3.2 and 3.4, if an rclf for a system with such a measurement constraint can be found, then this would guarantee the existence of an admissible control which simultaneously provides robust stability with respect to the disturbance w and input-to-state stability with respect to the disturbance d. To be precise, this would mean that there exist functions $\beta \in \mathcal{KL}$ and $\gamma \in \mathcal{K}$ such that solutions to the closed-loop system satisfy

$$|x(t)| \;\leq\; \beta(|x_0|, t) + \gamma(\|d\|_\infty) + \varepsilon \tag{6.23}$$

for all $t \geq 0$, every initial condition $x_0 \in \mathcal{X}$, every admissible disturbance w, and every measurement disturbance $d \in L_\infty$. Here $\varepsilon \geq 0$ is a constant which determines the size of the residual set Ω; it will be zero when the rclf satisfies the small control property (cf. Section 3.4.1).

Equipped with the measurement constraint (6.22), we can state our robust stabilization problem as follows: find a function $\rho \in \mathcal{K}_\infty$ such that the system $\Sigma = (f, U, W, Y)$, with f, U, W as above and Y as in (6.22), admits an rclf. The constructive solution to this problem is given in the remaining sections of this chapter.

6.2.2 Backstepping with measurement disturbances

We will use a recursive backstepping technique similar to the one outlined in Sections 5.2 and 5.3 to build an rclf for a strict feedback system with measurement constraint (6.22). The control law we obtain by means of this rclf will render the closed-loop system robustly stable with respect to the disturbance w as well as ISS with respect to the measurement disturbance d in (6.21).

Recall from Section 5.2 that a system

$$\dot{x} = F(x, w) + G(x, w)\, u \tag{6.24}$$

is in strict feedback form when F and G can be written as

$$F(x,w) = \begin{bmatrix} \phi_{11}(x,w) & \phi_{12}(x,w) & 0 & \dots & 0 \\ \phi_{21}(x,w) & \phi_{22}(x,w) & \phi_{23}(x,w) & \dots & 0 \\ \vdots & \vdots & \vdots & \ddots & \vdots \\ \phi_{n-1,1}(x,w) & \phi_{n-1,2}(x,w) & \phi_{n-1,3}(x,w) & \dots & \phi_{n-1,n}(x,w) \\ \phi_{n1}(x,w) & \phi_{n2}(x,w) & \phi_{n3}(x,w) & \dots & \phi_{nn}(x,w) \end{bmatrix} x + F(0,w)$$

$$G(x,w) = \begin{bmatrix} 0 \\ \vdots \\ 0 \\ \phi_{n,n+1}(x,w) \end{bmatrix} \tag{6.25}$$

for continuous scalar functions ϕ_{ij}. Each function ϕ_{ij} must depend only on w and the state components x_1 through x_i, namely,

$$\phi_{ij}(x,w) = \phi_{ij}(x_1, \dots, x_i, w) \tag{6.26}$$

for $1 \le i \le n$ and $1 \le j \le i+1$. Furthermore, we assume that

$$\phi_{i,i+1}(x_1, \ldots, x_i, w) \;\neq\; 0 \tag{6.27}$$

for all $x_1, \ldots, x_i \in I\!R$, for all $w \in B$, and for $1 \le i \le n$.

Our approach is to formulate a list of hypotheses H_1, H_2, ..., H_n, one for each of the n dimensions of the state space $\mathcal{X}$. We begin by showing that, if the last hypothesis H_n is true, then our robust stabilization problem has a solution. We then use a recursive argument to show that H_n is indeed true. First, we show in Section 6.2.3 that hypothesis H_1 is true. Finally, we show in Section 6.2.4 that, if hypothesis H_i is true for some integer i, then hypothesis H_{i+1} is also true.

To simplify notation, we let $\chi_i \in I\!R^i$ denote the first i components of the state vector x, that is, $\chi_i := [x_1 \;\ldots\; x_i]^{\mathrm{T}}$. In what follows, all time derivatives are with respect to the strict feedback system (6.24). Our entire list of hypotheses H_1, H_2, ..., H_n can be described by a single hypothesis H_i indexed by an integer $i \in \{1, \ldots, n\}$:

H_i: there exist functions V_i, α_i, λ_i, $\overline{\lambda}_i$, μ_i, $\overline{\nu}_i$, $\nu_i : I\!R^i \to I\!R$, a function $\Gamma_i \in \mathcal{K}_\infty$, and a scalar constant $\varepsilon_i > 0$ such that

(1) V_i and α_i are C^1, positive definite, and radially unbounded,

(2) λ_i, $\overline{\lambda}_i$, μ_i, $\overline{\nu}_i$, and ν_i are C^1, each of these functions is zero at $0 \in I\!R^i$, and for all $x \in \mathcal{X}$ such that $\chi_i \neq 0$ we have

$$\lambda_i(\chi_i) \;<\; \overline{\lambda}_i(\chi_i) \;<\; \mu_i(\chi_i) \;<\; \overline{\nu}_i(\chi_i) \;<\; \nu_i(\chi_i)$$

(3) for all $x \in \mathcal{X}$ and all $d_i \in I\!R^i$ such that $|\chi_i| \ge \Gamma_i(|d_i|) + \varepsilon_i$,

(a) $\lambda_i(\chi_i) \le x_{i+1} \le \nu_i(\chi_i) \implies \dot{V}_i \le -\alpha_i(\chi_i)$

(b) $\overline{\nu}_i(\chi_i) \;\ge\; \mu_i(\chi_i + d_i) + 2|\chi_i + d_i|$

(c) $\overline{\lambda}_i(\chi_i) \;\le\; \mu_i(\chi_i + d_i) - 2|\chi_i + d_i|$

When $i = n$, we interpret H_n(3a) by defining $x_{n+1} := u$.

Suppose that the last hypothesis H_n is true. If we choose the feedback law $u = \mu_n(y) = \mu_n(x + d)$, where the state measurement is corrupted by the additive disturbance d, then from H_n(2,3) we have

$$|x| \ge \Gamma_n(|d|) + \varepsilon_n \implies \dot{V}_n \le -\alpha_n(x) \tag{6.28}$$

for all x, $d \in \mathcal{X}$. If we let $\rho \in \mathcal{K}_\infty$ be such that $\rho(s) = \Gamma_n^{-1}(s - \varepsilon_n)$ for all $s \ge 2\varepsilon_n$, then (6.28) implies

$$\Big[\; |x| \ge 2\varepsilon_n \quad \text{and} \quad |d| \le \rho(|x|) \;\Big] \implies \dot{V}_n \le -\alpha_n(x) \tag{6.29}$$

It follows that V_n is an rclf for the strict feedback system (6.24) under the control constraint $U(x) \equiv \mathcal{U} = I\!R$, the disturbance constraint $W(x) \equiv B$, and the measurement constraint (6.22), that is, $Y(x) = x + \rho(|x|)B$. Furthermore, $\mu_n(y)$ is an admissible robustly stabilizing control associated with this rclf, and the resulting closed-loop system satisfies the ISS property (6.23). Because $\varepsilon_n > 0$, there will be a nontrivial residual set Ω; one would need to make the further assumption that $F(0, w) \equiv 0$ to eliminate this residual set and prove the small control property.

Our goal, therefore, is to show that hypothesis H_n is true and to construct the functions V_n and μ_n. We will accomplish this goal by proving the following two theorems:

Theorem 6.2 *Hypothesis* H_1 *is true.*

Theorem 6.3 *If hypothesis* H_i *is true, then hypothesis* H_{i+1} *is true.*

We will give constructive proofs of these theorems in the next sections. These proofs outline a systematic method for calculating the rclf V_n and the control law μ_n. The recursive nature of the construction is evident in the induction step in Theorem 6.3. Because the proof of Theorem 6.3 is long and technical, we will ignore all design flexibilities and concentrate on clarity and notational simplicity rather than achievable performance.

The proofs of these two theorems rely on the following technical lemma which is a consequence of the results on $C\mathcal{K}$-continuity in the Appendix:

Lemma 6.4 *If* $\omega : I\!R^i \to I\!R$ *is continuous, then there exists a* C^1 *nonzero function* $\zeta : I\!R^i \to I\!R_+$ *such that for all* χ_i, $d_i \in I\!R^i$ *such that* $|\chi_i| \geq 2|d_i|$ *we have*

$$\zeta(\chi_i) \;\geq\; \Big|\omega(\chi_i + d_i) - \omega(\chi_i)\Big| + 2|\chi_i + d_i| \tag{6.30}$$

$$\zeta(\chi_i + d_i) \;\geq\; \Big|\omega(\chi_i + d_i) - \omega(\chi_i)\Big| + 2|\chi_i + d_i| \tag{6.31}$$

Proof: It follows from Definition A.13 and Corollary A.15 that there exist $\gamma_\omega \in \mathcal{K}$ and a continuous function $\rho_\omega : I\!R^i \to I\!R_+$ such that

$$\Big|\omega(\chi_i + d_i) - \omega(\chi_i)\Big| \;\leq\; \rho_\omega(\chi_i) \cdot \gamma_\omega(|d_i|) \tag{6.32}$$

for all χ_i, $d_i \in I\!R^i$. Let ζ be any C^1 function such that

$$\zeta(\chi_i) \;\geq\; \rho_\omega(\chi_i) \cdot \gamma_\omega(|\chi_i|) + 3|\chi_i| \tag{6.33}$$

for all $\chi_i \in I\!R^i$. Suppose $|\chi_i| \geq 2|d_i|$; then from (6.32) and (6.33) we have

$$\begin{aligned}\varsigma(\chi_i) &\geq \rho_\omega(\chi_i)\cdot\gamma_\omega(|\chi_i|) + 3|\chi_i| \\ &\geq \rho_\omega(\chi_i)\cdot\gamma_\omega(|d_i|) + 2|\chi_i| + |\chi_i| \\ &\geq \left|\omega(\chi_i + d_i) - \omega(\chi_i)\right| + 2|\chi_i| + 2|d_i| \\ &\geq \left|\omega(\chi_i + d_i) - \omega(\chi_i)\right| + 2|\chi_i + d_i|\end{aligned}$$

which gives (6.30). Similarly we have

$$\begin{aligned}\varsigma(\chi_i + d_i) &\geq \rho_\omega(\chi_i + d_i)\cdot\gamma_\omega(|\chi_i + d_i|) + 3|\chi_i + d_i| \\ &\geq \rho_\omega(\chi_i + d_i)\cdot\gamma_\omega(|\chi_i| - |d_i|) + 3|\chi_i + d_i| \\ &\geq \rho_\omega(\chi_i + d_i)\cdot\gamma_\omega(|d_i|) + 2|\chi_i + d_i| \\ &\geq \left|\omega(\chi_i + d_i) - \omega(\chi_i)\right| + 2|\chi_i + d_i| \qquad (6.34)\end{aligned}$$

which gives (6.31). ■

6.2.3 Initialization step

We begin our construction by showing that hypothesis H_1 is true. We choose $V_1(x_1) = \alpha_1(x_1) = \frac{1}{2}x_1^2$ for all $x_1 \in I\!R$ and thus satisfy $H_1(1)$. Calculating the derivative $\dot{V}_1$ we obtain

$$\dot{V}_1 = -\alpha_1(x_1) + x_1\left[\tfrac{1}{2}x_1 + \phi_{11}(x_1, w)\,x_1 + \phi_{12}(x_1, w)\,x_2 + F_1(0, w)\right] \tag{6.35}$$

Recall from (6.27) that the function ϕ_{12} is never zero. From now on we will assume that ϕ_{12} takes on positive values; the analysis for the case in which ϕ_{12} takes on negative values is similar and involves simply reversing some inequalities below. As a result of ϕ_{12} being nonzero, there exist C^1 functions $\sigma, \varsigma : I\!R \to I\!R$ such that $\sigma(x_1) \leq \varsigma(x_1)$ and

$$\sigma(x_1) \leq -\frac{\frac{1}{2}x_1 + \phi_{11}(x_1, w)\,x_1 + F_1(0, w)}{\phi_{12}(x_1, w)} \tag{6.36}$$

$$\varsigma(x_1) \geq -\frac{\frac{1}{2}x_1 + \phi_{11}(x_1, w)\,x_1 + F_1(0, w)}{\phi_{12}(x_1, w)} \tag{6.37}$$

for all $x_1 \in I\!R$ and all $w \in B$. It follows from (6.35), (6.36), and (6.37) that

$$\left[\, x_1 \geq 0 \quad\text{and}\quad x_2 \leq \sigma(x_1) \,\right] \implies \dot{V}_1 \leq -\alpha_1(x_1) \tag{6.38}$$

$$\left[\, x_1 \leq 0 \quad\text{and}\quad x_2 \geq \varsigma(x_1) \,\right] \implies \dot{V}_1 \leq -\alpha_1(x_1) \tag{6.39}$$

We are now ready to construct the functions λ_1, $\overline{\lambda}_1$, μ_1, $\overline{\nu}_1$, and ν_1. Choose $\Gamma_1 \in \mathcal{K}_\infty$ such that $\Gamma_1(s) \geq 2s$ for all $s \geq 0$, and choose $\varepsilon_1 > 0$. Let $\sigma_1 : I\!R \to I\!R$ be the C^1 function defined by $\sigma_1(x_1) = \sigma(x_1) - 1 - \zeta(x_1)$ where the function ζ is from Lemma 6.4 with $\omega = \sigma - 1$. It follows from (6.31) that

$$\sigma(x_1) - 1 \;\geq\; \sigma_1(x_1 + d_1) + 2|x_1 + d_1| \tag{6.40}$$

whenever $|x_1| \geq \Gamma_1(|d_1|) + \varepsilon_1$. Next, let $\sigma_2 : I\!R \to I\!R$ be the C^1 function defined by $\sigma_2(x_1) = \sigma_1(x_1) - 1 - \zeta(x_1)$ where the (new) function ζ is from Lemma 6.4 with $\omega = \sigma_1$. It follows from (6.30) that

$$\sigma_2(x_1) + 1 \;\leq\; \sigma_1(x_1 + d_1) - 2|x_1 + d_1| \tag{6.41}$$

whenever $|x_1| \geq \Gamma_1(|d_1|) + \varepsilon_1$. By construction we have $\sigma_2 + 1 < \sigma_1 < \sigma - 1$ on $I\!R$. In a similar fashion we let $\varsigma_1 : I\!R \to I\!R$ be the C^1 function defined by $\varsigma_1(x_1) = \varsigma(x_1) + 1 + \zeta(x_1)$ where the (new) function ζ is from Lemma 6.4 with $\omega = \varsigma + 1$. It follows from (6.31) that

$$\varsigma(x_1) + 1 \;\leq\; \varsigma_1(x_1 + d_1) - 2|x_1 + d_1| \tag{6.42}$$

whenever $|x_1| \geq \Gamma_1(|d_1|) + \varepsilon_1$. We let $\varsigma_2 : I\!R \to I\!R$ be the C^1 function defined by $\varsigma_2(x_1) = \varsigma_1(x_1) + 1 + \zeta(x_1)$ where the (new) function ζ is from Lemma 6.4 with $\omega = \varsigma_1$. It follows from (6.30) that

$$\varsigma_2(x_1) - 1 \;\geq\; \varsigma_1(x_1 + d_1) + 2|x_1 + d_1| \tag{6.43}$$

whenever $|x_1| \geq \Gamma_1(|d_1|) + \varepsilon_1$. By construction we have $\varsigma + 1 < \varsigma_1 < \varsigma_2 - 1$ on $I\!R$. We now choose the functions λ_1, $\overline{\lambda}_1$, μ_1, $\overline{\nu}_1$, and ν_1 as follows:

$$\lambda_1(x_1) \;:=\; \begin{cases} \sigma_2(x_1) & \text{when } x_1 \geq \varepsilon_1 \\ \varsigma(x_1) & \text{when } x_1 \leq -\varepsilon_1 \\ \bigstar & \text{otherwise} \end{cases} \tag{6.44}$$

$$\mu_1(x_1) \;:=\; \begin{cases} \sigma_1(x_1) & \text{when } x_1 \geq \varepsilon_1 \\ \varsigma_1(x_1) & \text{when } x_1 \leq -\varepsilon_1 \\ \bigstar & \text{otherwise} \end{cases} \tag{6.45}$$

$$\nu_1(x_1) \;:=\; \begin{cases} \sigma(x_1) & \text{when } x_1 \geq \varepsilon_1 \\ \varsigma_2(x_1) & \text{when } x_1 \leq -\varepsilon_1 \\ \bigstar & \text{otherwise} \end{cases} \tag{6.46}$$

$$\overline{\lambda}_1(x_1) := \begin{cases} \lambda_1(x_1) + 1 & \text{when } |x_1| \geq \varepsilon_1 \\ \bigstar & \text{otherwise} \end{cases} \tag{6.47}$$

$$\overline{\nu}_1(x_1) := \begin{cases} \nu_1(x_1) - 1 & \text{when } |x_1| \geq \varepsilon_1 \\ \bigstar & \text{otherwise} \end{cases} \tag{6.48}$$

where each $\bigstar$ is chosen so that $H_1(2)$ is satisfied, that is, so that these functions are C^1, zero at $0 \in I\!R$, and such that $\lambda_1(x_1) < \overline{\lambda}_1(x_1) < \mu_1(x_1) < \overline{\nu}_1(x_1) < \nu_1(x_1)$ for all $x_1 \neq 0$. We have left to verify $H_1(3)$. Suppose $x \in \mathcal{X}$ and $d_1 \in I\!R$ are such that $|x_1| \geq \Gamma_1(|d_1|) + \varepsilon_1$. To verify $H_1(3a)$, suppose $\lambda_1(x_1) \leq x_2 \leq \nu_1(x_1)$. If $x_1 \geq \varepsilon_1$, then from (6.46) and (6.38) we have $\dot{V}_1 \leq -\alpha_1(x_1)$. If $x_1 \leq -\varepsilon_1$ then from (6.44) and (6.39) we again have $\dot{V}_1 \leq -\alpha_1(x_1)$, and we conclude that $H_1(3a)$ is true. $H_1(3b)$ and $H_1(3c)$ follow by construction from (6.40)–(6.48). This completes the proof of Theorem 6.2.

6.2.4 Recursion step

We now assume that hypothesis H_i is true and prove that hypothesis H_{i+1} is true. To simplify indices throughout the proof, we let $j = i + 1$ and $k = i + 2$. We define the following seven region in $I\!R^j$:

$$A^+ := \left\{ \chi_j \in I\!R^j : x_j > \nu_i(\chi_i) \right\} \tag{6.49}$$

$$A^0 := \left\{ \chi_j \in I\!R^j : \lambda_i(\chi_i) \leq x_j \leq \nu_i(\chi_i) \right\} \tag{6.50}$$

$$A^- := \left\{ \chi_j \in I\!R^j : x_j < \lambda_i(\chi_i) \right\} \tag{6.51}$$

$$D^+ := \left\{ \chi_j \in I\!R^j : x_j > \overline{\nu}_i(\chi_i) \right\} \tag{6.52}$$

$$D^- := \left\{ \chi_j \in I\!R^j : x_j < \overline{\lambda}_i(\chi_i) \right\} \tag{6.53}$$

$$E^+ := \left\{ \chi_j \in I\!R^j : x_j > \mu_i(\chi_i) + |\chi_i| \right\} \tag{6.54}$$

$$E^- := \left\{ \chi_j \in I\!R^j : x_j < \mu_i(\chi_i) - |\chi_i| \right\} \tag{6.55}$$

Note that $A^+ \cup A^0 \cup A^- = I\!R^j$, the sets A^+, A^0, and A^- are (pairwise) disjoint, $A^+ \subset D^+$, $A^- \subset D^-$, and $E^+ \cap E^- = \varnothing$. We now show that there exist a function $\Gamma_j \in \mathcal{K}_\infty$ and a constant $\overline{\varepsilon} > 0$ such that for all $\chi_j, d_j \in I\!R^j$ we have

$$\Big[\, |\chi_j| \geq \Gamma_j(|d_j|) + \overline{\varepsilon} \quad \text{and} \quad \chi_j \in D^+ \Big] \implies \chi_j + d_j \in E^+ \tag{6.56}$$

$$\Big[\, |\chi_j| \geq \Gamma_j(|d_j|) + \overline{\varepsilon} \quad \text{and} \quad \chi_j \in D^- \Big] \implies \chi_j + d_j \in E^- \tag{6.57}$$

To prove the existence of Γ_j and $\bar{\varepsilon}$, we first note that it follows from $\mathrm{H}_i(2)$ and Corollary A.15 that there exists a function $\gamma_\mu \in \mathcal{K}$ such that $|\mu_i(\chi_i)| \leq \gamma_\mu(|\chi_i|)$ for all $\chi_i \in I\!R^i$. Clearly we can find $\Gamma_j \in \mathcal{K}_\infty$ and $\bar{\varepsilon} > 0$ such that $\Gamma_j(s) \geq 2s$ and

$$\frac{\Gamma_j(s) + \bar{\varepsilon}}{\sqrt{2}} \geq \Gamma_i(s) + 2s + \varepsilon_i + \gamma_\mu(\Gamma_i(s) + 2s + \varepsilon_i) \tag{6.58}$$

for all $s \geq 0$. Let $\chi_j, d_j \in I\!R^j$ be such that $|\chi_j| \geq \Gamma_j(|d_j|) + \bar{\varepsilon}$. Then from (6.58) and the fact that $|\cdot| \leq \sqrt{2}|\cdot|_\infty$ on $I\!R^2$ we have

$$\begin{aligned} \max\Big\{|x_j|,\, |\chi_i|\Big\} \;&\geq\; \Gamma_i(|d_j|) + 2|d_j| + \varepsilon_i \\ &\quad + \gamma_\mu(\Gamma_i(|d_j|) + 2|d_j| + \varepsilon_i) \end{aligned} \tag{6.59}$$

Let $d_i \in I\!R^i$ denote the vector comprised of the first i components of $d_j \in I\!R^j$. We first consider the case in which $|\chi_i + d_i| < \Gamma_i(|d_i|) + |d_j| + \varepsilon_i$. Because $|d_i| \leq |d_j|$ we have $|\chi_i + d_i| < \Gamma_i(|d_j|) + |d_j| + \varepsilon_i$ which means the maximum in the left-hand side of (6.59) is equal to $|x_j|$. It follows from (6.59) that

$$\begin{aligned} |\chi_i + d_i| + |\mu_i(\chi_i + d_i)| \;&<\; \Gamma_i(|d_j|) + |d_j| + \varepsilon_i + \gamma_\mu(|\chi_i + d_i|) \\ &<\; \Gamma_i(|d_j|) + |d_j| + \varepsilon_i \\ &\qquad + \gamma_\mu(\Gamma_i(|d_j|) + 2|d_j| + \varepsilon_i) \\ &<\; |x_j| - |d_j| \\ &<\; |x_j + \delta| \end{aligned} \tag{6.60}$$

where $\delta \in I\!R$ denotes the last (j^{th}) component of d_j (note that $|\delta| \leq |d_j|$). Because $|x_j| \geq |\delta|$ from (6.59), if $x_j \geq 0$ then also $x_j + \delta \geq 0$, and it follows from (6.60) that $\chi_j + d_j \in E^+$. Similarly, if $x_j \leq 0$ then $\chi_j + d_j \in E^-$. Thus to prove (6.56) and (6.57) for the present case in which $|\chi_i + d_i| < \Gamma_i(|d_i|) + |d_j| + \varepsilon_i$, it suffices to show that

$$\chi_j \in D^+ \implies x_j \geq 0 \tag{6.61}$$

$$\chi_j \in D^- \implies x_j \leq 0 \tag{6.62}$$

To show (6.61), suppose $\chi_j \in D^+$ and $x_j < 0$. Then $\mu_i(\chi_i) < x_j < 0$ which means $|x_j| < |\mu_i(\chi_i)|$, and it follows from (6.59) that

$$|x_j| \;<\; \gamma_\mu(|\chi_i|)$$

$$
\begin{aligned}
&< \gamma_\mu(|\chi_i + d_i| + |d_i|) \\
&< \gamma_\mu(\Gamma_i(|d_j|) + 2|d_j| + \varepsilon_i) \\
&< |x_j| \qquad (6.63)
\end{aligned}
$$

which is a contradiction. A similar argument gives (6.62). We next consider the case in which $|\chi_i + d_i| \geq \Gamma_i(|d_i|) + |d_j| + \varepsilon_i$. In this case we have $|\chi_i| \geq \Gamma_i(|d_i|) + \varepsilon_i$, and if $\chi_j \in D^+$ then it follows from H_i(3b) that

$$
\begin{aligned}
x_j &> \overline{\nu}_i(\chi_i) \\
&> \mu_i(\chi_i + d_i) \; + \; 2|\chi_i + d_i| \\
&> \mu_i(\chi_i + d_i) \; + \; |\chi_i + d_i| \; + \; |d_j| \qquad (6.64)
\end{aligned}
$$

from which it follows that $\chi_j + d_j \in E^+$. A similar argument gives (6.57) from H_i(3c), and we conclude that both (6.56) and (6.57) are true.

We will employ the flattening technique introduced in Section 5.3.2 in the construction of our Lyapunov function. We define a Lyapunov function $V_j : I\!R^j \to I\!R_+$ flattened inside the region A^0 as follows:

$$
V_j(\chi_j) \;:=\; V_i(\chi_i) + \begin{cases} \frac{1}{2}\big[x_j \; - \; \nu_i(\chi_i)\big]^2 & \text{when } \; \chi_j \in A^+ \\ 0 & \text{when } \; \chi_j \in A^0 \\ \frac{1}{2}\big[x_j \; - \; \lambda_i(\chi_i)\big]^2 & \text{when } \; \chi_j \in A^- \end{cases} \qquad (6.65)
$$

This function V_j is C^1, positive definite, and radially unbounded as required by $\mathrm{H}_j(1)$. We will calculate the derivative of V_j in the three regions A^+, A^0, and A^-. In region A^0 we see from (6.65) that $\dot{V}_j = \dot{V}_i$, and it follows from H_i(3a) that

$$
\dot{V}_j \;\leq\; -\alpha_i(\chi_i) \qquad (6.66)
$$

whenever $\chi_j \in A^0$ and $|\chi_i| \geq \varepsilon_i$. Similarly, in region A^+ we have

$$
\begin{aligned}
\dot{V}_j \;=\; & \dot{V}_i\Big|_{x_j=\nu_i(\chi_i)} - \big[x_j \; - \; \nu_i(\chi_i)\big]^2 \\
& + \big[x_j \; - \; \nu_i(\chi_i)\big]\big[\phi_{jk}(\chi_j, w)\, x_k \; + \; \psi^+(\chi_j, w)\big] \qquad (6.67)
\end{aligned}
$$

and in region A^- we have

$$
\begin{aligned}
\dot{V}_j \;=\; & \dot{V}_i\Big|_{x_j=\lambda_i(\chi_i)} - \big[x_j \; - \; \lambda_i(\chi_i)\big]^2 \\
& + \big[x_j \; - \; \lambda_i(\chi_i)\big]\big[\phi_{jk}(\chi_j, w)\, x_k \; + \; \psi^-(\chi_j, w)\big] \qquad (6.68)
\end{aligned}
$$

where the continuous functions ψ^+ and ψ^- are given by

$$\begin{aligned}
\psi^+(\chi_j, w) \;\; &:= \;\; \frac{\partial V_i}{\partial x_i}\,\phi_{ij}(\chi_i, w) + \big[x_j - \nu_i(\chi_i)\big] \\
&\quad + F_j(0,w) + \sum_{\ell=1}^{j} \phi_{j\ell}(\chi_j, w)\, x_\ell \\
&\quad - \sum_{p=1}^{i} \frac{\partial \nu_i}{\partial x_p}\left[F_p(0,w) + \sum_{\ell=1}^{p+1} \phi_{p\ell}(\chi_p, w)\, x_\ell\right] \qquad (6.69) \\
\psi^-(\chi_j, w) \;\; &:= \;\; \frac{\partial V_i}{\partial x_i}\,\phi_{ij}(\chi_i, w) + \big[x_j - \lambda_i(\chi_i)\big] \\
&\quad + F_j(0,w) + \sum_{\ell=1}^{j} \phi_{j\ell}(\chi_j, w)\, x_\ell \\
&\quad - \sum_{p=1}^{i} \frac{\partial \lambda_i}{\partial x_p}\left[F_p(0,w) + \sum_{\ell=1}^{p+1} \phi_{p\ell}(\chi_p, w)\, x_\ell\right] \qquad (6.70)
\end{aligned}$$

We define the function $\alpha_j : I\!R^j \to I\!R_+$ as follows:

$$\alpha_j(\chi_j) \;\; := \;\; \alpha_i(\chi_i) + \begin{cases} \frac{1}{2}\big[x_j - \nu_i(\chi_i)\big]^2 & \text{when } \chi_j \in A^+ \\ 0 & \text{when } \chi_j \in A^0 \\ \frac{1}{2}\big[x_j - \lambda_i(\chi_i)\big]^2 & \text{when } \chi_j \in A^- \end{cases} \qquad (6.71)$$

This function α_j is C^1, positive definite, and radially unbounded as required by $\mathrm{H}_j(1)$. One can use (6.66), (6.67), and (6.68) to show that there exists a constant $\varepsilon_j \geq \bar{\varepsilon}$ such that whenever $|\chi_j| \geq \varepsilon_j$ we have

$$\dot{V}_j \;\leq\; -\alpha_j(\chi_j) + \begin{cases} \big[x_j - \nu_i(\chi_i)\big]\big[\phi_{jk}(\chi_j, w)\, x_k + \psi^+(\chi_j, w)\big] \\ \qquad\qquad \text{when } \chi_j \in A^+ \\ 0 \qquad\qquad \text{when } \chi_j \in A^0 \\ \big[x_j - \lambda_i(\chi_i)\big]\big[\phi_{jk}(\chi_j, w)\, x_k + \psi^-(\chi_j, w)\big] \\ \qquad\qquad \text{when } \chi_j \in A^- \end{cases} \qquad (6.72)$$

Recall from (6.27) that the function ϕ_{jk} is never zero. From now on we will assume that ϕ_{jk} takes on positive values; the analysis for the case in which ϕ_{jk} takes on negative values is similar and involves simply reversing some inequalities below. As a result of ϕ_{jk} being nonzero, there exist C^1 functions $\sigma, \varsigma : I\!R^j \to I\!R$ such that $\sigma(\chi_j) \leq \varsigma(\chi_j)$ and

$$\sigma(\chi_j) \;\leq\; -\frac{\psi^+(\chi_j, w)}{\phi_{jk}(\chi_j, w)} \qquad (6.73)$$

$$\varsigma(\chi_j) \;\geq\; -\frac{\psi^-(\chi_j, w)}{\phi_{jk}(\chi_j, w)} \tag{6.74}$$

for all $\chi_j \in \mathbb{R}^j$ and all $w \in B$. It now follows from (6.72), (6.73), and (6.74) that

$$\Big[\, |\chi_j| \geq \varepsilon_j, \quad \chi_j \in A^+, \quad \text{and} \quad x_k \leq \sigma(\chi_j) \Big] \implies \dot{V}_j \leq -\alpha_j(\chi_j) \tag{6.75}$$

$$\Big[\, |\chi_j| \geq \varepsilon_j \quad \text{and} \quad \chi_j \in A^0 \Big] \implies \dot{V}_j \leq -\alpha_j(\chi_j) \tag{6.76}$$

$$\Big[\, |\chi_j| \geq \varepsilon_j, \quad \chi_j \in A^-, \quad \text{and} \quad x_k \geq \varsigma(\chi_j) \Big] \implies \dot{V}_j \leq -\alpha_j(\chi_j) \tag{6.77}$$

We are now ready to begin the construction of the functions λ_j, $\overline{\lambda}_j$, μ_j, $\overline{\nu}_j$, and ν_j. Let $\sigma_1 : \mathbb{R}^j \to \mathbb{R}$ be the C^1 function defined by $\sigma_1(\chi_j) = \sigma(\chi_j) - 1 - \zeta(\chi_j)$ where the function ζ is from Lemma 6.4 with $\omega = \sigma - 1$. It follows from (6.31) that

$$\sigma(\chi_j) - 1 \;\geq\; \sigma_1(\chi_j + d_j) + 2|\chi_j + d_j| \tag{6.78}$$

whenever $|\chi_j| \geq \Gamma_j(|d_j|) + \varepsilon_j$. Let $\sigma_2 : \mathbb{R}^j \to \mathbb{R}$ be the C^1 function defined by $\sigma_2(\chi_j) = \sigma_1(\chi_j) - 1 - \zeta(\chi_j)$ where the (new) function ζ is from Lemma 6.4 with $\omega = \sigma_1$. It follows from (6.30) that

$$\sigma_2(\chi_j) + 1 \;\leq\; \sigma_1(\chi_j + d_j) - 2|\chi_j + d_j| \tag{6.79}$$

whenever $|\chi_j| \geq \Gamma_j(|d_j|) + \varepsilon_j$. By construction we have $\sigma_2 + 1 < \sigma_1 < \sigma - 1$ on $\mathbb{R}^j$. We let $\varsigma_1 : \mathbb{R}^j \to \mathbb{R}$ be the C^1 function defined by $\varsigma_1(\chi_j) = \varsigma(\chi_j) + 1 + \zeta(\chi_j)$ where the (new) function ζ is from Lemma 6.4 with $\omega = \varsigma + 1$. It follows from (6.31) that

$$\varsigma(\chi_j) + 1 \;\leq\; \varsigma_1(\chi_j + d_j) - 2|\chi_j + d_j| \tag{6.80}$$

whenever $|\chi_j| \geq \Gamma_j(|d_j|) + \varepsilon_j$. We let $\varsigma_2 : \mathbb{R}^j \to \mathbb{R}$ be the C^1 function defined by $\varsigma_2(\chi_j) = \varsigma_1(\chi_j) + 1 + \zeta(\chi_j)$ where the (new) function ζ is from Lemma 6.4 with $\omega = \varsigma_1$. It follows from (6.30) that

$$\varsigma_2(\chi_j) - 1 \;\geq\; \varsigma_1(\chi_j + d_j) + 2|\chi_j + d_j| \tag{6.81}$$

whenever $|\chi_j| \geq \Gamma_j(|d_j|) + \varepsilon_j$. By construction we have $\varsigma + 1 < \varsigma_1 < \varsigma_2 - 1$ on $\mathbb{R}^j$. We now choose the functions λ_j, $\overline{\lambda}_j$, μ_j, $\overline{\nu}_j$, and ν_j as follows:

$$\lambda_j(\chi_j) \;:=\; \begin{cases} \sigma_2(\chi_j) & \text{when } \chi_j \notin D^- \text{ and } |\chi_j| \geq \varepsilon_j \\ \varsigma(\chi_j) & \text{when } \chi_j \in A^- \text{ and } |\chi_j| \geq \varepsilon_j \\ \bigstar & \text{otherwise} \end{cases} \tag{6.82}$$

$$\mu_j(\chi_j) \;:=\; \begin{cases} \sigma_1(\chi_j) & \text{when } \chi_j \in E^+ \text{ and } |\chi_j| \geq \varepsilon_j \\ \varsigma_1(\chi_j) & \text{when } \chi_j \in E^- \text{ and } |\chi_j| \geq \varepsilon_j \\ \bigstar & \text{otherwise} \end{cases} \tag{6.83}$$

$$\nu_j(\chi_j) \;:=\; \begin{cases} \sigma(\chi_j) & \text{when } \chi_j \in A^+ \text{ and } |\chi_j| \geq \varepsilon_j \\ \varsigma_2(\chi_j) & \text{when } \chi_j \notin D^+ \text{ and } |\chi_j| \geq \varepsilon_j \\ \bigstar & \text{otherwise} \end{cases} \tag{6.84}$$

$$\overline{\lambda}_j(\chi_j) \;:=\; \begin{cases} \lambda_j(\chi_j) + 1 & \text{when } |\chi_j| \geq \varepsilon_j \\ \bigstar & \text{otherwise} \end{cases} \tag{6.85}$$

$$\overline{\nu}_j(\chi_j) \;:=\; \begin{cases} \nu_j(\chi_j) - 1 & \text{when } |\chi_j| \geq \varepsilon_j \\ \bigstar & \text{otherwise} \end{cases} \tag{6.86}$$

where each $\bigstar$ denotes a part of the respective function about which we are not yet concerned. Note that these definitions make sense because the pairs of closed sets $(I\!R^j \backslash D^-, \overline{A^-})$, $(\overline{E^+}, \overline{E^-})$, and $(\overline{A^+}, I\!R^j \backslash D^+)$ are disjoint relative to the set $I\!R^j \backslash \varepsilon_j B$. Note also from (6.56) and (6.57) that $D^+ \subset E^+$ and $D^- \subset E^-$ relative to $I\!R^j \backslash \varepsilon_j B$. As a result, one can easily verify that, for an appropriate choice of each $\bigstar$, these functions can be made to satisfy $\mathrm{H}_j(2)$, that is, they can be made C^1, zero at $0 \in I\!R^j$, and such that $\lambda_j(\chi_j) < \overline{\lambda}_j(\chi_j) < \mu_j(\chi_j) < \overline{\nu}_j(\chi_j) < \nu_j(\chi_j)$ for all $\chi_j \neq 0$. We can assume further from (6.82)–(6.84) that each $\bigstar$ is chosen so that

$$\sigma(\chi_j) \;\leq\; \nu_j(\chi_j) \;\leq\; \varsigma_2(\chi_j) \tag{6.87}$$

$$\sigma_1(\chi_j) \;\leq\; \mu_j(\chi_j) \;\leq\; \varsigma_1(\chi_j) \tag{6.88}$$

$$\sigma_2(\chi_j) \;\leq\; \lambda_j(\chi_j) \;\leq\; \varsigma(\chi_j) \tag{6.89}$$

whenever $|\chi_j| \geq \varepsilon_j$. The above construction of the functions λ_j, $\overline{\lambda}_j$, μ_j, $\overline{\nu}_j$, and ν_j in (6.82)–(6.86) is illustrated conceptually in Figure 6.2.

We have left to verify $\mathrm{H}_j(3)$. Suppose $x \in \mathcal{X}$ and $d_j \in I\!R^j$ are such that $|\chi_j| \geq \Gamma_j(|d_j|) + \varepsilon_j$ (in particular $|\chi_j| \geq \varepsilon_j$ and $|\chi_j + d_j| \geq \varepsilon_j$). To verify $\mathrm{H}_j(3\mathrm{a})$, suppose $\lambda_j(\chi_j) \leq x_k \leq \nu_j(\chi_j)$. If $\chi_j \in A^0$, then from (6.76) we have $\dot{V}_j \leq -\alpha_j(\chi_j)$ as desired. If $\chi_j \in A^+$, then from (6.84) and (6.75) we again have $\dot{V}_j \leq -\alpha_j(\chi_j)$. Finally, if $\chi_j \in A^-$ then from (6.82) and (6.77) we again have $\dot{V}_j \leq -\alpha_j(\chi_j)$, and we conclude that $\mathrm{H}_j(3\mathrm{a})$ is true. To verify $\mathrm{H}_j(3\mathrm{b})$, we must show that

$$\overline{\nu}_j(\chi_j) \;\geq\; \mu_j(\chi_j + d_j) + 2|\chi_j + d_j| \tag{6.90}$$

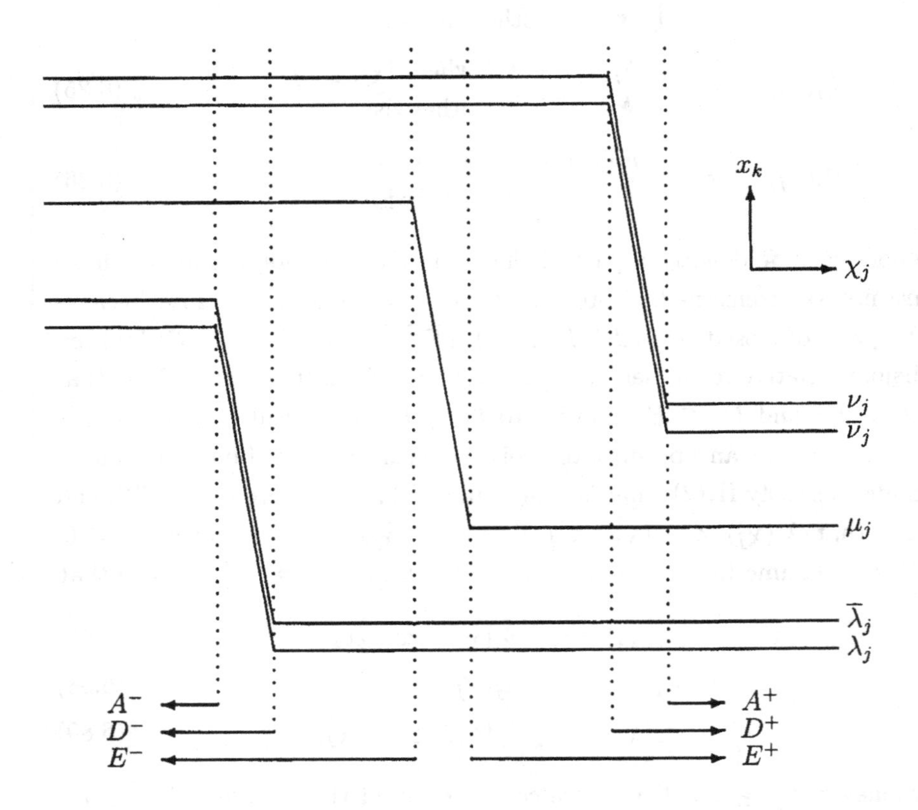

Figure 6.2: Construction of the functions λ_j, $\overline{\lambda}_j$, μ_j, $\overline{\nu}_j$, and ν_j. The horizontal axis represents the χ_j-space partitioned into sets $A^{\pm}$, $D^{\pm}$, and $E^{\pm}$. The vertical axis represents the x_k-space. The named functions are constructed by patching together pieces on different partitions as shown.

If $\chi_j \notin D^+$, then it follows from (6.84), (6.86), (6.81), and (6.88) that

$$\begin{aligned} \overline{\nu}_j(\chi_j) &= \varsigma_2(\chi_j) - 1 \\ &\geq \varsigma_1(\chi_j + d_j) + 2|\chi_j + d_j| \\ &\geq \mu_j(\chi_j + d_j) + 2|\chi_j + d_j| \end{aligned} \tag{6.91}$$

which gives (6.90). If $\chi_j \in D^+$, then from (6.56) we have $\chi_j + d_j \in E^+$, and it follows from (6.86), (6.87), (6.78), and (6.83) that

$$\begin{aligned} \overline{\nu}_j(\chi_j) &\geq \sigma(\chi_j) - 1 \\ &\geq \sigma_1(\chi_j + d_j) + 2|\chi_j + d_j| \\ &\geq \mu_j(\chi_j + d_j) + 2|\chi_j + d_j| \end{aligned} \tag{6.92}$$

which also gives (6.90). Finally, to verify H$_j$(3c) we must show that

$$\overline{\lambda}_j(\chi_j) \leq \mu_j(\chi_j + d_j) - 2|\chi_j + d_j| \tag{6.93}$$

If $\chi_j \notin D^-$, then it follows from (6.82), (6.85), (6.79), and (6.88) that

$$\begin{aligned} \overline{\lambda}_j(\chi_j) &= \sigma_2(\chi_j) + 1 \\ &\leq \sigma_1(\chi_j + d_j) - 2|\chi_j + d_j| \\ &\leq \mu_j(\chi_j + d_j) - 2|\chi_j + d_j| \end{aligned} \tag{6.94}$$

which gives (6.93). If $\chi_j \in D^-$, then from (6.57) we have $\chi_j + d_j \in E^-$, and it follows from (6.85), (6.89), (6.80), and (6.83) that

$$\begin{aligned} \overline{\lambda}_j(\chi_j) &\leq \varsigma(\chi_j) + 1 \\ &\leq \varsigma_1(\chi_j + d_j) - 2|\chi_j + d_j| \\ &\leq \mu_j(\chi_j + d_j) - 2|\chi_j + d_j| \end{aligned} \tag{6.95}$$

which also gives (6.93). This completes the proof of Theorem 6.3.

6.2.5 Design procedure and example

The design procedure outlined in the previous sections can be summarized as a list of tasks to perform at each step i in the recursion, where i takes values sequentially from 0 to $n-1$. The tasks for the initialization step $i = 0$ are as follows:

Step 0_1: define the function $V_1 = \frac{1}{2}x_1^2$
Step 0_2: calculate the functions σ and ς from (6.36) and (6.37)
Step 0_3: use Lemma 6.4 to construct functions σ_1, σ_2, ς_1, and ς_2 that satisfy (6.40)–(6.43)
Step 0_4: construct λ_1, $\overline{\lambda}_1$, μ_1, $\overline{\nu}_1$, and ν_1 as in (6.44)–(6.48)

If $n = 1$, then the design is complete: V_1 is the desired rclf and μ_1 is the desired control law. If $n > 1$, then we must continue the design in the recursive fashion outlined in previous sections. The list of tasks to perform at each step i is as follows, where $j := i + 1$:

Step i_1: construct the function V_j as in (6.65) and (6.71)
Step i_2: calculate the functions σ and ς from (6.73) and (6.74)
Step i_3: use Lemma 6.4 to construct functions σ_1, σ_2, ς_1, and ς_2 that satisfy (6.78)–(6.81)
Step i_4: construct λ_j, $\overline{\lambda}_j$, μ_j, $\overline{\nu}_j$, and ν_j as in (6.82)–(6.86)

At the completion of step $i = n - 1$, we will have our desired rclf V_n and control law μ_n.

We now illustrate this design procedure for the system

$$\begin{aligned} \dot{x}_1 &= x_2 + wx_1^2 + w \qquad (6.96) \\ \dot{x}_2 &= u \\ y &= x + d \qquad (6.97) \end{aligned}$$

where the disturbance w takes values in the interval $[-1, 1]$ and the measurement disturbance d is bounded. This system is in strict feedback form with $\phi_{11} = wx_1$, $\phi_{12} = \phi_{23} = 1$, and $F(0, w) = [w\ \ 0]^{\mathrm{T}}$.

We first perform the list of tasks for $i = 0$. We assign $V_1 = \frac{1}{2}x_1^2$ and choose the functions σ and ς according to (6.36) and (6.37):

$$\begin{aligned} \sigma(x_1) &= -\tfrac{1}{2}x_1 - x_1^2 - 1 \qquad (6.98) \\ \varsigma(x_1) &= -\tfrac{1}{2}x_1 + x_1^2 + 1 \qquad (6.99) \end{aligned}$$

The next task is to construct the functions σ_1, σ_2, ς_1, and ς_2 using Lemma 6.4. To construct σ_1, we define $\omega(x_1) = \sigma(x_1) - 1$ and calculate ρ_ω and γ_ω in (6.32):

$$\begin{aligned} \left|\omega(x_1 + d_1) - \omega(x_1)\right| &\le \left|-\tfrac{1}{2}(x_1 + d_1) - (x_1 + d_1)^2 + \tfrac{1}{2}x_1 + x_1^2\right| \\ &\le \left|(x_1 + d_1)^2 - x_1^2\right| + \tfrac{1}{2}|d_1| \end{aligned}$$

$$\begin{aligned} &\leq \; 2|x_1||d_1| + d_1^2 + \tfrac{1}{2}|d_1| \\ &\leq \; \big[2|x_1| + 1\big]\big[|d_1| + d_1^2\big] \end{aligned} \tag{6.100}$$

Following (6.33), we choose ζ to be a C^1 function such that

$$\zeta(x_1) \;\geq\; \big[2|x_1| + 1\big]\big[|x_1| + x_1^2\big] + 3|x_1| \tag{6.101}$$

for all $x_1 \in I\!R$, and we assign $\sigma_1(x_1) = \sigma(x_1) - 1 - \zeta(x_1)$. Using this choice for σ_1, we again apply Lemma 6.4, this time with $\omega(x_1) = \sigma_1(x_1)$. The resulting function ζ, whose calculation we omit for brevity, defines the function $\sigma_2(x_1) = \sigma_1(x_1) - 1 - \zeta(x_1)$. The functions ς_1 and ς_2 are computed in an analogous manner. Finally, we patch these six functions σ, σ_1, σ_2, ς, ς_1, and ς_2 together as in (6.44)–(6.48) to obtain the functions λ_1, $\overline{\lambda}_1$, μ_1, $\overline{\nu}_1$, and ν_1. Here $\varepsilon_1 > 0$ is a design parameter.

We now perform a similar list of tasks for $i = 1$. The function V_2, which will be the complete rclf for our system (6.96)–(6.97), is given by (6.65), namely,

$$V_2(x_1, x_2) \;=\; \tfrac{1}{2}x_1^2 + \begin{cases} \tfrac{1}{2}\big[x_2 - \nu_1(x_1)\big]^2 & \text{when } x_2 > \nu_1(x_1) \\ 0 & \text{when } \lambda_1(x_1) \leq x_2 \leq \nu_1(x_1) \\ \tfrac{1}{2}\big[x_2 - \lambda_1(x_1)\big]^2 & \text{when } x_2 < \lambda_1(x_1) \end{cases}$$

We have left to construct the associated control law μ_2. We begin by calculating the functions ψ^+ and ψ^- as in (6.69)–(6.70); this involves computing the derivatives of the functions V_1, ν_1, and λ_1. We then choose (new) C^1 functions σ and ς such that $\sigma(x_1, x_2) \leq \varsigma(x_1, x_2)$ and

$$\sigma(x_1, x_2) \;\leq\; -\psi^+(x_1, x_2, w) \tag{6.102}$$

$$\varsigma(x_1, x_2) \;\geq\; -\psi^-(x_1, x_2, w) \tag{6.103}$$

for all $x_1, x_2 \in I\!R$ and all $w \in [-1, 1]$. Two applications of Lemma 6.4 yield the functions σ_1 and ς_1, which when patched together form our control law μ_2 as in (6.83).

The closed-loop system with the control law μ_2 is

$$\dot{x}_1 \;=\; x_2 + w x_1^2 + w \tag{6.104}$$

$$\dot{x}_2 \;=\; \mu_2(x + d) \tag{6.105}$$

where $d \in I\!R^2$ represents the bounded state measurement disturbance. This system is (robustly) ISS with respect to the disturbance d, namely, every state trajectory will converge to some compact set whose size depends on the size of d.

This design procedure is less explicit than the backstepping procedures of Chapter 5 because of Step i_3. The source of difficulty here is in the application of Lemma 6.4: we have no explicit formula for the functions ρ_ω and γ_ω in (6.32) in terms of the given function ω. We showed in (6.100) how to find these functions for a particular choice for ω, and similar algebraic manipulations will yield ρ_ω and γ_ω for any polynomial ω. However, the calculation of ρ_ω and γ_ω may be tedious in general.

6.3 Summary

In this chapter we considered systems whose state measurements are corrupted by additive disturbances. We showed that these exogenous disturbances can destroy global stability and, in the case of memoryless time-invariant feedback, even global stabilizability. Such phenomena are inherently nonlinear because exogenous disturbances do not affect the stability of linear systems.

Using the flattened Lyapunov functions developed in Chapter 5, we showed that systems in strict feedback form remain globally robustly stabilizable in the presence of additive state measurement disturbances. We presented a recursive backstepping procedure for constructing the rclf and stabilizing control law. At the time of this writing, we know of no larger class of nonlinear systems for which global robustness to state measurement disturbances is possible.

Chapter 7

Dynamic Partial State Feedback

The controllers we have designed thus far have been static (memoryless) and have employed state feedback either perfect or corrupted by additive measurement disturbances. Potential advantages of dynamic over static feedback have been explored in the nonlinear control literature, and several paradigms for dynamic feedback design have been introduced. Among them, paradigms for dynamic feedback linearization and disturbance decoupling [60, 111] are being developed elsewhere and will not be pursued here.

Another dynamic feedback paradigm is the adaptive control of nonlinear systems with unknown constant parameters [85]. Although not commonly perceived as such, this adaptive control paradigm is a special case of a nonlinear observer paradigm in which the various filters and parameter update laws comprise a type of "observer" for the unknown parameter vector, regarded now as an unmeasured state variable. In this chapter we adopt this nonlinear observer point of view and show how globally stabilizing dynamic feedback controllers can be constructed for larger classes of systems without the need for full state information. We thereby develop a controller design procedure for a new class of *extended strict feedback systems.*

The class of extended strict feedback systems is introduced in Section 7.1. Such systems have a strict feedback structure (similar to that described in Section 5.2) in which the unmeasured states enter in an affine manner. A dynamic backstepping controller design for this class of systems is presented in Section 7.2. In Section 7.3 we apply this design to a nonlinear mechanical system.

7.1 Nonlinear observer paradigm

To avoid cumbersome notation, we restrict our presentation to systems with no uncertainty. Using the tools of Chapters 5 and 6, similar results can be obtained for classes of systems with uncertainties.

7.1.1 Extended strict feedback systems

We consider single-input nonlinear systems of the form

$$\dot{x} = A(x) + B(x)\,u \tag{7.1}$$

$$z = c(x) \tag{7.2}$$

$$y = C(x) \tag{7.3}$$

with two output variables z and y. The scalar *tracking output* z is the variable we wish to control using feedback from the (vector) *measurement output* y. We restrict our attention those systems which can be converted via smooth global state and measurement output transformations into the following form:

$$\begin{array}{|rcl|c|}
\hline
\dot{\eta} & = & F(\zeta,\xi_1)\,\eta + G(\zeta,\xi_1) & \eta \in I\!R^p \\
\hline
\dot{\zeta} & = & f(\zeta) + g(\zeta)\,\xi_1 & \zeta \in I\!R^q \\
z & = & h(\zeta) + k(\zeta)\,\xi_1 & z \in I\!R \\
\hline
\dot{\xi}_1 & = & \xi_2 + \phi_1(\zeta,\xi_1)\,\eta & \\
\dot{\xi}_2 & = & \xi_3 + \phi_2(\zeta,\xi_1,\xi_2)\,\eta & \\
 & \vdots & & \xi_i \in I\!R \\
\dot{\xi}_{n-1} & = & \xi_n + \phi_{n-1}(\zeta,\xi_1,\ldots,\xi_{n-1})\,\eta & \\
\dot{\xi}_n & = & \phi_0(\zeta,\xi) + \phi_n(\zeta,\xi)\,\eta + \phi_u(\zeta,\xi)\,u & u \in I\!R \\
\hline
y & = & \begin{bmatrix} \zeta \\ \xi \end{bmatrix} & y \in I\!R^{q+n} \\
\hline
\end{array} \tag{7.4}$$

where all functions are smooth and $\xi := [\xi_1 \ldots \xi_n]^{\mathrm{T}} \in I\!R^n$. We assume that the coefficient $\phi_u(\zeta,\xi)$ of the control variable u is nonzero for all

$(\zeta, \xi) \in I\!R^{q+n}$. The system (7.4) consists of three interconnected subsystems: the η-subsystem, the ζ-subsystem, and the ξ-subsystem. The tracking output z appears as the output of the ζ-subsystem. The measurement output y, which is available to the controller, consists of the states ζ and ξ. The η states, which are not measured, enter the η- and ξ-subsystems in an affine fashion and do not enter the ζ-subsystem. The ξ-subsystem is in a form similar to the strict feedback form of Section 5.2, and it is reminiscent of the parametric strict feedback form of adaptive backstepping [72].

Our control objective is to drive the tracking output z asymptotically to a desired bounded reference trajectory $z_r(t)$ from any initial condition while maintaining the boundedness of η, ζ, ξ, u, and any internal controller states. Our dynamic controller will have access to the measurement output y (partial state feedback) as well as the reference $z_r(t)$ and its derivatives. Figure 7.1 illustrates the closed-loop system structure. The dynamic controller will be of the form

$$u = \mu_n(\zeta, \xi, \omega, t) \tag{7.5}$$

$$\dot{\omega} = \Omega_n(\zeta, \xi, \omega, t) \qquad \omega \in I\!R^p \tag{7.6}$$

where μ_n and Ω_n are smooth functions whose time dependencies occur solely through the reference trajectory $z_r(t)$ and its derivatives. The dynamic order of our controller is the same as the order of the η-subsystem. The observer paradigm will become more clear later when we interpret the dynamics (7.6) as nonlinear observer for the unmeasured state η. For this reason we will refer to the variable $\sigma := \eta - \omega$ as the *observer error*.

7.1.2 Assumptions and system structure

We will establish the existence of a controller (7.5)–(7.6) which meets our control objective under the following assumptions on the η- and ζ-subsystems and the reference signals.

Our assumptions on the η-subsystem $\dot{\eta} = F(\zeta, \xi_1)\,\eta + G(\zeta, \xi_1)$ are:

H1: There exists a symmetric positive definite matrix P such that for all $(\zeta, \xi_1) \in I\!R^{q+1}$,

$$F^{\mathrm{T}}(\zeta, \xi_1)\,P + PF(\zeta, \xi_1) \;\leq\; 0 \tag{7.7}$$

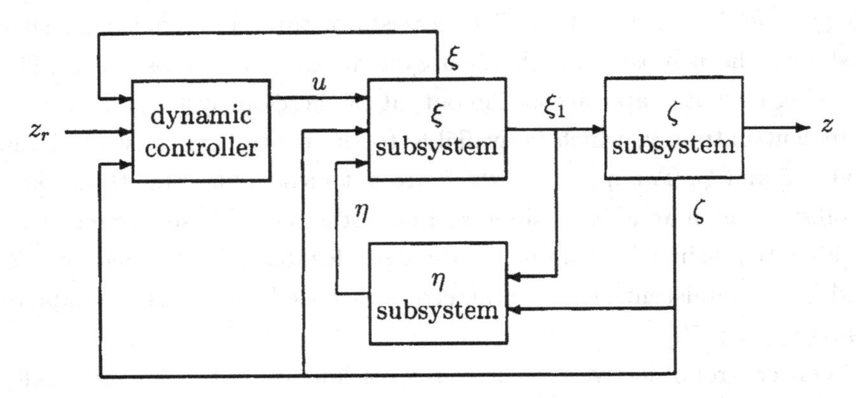

Figure 7.1: Extended strict feedback system.

H2: For every $\eta_0 \in I\!R^p$, every $\zeta(t) \in L^q_\infty$, and every $\xi_1(t) \in L_\infty$, the solution $\eta(t)$ starting from the initial condition η_0 is bounded, that is, $\eta(t) \in L^p_\infty$.

If F is a constant matrix, then it follows from [104, Theorem 4] that assumption H1 is a consequence of assumption H2. In this case, assumption H1 is equivalent to the assumption that the linear system $\dot{x} = Fx$ is stable in the sense of Lyapunov but not necessarily asymptotically stable.

Our assumptions on the ζ-subsystem $[f, g, h, k]$ with input ξ_1 are:

H3: The SISO system $[f, g, h, k]$ is globally input/output linearizable with uniform relative degree $r \in \{0, \ldots, q\}$.

H4: If $\zeta(t)$ is the solution on an interval $[0, t_f)$ for some $t_f \in (0, \infty]$, some initial condition $\zeta_0 \in I\!R^q$, and some input $\xi_1(t) \in L_{\infty e}[0, t_f)$, and if the resulting output $z(t)$ and its first r derivatives are bounded on $[0, t_f)$, then $\zeta(t)$ is also bounded on $[0, t_f)$, that is, $\zeta(t) \in L^q_\infty[0, t_f)$.

Finally, our assumption on the reference signals is:

H5: The reference trajectory $z_r(t)$ and its first $r + n$ derivatives are bounded.

We now interpret the structure of the extended strict feedback system (7.4) and the roles of assumptions H1–H4.

Global input/output linearizability. It is clear from assumption H3 and the strict feedback structure of the ξ-subsystem that the whole system (7.4) has a uniform relative degree $r+n$ with respect to the tracking output z and is therefore globally input/output linearizable via *full state feedback.* To be precise, there exists a feedback transformation $u = \alpha(\eta, \zeta, \xi) + \beta(\zeta, \xi)\, v$ such that the tracking output z is given by the chain of integrators $z^{(r+n)} = v$. However, this feedback transformation is not implementable because η is not available for measurement.

Inverse dynamics. The inverse dynamics of the system (7.4) are defined as the dynamics which remain after the (non-implementable) transformation into the chain of integrators $z^{(r+n)} = v$. These dynamics include the η-subsystem and, if $r < q$, all or part of the ζ-subsystem. Assumptions H2 and H4 together state that these inverse dynamics are globally bounded-input/bounded-state (BIBS) stable. This assumption is necessary if we want to track *arbitrary* bounded reference trajectories while maintaining internal boundedness. It is not the same as the minimum-phase assumption commonly used in the stabilization problem, which requires the inverse dynamics to be zero-input *asymptotically* stable but not necessarily BIBS stable. Also, our BIBS assumption on the inverse dynamics is weaker than the input-to-state stability assumption used in [143] in the context of the stabilization of uncertain systems. The key feature of our assumption is that we do not require the η- or ζ-subsystems to be zero-input *asymptotically* stable. These subsystems can therefore include disturbance exogenerators (considered in [62] for local output regulation) or unknown constant parameters. In fact, if we consider the special case where η is an unknown constant parameter with trivial dynamics $\dot{\eta} = 0$, then Theorem 7.1 below reduces to a theorem proved in [84] in the context of adaptive nonlinear control (see also [85]).

Inputs to inverse dynamics. In the case $n \geq 2$, the states $(\xi_2, \ldots, \xi_n)$ are not allowed to appear as inputs to the inverse dynamics. Such restrictions on the connection between the inverse dynamics and the linearizable dynamics are commonly assumed to prevent the destabilizing effects of peaking [140].

Structural restrictions regarding η. Our most restrictive assumption is that the unmeasured state variable η appears linearly in the system equations, and that the ξ-subsystem has a lower triangular strict feedback structure with respect to η. This assumption of linearity in the unmeasured states has also been exploited in [117, 116, 13] for stabilization and in [100, 73] for tracking.

Let us now examine some simple systems and determine whether or not they can be put into extended strict feedback form (7.4) while satisfying H1–H4. The first system is

$$\begin{aligned} \dot{x}_1 &= -x_1 + x_2 \\ \dot{x}_2 &= x_1^3 + u \\ z &= x_2 \\ y &= x_2 \end{aligned} \tag{7.8}$$

Here x_1 is an unmeasured state, so we assign $\eta := x_1$ when writing this system in the form (7.4). However, we see that the system is not affine in this unmeasured state because of the x_1^3 term in (7.8), and we conclude that it is not in extended strict feedback form.

We next consider the system

$$\begin{aligned} \dot{x}_1 &= -x_1 + x_2^2 + a x_3^4 \\ \dot{x}_2 &= x_3 + x_1 x_2^3 \\ \dot{x}_3 &= x_1 x_3 + u \\ z &= x_2 \\ y &= \begin{bmatrix} x_2 \\ x_3 \end{bmatrix} \end{aligned} \tag{7.9}$$

where a is a constant parameter. This system is affine in the unmeasured state $\eta := x_1$. Because x_1 appears in (7.9), there will be no ζ variable and we thus assign $\xi_1 := x_2$ and $\xi_2 := x_3$. If $a = 0$, then this system is in extended strict feedback form and satisfies H1–H4. If a is nonzero, however, then ξ_2 drives the η-subsystem which is prohibited in (7.4).

Finally, we consider the system

$$\begin{aligned} \dot{x}_1 &= -ax_1 + x_2^2 + x_3^4 \\ \dot{x}_2 &= x_3 + x_2^3 \\ \dot{x}_3 &= x_1x_3 + u \\ z &= x_2 \\ y &= \begin{bmatrix} x_2 \\ x_3 \end{bmatrix} \end{aligned} \tag{7.10}$$

where a is a constant parameter. Again we assign $\eta := x_1$, but this time x_1 does not appear in (7.10) which means we may introduce the ζ variable as $\zeta := x_2$ and then assign $\xi_1 := x_3$. If $a > 0$, then this system is in extended strict feedback form and satisfies H1–H4. If $a \leq 0$, however, then the η-subsystem is not BIBS stable and thus violates H1–H2.

7.2 Controller design

We have introduced a class of extended strict feedback systems and discussed our assumptions H1–H5. Our main result, which we report in Section 7.2.1, states that a dynamic feedback controller exists which achieves global asymptotic tracking for our system (7.4). We prove this result in Sections 7.2.2–7.2.6 by constructing such a controller together with an associated Lyapunov function.

7.2.1 Main result

The main result of this chapter is:

Theorem 7.1 *Under assumptions* H1–H5, *there exists a dynamic controller* (7.5)-(7.6) *for the system* (7.4) *which achieves global asymptotic tracking with internal boundedness: for any initial condition of the closed-loop system, all states are bounded for all* $t \geq 0$ *and* $z(t) \to z_r(t)$ *as* $t \to \infty$. *If in addition the inequality* (7.7) *in assumption* H1 *is strict, then the observer error* $\sigma(t) = \eta(t) - \omega(t)$ *converges to zero as* $t \to \infty$.

We split the controller design for the system (7.4) into two parts. In Section 7.2.2 we do the design for the case $n = 1$. Then in Section 7.2.5 we do the design for the case $n \geq 2$ by recursively applying the backstepping lemma given in Section 7.2.4. Finally, in Section 7.2.6 we analyze the closed-loop system and give the proof of Theorem 7.1.

7.2.2 Controller design for $n = 1$

If $n = 1$, then the system (7.4) is

$$\begin{array}{|rcl|}\hline \dot{\eta} & = & F(\zeta,\xi_1)\,\eta + G(\zeta,\xi_1) \\ \hline \dot{\zeta} & = & f(\zeta) + g(\zeta)\,\xi_1 \\ z & = & h(\zeta) + k(\zeta)\,\xi_1 \\ \hline \dot{\xi}_1 & = & \phi_0(\zeta,\xi_1) + \phi_1(\zeta,\xi_1)\,\eta + \phi_u(\zeta,\xi_1)\,u \\ \hline y & = & \begin{bmatrix} \zeta \\ \xi_1 \end{bmatrix} \\ \hline \end{array} \tag{7.11}$$

We first pretend that ξ_1 is the control variable for the ζ-subsystem

$$\dot{\zeta} = f(\zeta) + g(\zeta)\,\xi_1 \tag{7.12}$$

$$z = h(\zeta) + k(\zeta)\,\xi_1 \tag{7.13}$$

and construct a static tracking controller $\xi_1 = \mu_0(\zeta, t)$ for this subsystem. Because the unmeasured variable η does not enter these equations, this is a full-state feedback design and we can therefore use the results on input/output linearization in [60, Section 4.5].

By assumption H3, the system (7.12)–(7.13) has uniform relative degree r. We will assume for the moment that $r \geq 1$, in which case $k(\zeta) \equiv 0$. Later we will give the modifications needed for the case $r = 0$. We define the output tracking error $e := z - z_d(t)$ and evaluate its first r derivatives:

$$e^{(i)} = L_f^i h(\zeta) - z_d^{(i)}(t) \qquad i \in \{0, \ldots, r-1\} \tag{7.14}$$

$$e^{(r)} = L_f^r h(\zeta) + L_g L_f^{r-1} h(\zeta) \cdot \xi_1 - z_d^{(r)}(t) \tag{7.15}$$

$$\bar{e} := [e\ \dot{e}\ \ldots\ e^{(r-1)}]^{\mathrm{T}} \tag{7.16}$$

where by assumption $L_g L_f^{r-1} h(\zeta) \neq 0$ for all $\zeta \in I\!R^q$. Here $\bar{e} \in I\!R^r$ is the vector containing the tracking error e and its first $(r-1)$ derivatives. A global tracking controller $\xi_1 = \mu_0(\zeta, t)$ for the ζ-subsystem (7.12)–(7.13) is

$$\mu_0(\zeta, t) := \frac{1}{L_g L_f^{r-1} h(\zeta)} \left[-L_f^r h(\zeta) + z_d^{(r)}(t) - \sum_{i=0}^{r-1} c_i\, e^{(i)} \right] \tag{7.17}$$

where the constants c_i are such that the polynomial $s^r + c_{r-1}\, s^{r-1} + \ldots + c_0$ is Hurwitz. Using (7.17) we can rewrite (7.15) as

$$e^{(r)} \;=\; -\sum_{i=0}^{r-1} c_i\, e^{(i)} + L_g L_f^{r-1} h(\zeta) \left[\xi_1 - \mu_0(\zeta, t)\right] \tag{7.18}$$

We now see that if we could implement the controller $\xi_1 = \mu_0(\zeta, t)$, then the tracking error e would converge to zero from any initial condition. However, this controller is only conceptual: it cannot be implemented because the state ξ_1 is not a control input but is governed by its own dynamics. Nevertheless, we can use this controller to rewrite the system (7.14)–(7.15) in the form of an asymptotically stable system driven by the error variable

$$\rho_0 \;:=\; L_g L_f^{r-1} h(\zeta) \left[\xi_1 - \mu_0(\zeta, t)\right] \tag{7.19}$$

In this error variable ρ_0, the crucial factor is difference between the actual input ξ_1 to the ζ-subsystem and the conceptual control signal $\mu_0(\zeta(t), t)$. In state space form, the equation (7.18) becomes

$$\dot{\bar{e}} \;=\; A_e\, \bar{e} + B_e\, \rho_0 \tag{7.20}$$

where the constant matrices A_e and B_e are given by

$$A_e := \begin{bmatrix} 0 & 1 & & \\ \vdots & & \ddots & \\ 0 & & & 1 \\ -c_0 & -c_1 & \cdots & -c_{r-1} \end{bmatrix} \qquad B_e := \begin{bmatrix} 0 \\ \vdots \\ 0 \\ 1 \end{bmatrix} \tag{7.21}$$

The factor $L_g L_f^{r-1} h(\zeta)$ in our definition (7.19) of the error variable ρ_0 allows us to conclude from (7.20) that $\bar{e}$ is bounded whenever ρ_0 is bounded. Now A_e is Hurwitz by the choice of the constants c_i, and so there exist symmetric positive definite matrices P_e and Q_e such that

$$A_e^{\mathrm{T}} P_e + P_e A_e \;=\; -Q_e \tag{7.22}$$

We next design a dynamic controller for the complete system (7.11) with $n = 1$. This is a partial state feedback problem in which the unmeasured variable η is *matched* with the control variable u. We let $\omega \in I\!R^p$ denote the state of our controller, we define $\sigma := \eta - \omega$, and we employ

$$V_1(\bar{e}, \rho_0, \sigma) \;:=\; \bar{e}^{\mathrm{T}} P_e \bar{e} + d_1 \rho_0^2 + \sigma^{\mathrm{T}} P \sigma \tag{7.23}$$

as our Lyapunov function for the closed-loop system. Here P is as in H1 and $d_1 > 0$ is a constant design parameter. We define $\rho_1 := [\bar{e}^{\mathrm{T}}\ \rho_0]^{\mathrm{T}}$ and $P_1 := \mathrm{diag}(P_e,\, d_1)$, and we rewrite (7.23) in more compact notation as

$$V_1(\rho_1,\, \sigma) \quad := \quad {\rho_1}^{\mathrm{T}} P_1\, \rho_1 + \sigma^{\mathrm{T}} P \sigma \tag{7.24}$$

We next calculate the derivative $\dot{V}_1$ along solutions of the system (7.11):

$$\dot{V}_1 = -\bar{e}^{\mathrm{T}} Q_e \bar{e} + 2 d_1 \rho_0 \left[a_1(\zeta,\xi_1)\,\eta + a_2(\zeta,\xi_1)\,u + a_3(\zeta,\xi_1,t)\right] + 2(\dot{\eta} - \dot{\omega})^{\mathrm{T}} P \sigma \tag{7.25}$$

where the functions a_i are smooth and explicitly known from (7.11) and (7.20). For example, the functions a_1 and a_2 are given by the formulas

$$a_1(\zeta,\xi_1) = \frac{\partial \rho_0}{\partial \xi_1}\, \phi_1(\zeta,\xi_1) \tag{7.26}$$

$$a_2(\zeta,\xi_1) = \frac{\partial \rho_0}{\partial \xi_1}\, \phi_u(\zeta,\xi_1) \tag{7.27}$$

Note that $a_2(\zeta,\xi_1) \neq 0$ for all $(\zeta,\xi_1) \in I\!R^{q+1}$. We now look for smooth functions μ_1 and Ω_1 such that the dynamic feedback

$$u = \mu_1(\zeta,\xi_1,\omega,t) \tag{7.28}$$

$$\dot{\omega} = \Omega_1(\zeta,\xi_1,\omega,t) \tag{7.29}$$

makes the Lyapunov derivative in (7.25) negative. We first choose a certainty equivalence control law

$$\mu_1(\zeta,\xi_1,\omega,t) := \frac{1}{a_2(\zeta,\xi_1)} \left[-m_1\rho_0 - a_1(\zeta,\xi_1)\,\omega - a_3(\zeta,\xi_1,t)\right] \tag{7.30}$$

which upon substitution into (7.25) yields

$$\dot{V}_1\Big|_{u=\mu_1} = -\bar{e}^{\mathrm{T}} Q_e \bar{e} - 2 d_1 m_1 \rho_0^2 + 2 d_1 \rho_0\, a_1(\zeta,\xi_1)\,\sigma + 2(\dot{\eta} - \dot{\omega})^{\mathrm{T}} P \sigma \tag{7.31}$$

where $m_1 > 0$ is a constant design parameter. By "certainty equivalence" we mean that the control law μ_1 is obtained by first cancelling the term $a_1(\zeta,\xi_1)\,\eta$ in (7.25) and then simply replacing the unmeasured variable η

with its "estimate" ω. We next choose the controller dynamics $\dot{\omega}$ to cancel the term $2d_1\rho_0\, a_1(\zeta,\xi_1)\,\sigma$ as well as the terms coming from $\dot{\eta}$:

$$\Omega_1(\zeta,\xi_1,\omega,t) \;:=\; F(\zeta,\xi_1)\,\omega \;+\; G(\zeta,\xi_1) \;+\; d_1\rho_0\,P^{-1}a_1^{\mathrm{T}}(\zeta,\xi_1) \tag{7.32}$$

Substituting (7.32) for $\dot{\omega}$ in (7.31) yields

$$\dot{V}_1\Big|_{\substack{u=\mu_1\\ \dot{\omega}=\Omega_1}} \;=\; -W_1(\rho_1) \;+\; \sigma^{\mathrm{T}}\left[F^{\mathrm{T}}(\zeta,\xi_1)\,P \;+\; PF(\zeta,\xi_1)\right]\sigma \tag{7.33}$$

where $W_1(\rho_1) := \bar{e}^{\mathrm{T}}Q_e\bar{e}+2d_1m_1\rho_0^2$ is a positive definite quadratic function of the variable $\rho_1 := [\bar{e}^{\mathrm{T}}\ \rho_0]^{\mathrm{T}}$.

The controller dynamics $\dot{\omega} = \Omega_1$, with Ω_1 defined in (7.32), illustrate the nonlinear observer paradigm. These dynamics consist of a copy of the η-subsystem plus a nonlinear driving term, and can be interpreted as a type of nonlinear observer for the unmeasured state η. The driving term is *not* an output error as is usual in observer designs, but is instead a weighted difference between $\mu_0(\zeta,t)$ and the variable ξ_1 as can be seen from (7.19).

We pause for a moment to point out another important feature of this approach. The usual combination of a certainty equivalence control law and a separately designed exponentially convergent *linear* observer for η would fail here; in general, such a combination would not be able to prevent finite escape times, let alone achieve the control objective. A situation of this type occurs already with the simple extended strict feedback system

$$\dot{\eta} \;=\; 0 \tag{7.34}$$

$$\dot{\xi}_1 \;=\; (1+\xi_1^2)\eta \;+\; u \tag{7.35}$$

Standard adaptive control techniques can be used to construct dynamics for $\dot{\omega}$ so that the variable ω converges exponentially to the unknown value of η from any initial condition. If we were to use ω in a certainty equivalence control law $u = -(1+\xi_1^2)\omega - \xi_1$, then the closed-loop ξ_1-subsystem would become

$$\dot{\xi}_1 \;=\; -\xi_1 \;+\; \xi_1^2(\eta-\omega) \tag{7.36}$$

It is easy to show that this system exhibits finite escape times from some initial conditions even though the observer error $\sigma := \eta - \omega$ converges exponentially to zero.

In contrast, our nonlinear observer (7.32) and control law (7.30) will work together to guarantee the boundedness of the observer error σ, and, if the inequality (7.7) is strict, the convergence of σ to zero. Our interpretation of the controller dynamics (7.32) as an observer will carry over to the cases $n \geq 2$, but the certainty equivalence form of our control law (7.30) is particular to the case $n = 1$.

We next give the appropriate modifications in the design when $r = 0$. In this case we have $k(\zeta) \neq 0$ for all $\zeta \in \mathbb{R}^q$, and instead of (7.17) we define

$$\mu_0(\zeta, t) \quad := \quad \frac{1}{k(\zeta)} \left[-h(\zeta) + z_d(t) \right] \tag{7.37}$$

Instead of (7.19) we define

$$\rho_0 := k(\zeta) \left[\xi_1 - \mu_0(\zeta, t) \right] \tag{7.38}$$

and instead of (7.20) we have simply

$$e \;=\; \rho_0 \tag{7.39}$$

We again use the Lyapunov function (7.24) with the modified definitions $\rho_1 := \rho_0$ and $P_1 := d_1$. The design proceeds exactly as above, and in the end we obtain (7.33) with $W_1(\rho_1) := 2d_1 m_1 \rho_0^2$. This completes the controller design for $n = 1$.

7.2.3 Conceptual controllers and derivatives

We now introduce some terminology to illuminate the ideas behind the controller design for $n \geq 2$. Recall that we designed a control law $\xi_1 = \mu_0(\zeta, t)$ for the ζ-subsystem (7.12)–(7.13), pretending that ξ_1 was a control variable. This *conceptual controller* was not implementable, of course, because ξ_1 was not a control variable. To see what we achieved by designing this conceptual controller $\mu_0(\zeta, t)$, consider the first term $\bar{e}^{\mathrm{T}} P_e \bar{e}$ of V_1 in (7.23) as a Lyapunov function

$$V_0(\bar{e}) \quad := \quad \bar{e}^{\mathrm{T}} P_e \bar{e} \tag{7.40}$$

for the system (7.20). The derivative of V_0 along solutions of (7.20) is

$$\dot{V}_0 = -\bar{e}^{\mathrm{T}} Q_e \bar{e} + 2\bar{e}^{\mathrm{T}} P_e B_e \rho_0 \tag{7.41}$$

Suppose now that we evaluate $\dot{V}_0$ only at points where $\xi_1 = \mu_0(\zeta, t)$; at such points we have from (7.19) that $\rho_0 = 0$ and thus

$$\dot{V}_0\Big|_{\xi_1 = \mu_0} = -\bar{e}^{\mathrm{T}} Q_e \bar{e} \tag{7.42}$$

This *conceptual derivative* (7.42) is not the true derivative $\dot{V}_0$ along solutions of the system (7.20), but instead represents the derivative (7.41) evaluated along the lower-dimensional submanifold $\xi_1 = \mu_0(\zeta, t)$ where the conceptual controller μ_0 agrees with the variable ξ_1. Our conceptual controller μ_0 was designed to make this conceptual derivative negative.

Let us now suppose that $n \geq 2$. In this case, the control u does not appear in the equation for $\dot{\xi}_1$, but the variable ξ_2 does. If we pretend that ξ_2 is the control variable, we can perform the calculations in Section 7.2.2 and obtain a *dynamic* conceptual controller

$$\xi_2 = \mu_1(\zeta, \xi_1, \omega, t) \tag{7.43}$$

$$\dot{\omega} = \Omega_1(\zeta, \xi_1, \omega, t) \tag{7.44}$$

It follows from (7.33) that this new conceptual controller yields the conceptual derivative

$$\dot{V}_1\Big|_{\substack{\xi_2 = \mu_1 \\ \dot{\omega} = \Omega_1}} = -W_1(\rho_1) + \sigma^{\mathrm{T}}\left[F^{\mathrm{T}}(\zeta, \xi_1)\, P + PF(\zeta, \xi_1)\right]\sigma \tag{7.45}$$

The key observation here is that the conceptual derivative (7.45) for the case $n \geq 2$ is identical to the actual derivative (7.33) for the case $n = 1$, but with the state variable ξ_2 appearing instead of the control variable u.

Our knowledge of the conceptual controller (7.43)–(7.44) and the corresponding conceptual derivative (7.45) will now be used in the construction of an *actual* controller. This will be a recursive backstepping construction. We first consider the case $n = 2$ and design an actual controller

$$u = \mu_2(\zeta, \xi_1, \xi_2, \omega, t) \tag{7.46}$$

$$\dot{\omega} = \Omega_2(\zeta, \xi_1, \xi_2, \omega, t) \tag{7.47}$$

This backstepping construction will be given in the next section. When we enlarge the system to the case $n = 3$, the actual controller (7.46)–(7.47) for $n = 2$ is demoted to the status of conceptual controller:

$$\xi_3 = \mu_2(\zeta, \xi_1, \xi_2, \omega, t) \tag{7.48}$$

$$\dot{\omega} = \Omega_2(\zeta, \xi_1, \xi_2, \omega, t) \tag{7.49}$$

We then perform the backstepping through the ξ_3-integrator to obtain an actual controller for the $n = 3$ system. For $n \geq 4$, we proceed in this manner until the control variable u appears. In each step of the design, the actual controller for the case $n = i$ is demoted to the status of conceptual controller for the case $n = i + 1$. The backstepping construction is the same at each step, and we now present it in the form of a lemma.

7.2.4 Backstepping lemma

The following backstepping construction, based on the adaptive control results of [84], is a main ingredient in our proof of Theorem 7.1 for $n \geq 2$. Consider the system

$$\dot{\chi}_1 = \alpha_1(\chi_1) + \beta_1(\chi_1)\,\eta + \gamma_1(\chi_1)\,\chi_2 \qquad \chi_1 \in I\!R^m \tag{7.50}$$

$$\dot{\chi}_2 = \alpha_2(\chi_1, \chi_2) + \beta_2(\chi_1, \chi_2)\,\eta + \gamma_2(\chi_1, \chi_2)\,u \qquad \chi_2 \in I\!R \tag{7.51}$$

where all functions are smooth, $u \in I\!R$ is a control variable, and $\eta \in I\!R^p$ is an unmeasured state variable with some dynamics $\dot{\eta}$. We assume $\gamma_2(\chi_1, \chi_2) \neq 0$ for all $(\chi_1, \chi_2) \in I\!R^{m+1}$. Suppose that we have designed a smooth p^{th}-order dynamic partial state feedback controller for the χ_1-subsystem (7.50), pretending that χ_2 is the control variable. Let us denote this conceptual controller by

$$\chi_2 = \mu_c(\chi_1, \omega) \tag{7.52}$$

$$\dot{\omega} = \Omega_c(\chi_1, \omega) \qquad \omega \in I\!R^p \tag{7.53}$$

where μ_c and Ω_c are smooth functions. Suppose also that we have an associated conceptual Lyapunov function

$$V_c(\chi_1, \omega, \eta) \;:=\; U(\chi_1, \omega) + (\eta - \omega)^{\mathrm{T}} P(\eta - \omega) \tag{7.54}$$

where $U(\chi_1, \omega)$ is a given smooth function and P is a constant symmetric positive definite matrix. The conceptual derivative of V_c is represented by the notation

$$\dot{V}_c\Big|_{\substack{\chi_2 = \mu_c \\ \dot{\omega} = \Omega_c}} \tag{7.55}$$

We will use this conceptual controller to construct an actual controller for the complete system (7.50)–(7.51). This actual controller will be a smooth p^{th}-order dynamic partial state feedback controller of the form

$$u = \mu_a(\chi_1, \chi_2, \omega) \tag{7.56}$$

$$\dot{\omega} = \Omega_a(\chi_1, \chi_2, \omega) \qquad \omega \in I\!R^p \tag{7.57}$$

where μ_a and Ω_a are smooth functions. Our goal is to find functions μ_a and Ω_a such that the actual derivative of the new Lyapunov function

$$V_a(\chi_1, \chi_2, \omega, \eta) := V_c(\chi_1, \omega, \eta) + d\left[\chi_2 - \mu_c(\chi_1, \omega)\right]^2 \tag{7.58}$$

along closed-loop trajectories is *less* than the conceptual derivative (7.55). We obtain the new Lyapunov function (7.58) by adding to (7.54) the square of the error between the conceptual control law $\mu_c(\chi_1, \omega)$ and the state variable χ_2. Here $d > 0$ is a constant design parameter.

Lemma 7.2 *There exist smooth functions $\mu_a(\chi_1, \chi_2, \omega)$ and $\Omega_a(\chi_1, \chi_2, \omega)$ such that*

$$\dot{V}_a\Big|_{\substack{u = \mu_a \\ \dot{\omega} = \Omega_a}} = \dot{V}_c\Big|_{\substack{\chi_2 = \mu_c \\ \dot{\omega} = \Omega_c}} - 2dm\left[\chi_2 - \mu_c(\chi_1, \omega)\right]^2 \tag{7.59}$$

where $m > 0$ is a constant design parameter.

Proof: We calculate $\dot{V}_c$ along trajectories of the system (7.50)–(7.51):

$$\begin{aligned}
\dot{V}_c &= \frac{\partial U}{\partial \chi_1}\left[\alpha_1(\chi_1) + \beta_1(\chi_1)\,\eta + \gamma_1(\chi_1)\,\chi_2\right] \\
&\quad + \frac{\partial U}{\partial \omega}\dot{\omega} + 2(\eta - \omega)^{\mathrm{T}} P\,(\dot{\eta} - \dot{\omega}) \\
&= \frac{\partial U}{\partial \chi_1}\left[\alpha_1(\chi_1) + \beta_1(\chi_1)\,\eta + \gamma_1(\chi_1)\,\mu_c(\chi_1, \omega)\right] + \frac{\partial U}{\partial \omega}\Omega_c(\chi_1, \omega)
\end{aligned}$$

$$
\begin{aligned}
&+ 2(\eta - \omega)^{\mathrm{T}} P \left[\dot{\eta} - \Omega_c(\chi_1, \omega)\right] \\
&+ \frac{\partial U}{\partial \chi_1} \gamma_1(\chi_1) \left[\chi_2 - \mu_c(\chi_1, \omega)\right] \\
&+ \left[\frac{\partial U}{\partial \omega} - 2(\eta - \omega)^{\mathrm{T}} P\right] \left[\dot{\omega} - \Omega_c(\chi_1, \omega)\right]
\end{aligned} \tag{7.60}
$$

$$
\begin{aligned}
= \; & \dot{V}_c\Big|_{\substack{\chi_2 = \mu_c \\ \dot{\omega} = \Omega_c}} + \frac{\partial U}{\partial \chi_1} \gamma_1(\chi_1) \left[\chi_2 - \mu_c(\chi_1, \omega)\right] \\
&+ \left[\frac{\partial U}{\partial \omega} + 2\omega^{\mathrm{T}} P\right] \left[\dot{\omega} - \Omega_c(\chi_1, \omega)\right] - 2\eta^{\mathrm{T}} P \left[\dot{\omega} - \Omega_c(\chi_1, \omega)\right]
\end{aligned} \tag{7.61}
$$

Note that the first three terms in (7.60) constitute the conceptual derivative (7.55) in (7.61). Introducing the error variable

$$
\rho \;:=\; \chi_2 - \mu_c(\chi_1, \omega) \tag{7.62}
$$

and calculating the derivative $\dot{V}_a$ of (7.58), we obtain

$$
\dot{V}_a \;=\; \dot{V}_c + \frac{d}{dt}\left(d\rho^2\right) \tag{7.63}
$$

where

$$
\begin{aligned}
\frac{d}{dt}\left(d\rho^2\right) \;=\; & 2d\rho \left[\alpha(\chi_1, \chi_2, \omega) + \beta(\chi_1, \chi_2, \omega)\, \eta \right. \\
& \left. + \gamma_2(\chi_1, \chi_2)\, u + \delta(\chi_1, \omega)\, \dot{\omega}\right]
\end{aligned} \tag{7.64}
$$

The functions α, β, and δ are explicitly known from (7.50)–(7.51) and (7.62). From (7.63), (7.61), and (7.64) we have

$$
\begin{aligned}
\dot{V}_a \;=\; & \dot{V}_c\Big|_{\substack{\chi_2 = \mu_c \\ \dot{\omega} = \Omega_c}} + \left[\frac{\partial U}{\partial \omega} + 2\omega^{\mathrm{T}} P\right] \left[\dot{\omega} - \Omega_c(\chi_1, \omega)\right] \\
& - 2\eta^{\mathrm{T}} P \left[\dot{\omega} - \Omega_c(\chi_1, \omega)\right] \\
& + 2d\rho \left[\frac{1}{2d} \frac{\partial U}{\partial \chi_1} \gamma_1(\chi_1) + \alpha(\chi_1, \chi_2, \omega) + \beta(\chi_1, \chi_2, \omega)\, \eta \right. \\
& \left. + \gamma_2(\chi_1, \chi_2)\, u + \delta(\chi_1, \omega)\, \dot{\omega}\right]
\end{aligned} \tag{7.65}
$$

Because η is not measured, the control u cannot be used to cancel the η-terms in (7.65). However, what has been achieved by this construction is that η appears as a factor multiplying a quantity containing $\dot{\omega}$. We can thus eliminate η from (7.65) by choosing our controller dynamics $\dot{\omega} = \Omega_a$ to set the quantity multiplying η to zero:

$$\Omega_a(\chi_1, \chi_2, \omega) \quad := \quad \Omega_c(\chi_1, \omega) + d\rho P^{-1}\beta^{\mathrm{T}}(\chi_1, \chi_2, \omega) \tag{7.66}$$

Substituting (7.66) for $\dot{\omega}$ in (7.65) yields

$$\dot{V}_a\Big|_{\dot{\omega} = \Omega_a} \quad = \quad \dot{V}_c\Big|_{\substack{\chi_2 = \mu_c \\ \dot{\omega} = \Omega_c}} + 2d\rho\left[\psi(\chi_1, \chi_2, \omega) + \gamma_2(\chi_1, \chi_2)\, u\right] \tag{7.67}$$

where

$$\psi(\chi_1, \chi_2, \omega) := \frac{1}{2}\frac{\partial U}{\partial \omega}P^{-1}\beta^{\mathrm{T}}(\chi_1, \chi_2, \omega) + \beta(\chi_1, \chi_2, \omega)\,\omega + \frac{1}{2d}\frac{\partial U}{\partial \chi_1}\gamma_1(\chi_1)$$
$$+ \alpha(\chi_1, \chi_2, \omega) + \delta(\chi_1, \omega)\,\Omega_a(\chi_1, \chi_2, \omega) \tag{7.68}$$

We now see from (7.67) that to achieve (7.59) we must choose

$$\mu_a(\chi_1, \chi_2, \omega) \quad := \quad \frac{1}{\gamma_2(\chi_1, \chi_2)}\left[-m\rho - \psi(\chi_1, \chi_2, \omega)\right] \tag{7.69}$$

We substitute μ_a for u in (7.67) and achieve the desired relationship (7.59) between the actual and conceptual Lyapunov derivatives. ∎

7.2.5 Controller design for $n \geq 2$

We use the controller designed in Section 7.2.2 as a conceptual controller for the (ζ, ξ_1)-subsystem, pretending that ξ_2 is the control variable. This conceptual controller (7.43)–(7.44) achieves the conceptual derivative (7.45). Suppose first that $n = 2$, and apply Lemma 7.2 with $\chi_1 = [t \;\; \zeta^{\mathrm{T}} \;\; \xi_1]^{\mathrm{T}}$, $\chi_2 = \xi_2$, $\mu_c = \mu_1$, $\Omega_c = \Omega_1$, and $V_c = V_1$. We obtain smooth functions $\mu_2(\zeta, \xi_1, \xi_2, \omega, t)$ $(= \mu_a)$ and $\Omega_2(\zeta, \xi_1, \xi_2, \omega, t)$ $(= \Omega_a)$ such that the derivative of the new Lyapunov function

$$V_2(\rho_1, \rho_2, \sigma) \quad := \quad V_1(\rho_1, \sigma) + d_2\rho_2^2 \tag{7.70}$$

(where $\rho_2 := \xi_2 - \mu_1(\zeta, \xi_1, \omega, t)$ and $d_2 > 0$ is a design parameter) satisfies

$$\dot{V}_2\Big|_{\substack{u = \mu_2 \\ \dot{\omega} = \Omega_2}} \quad = \quad \dot{V}_1\Big|_{\substack{\xi_2 = \mu_1 \\ \dot{\omega} = \Omega_1}} - 2d_2 m_2 \rho_2^2 \tag{7.71}$$

where $m_2 > 0$ is a design parameter. Our actual controller for $n=2$ is

$$u = \mu_2(\zeta, \xi_1, \xi_2, \omega, t) \tag{7.72}$$

$$\dot{\omega} = \Omega_2(\zeta, \xi_1, \xi_2, \omega, t) \tag{7.73}$$

If $n = 3$, then we demote the controller (7.72)–(7.73) to the status of conceptual controller, and the derivative (7.71) becomes the conceptual derivative

$$\dot{V}_2\Big|_{\substack{\xi_3 = \mu_2 \\ \dot{\omega} = \Omega_2}} = \dot{V}_1\Big|_{\substack{\xi_2 = \mu_1 \\ \dot{\omega} = \Omega_1}} - 2d_2 m_2 \rho_2^2 \tag{7.74}$$

where we have replaced the control u with the state variable ξ_3. We again apply Lemma 7.2, this time with $\chi_1 = [t \; \zeta^{\mathrm{T}} \; \xi_1 \; \xi_2]^{\mathrm{T}}$, $\chi_2 = \xi_3$, $\mu_c = \mu_2$, $\Omega_c = \Omega_2$, and $V_c = V_2$. We obtain smooth functions $\mu_3(\zeta, \xi_1, \xi_2, \xi_3, \omega, t)$ $(= \mu_a)$ and $\Omega_3(\zeta, \xi_1, \xi_2, \xi_3, \omega, t)$ $(= \Omega_a)$ such that the derivative of the new Lyapunov function

$$V_3(\rho_1, \rho_2, \rho_3, \sigma) \;:=\; V_2(\rho_1, \rho_2, \sigma) + d_3 \rho_3^2 \tag{7.75}$$

(where $\rho_3 := \xi_3 - \mu_2(\zeta, \xi_1, \xi_2, \omega, t)$ and $d_3 > 0$ is a design parameter) satisfies

$$\dot{V}_3\Big|_{\substack{u = \mu_3 \\ \dot{\omega} = \Omega_3}} = \dot{V}_1\Big|_{\substack{\xi_2 = \mu_1 \\ \dot{\omega} = \Omega_1}} - 2d_2 m_2 \rho_2^2 - 2d_3 m_3 \rho_3^2 \tag{7.76}$$

where $m_3 > 0$ is a design parameter. Our actual controller for $n=3$ is

$$u = \mu_3(\zeta, \xi_1, \xi_2, \xi_3, \omega, t) \tag{7.77}$$

$$\dot{\omega} = \Omega_3(\zeta, \xi_1, \xi_2, \xi_3, \omega, t) \tag{7.78}$$

If $n \geq 4$, then we continue the recursive design, applying Lemma 7.2 again for each additional integrator $(\xi_4, \ldots \xi_n)$. Upon the $(i-1)^{st}$ application of this backstepping lemma, for $i \in \{2, \ldots, n\}$, we obtain smooth functions $\mu_i(\zeta, \xi_1, \ldots, \xi_i, \omega, t)$ and $\Omega_i(\zeta, \xi_1, \ldots, \xi_i, \omega, t)$. This recursive construction accomplishes the following: if we define $P_n := \mathrm{diag}(P_e, d_1, \ldots, d_n)$ for design parameters $d_i > 0$, and

$$V_n(\rho, \sigma) \;:=\; \rho^{\mathrm{T}} P_n \, \rho + \sigma^{\mathrm{T}} P \sigma \tag{7.79}$$

where $\rho := [\rho_1{}^{\mathrm{T}} \; \rho_2 \; \dots \; \rho_n]^{\mathrm{T}}$ with $\rho_{i+1} := \xi_{i+1} - \mu_i(\zeta, \xi_1, \dots, \xi_i, \omega, t)$ for $i \in \{1, \dots, n-1\}$, then we obtain

$$\dot{V}_n\Big|_{\substack{u = \mu_n \\ \dot{\omega} = \Omega_n}} = -W_n(\rho) + \sigma^{\mathrm{T}} [F^{\mathrm{T}}(\zeta, \xi_1) P + PF(\zeta, \xi_1)] \sigma \tag{7.80}$$

where $W_n(\rho)$ is a positive definite quadratic function of ρ. For example, if $n = 3$ then (7.80) follows from (7.45) and (7.76). Note that that the expressions (7.79) and (7.80) are also valid for $n = 1$ if for this case we define $\rho := \rho_1$. In any case, our actual p^{th}-order dynamic partial state feedback controller for the extended strict feedback system (7.4) is

$$u = \mu_n(\zeta, \xi_1, \dots, \xi_n, \omega, t) \tag{7.81}$$

$$\dot{\omega} = \Omega_n(\zeta, \xi_1, \dots, \xi_n, \omega, t) \tag{7.82}$$

The functions Ω_i in this design are modifications of the *tuning functions* recently developed in [84, 85] for adaptive nonlinear control.

7.2.6 Proof of the main result

It follows from H1 and (7.80) that the dynamic controller (7.81)–(7.82) for the system (7.4) achieves the Lyapunov derivative

$$\dot{V}_n \le -W_n(\rho) \tag{7.83}$$

along closed-loop trajectories. Let $[0, t_f)$ denote the maximal interval of existence of a closed-loop trajectory from some initial condition, where $t_f \in (0, \infty]$. It follows from (7.79) and (7.83) that $\rho(t)$ and $\sigma(t)$ are bounded on $[0, t_f)$. In particular, $\rho_1(t)$ is bounded on $[0, t_f)$. It then follows from H5 and (7.20) or (7.39) that $z(t), \dots, z^{(r)}(t)$ are bounded on $[0, t_f)$. Having established the boundedness of $z(t)$ and its first r derivatives, we conclude from H4 that $\zeta(t)$ is bounded on $[0, t_f)$. Therefore $\mu_0(\zeta, t)$ is bounded, and we conclude from (7.19) or (7.38) that $\xi_1(t)$ is bounded on $[0, t_f)$. It follows from H2 that $\eta(t)$ is bounded on $[0, t_f)$, which further implies that $\omega(t) = \eta(t) - \sigma(t)$ is bounded on $[0, t_f)$. Finally, if $n \ge 2$ then we conclude from H5 and the definition of each ρ_i that $\xi_i(t)$ is bounded on $[0, t_f)$ for $i \in \{2, \dots, n\}$. Because all states are bounded on $[0, t_f)$ we have $t_f = \infty$, and we conclude from (7.83) and standard arguments that $\rho(t) \to 0$ as $t \to \infty$. From this it follows that $z(t) \to z_d(t)$ as $t \to \infty$.

Suppose now that the inequality (7.7) is strict for all $(\zeta, \xi_1) \in I\!R^{q+1}$. Because $\zeta(t)$ and $\xi_1(t)$ are bounded, there exists $\varepsilon > 0$ such that all eigenvalues of the matrix $F^{\mathrm{T}}(\zeta, \xi_1)\, P + PF(\zeta, \xi_1)$ are less than $-\varepsilon$ for all $t \geq 0$. It follows from (7.80) that

$$\dot{V}_n \;\leq\; -W_n(\rho) \;-\; \varepsilon\|\sigma\|^2 \tag{7.84}$$

along closed-loop trajectories, and we conclude that $\sigma(t) \to 0$ as $t \to \infty$. This completes the proof of Theorem 7.1.

7.3 Design example

The procedure developed in the preceding section will now be applied to design a controller for the arm/rotor/platform (ARP) system in Figure 7.2. The robot arm **A** is driven by the motor rotor **R** to which it is connected by a flexible joint. The motor stator with rotor bearings is mounted on the platform **P** which in turn is flexibly connected to the fixed base **B**. The platform/base connection admits both rotational and translational motion. The parameters for the ARP system (masses, lengths, moments of inertia, spring constants, and friction coefficients in *mks* units) are listed in Table 7.1.

The design task is to make the position of the arm relative to the platform track a given command signal with high precision. This task is difficult because of the flexibility on both sides of the rotor (in the arm/rotor joint and in the platform/base connection). The only variables available for feedback are the angular positions and velocities of the arm and rotor relative to the platform.

In Section 7.3.1 we develop a "truth model" of the ARP system, which we use for simulations, together with a simplified "design model" which we use for controller design. We begin the design of our controller in Section 7.3.2 by constructing a full-state feedback control law which achieves our tracking objective. This control law cannot be implemented because not all state variables are measured; nevertheless, it will serve as a starting point for the partial state feedback design we present in Section 7.3.3.

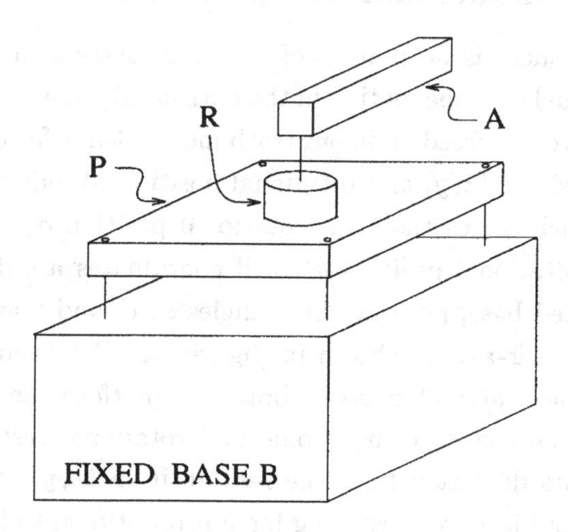

Figure 7.2: A robot arm **A** is attached with a flexible joint to the rotor **R** of a motor mounted on the platform **P**. The platform **P** is attached with flexible mounts to the fixed base **B**.

M	5.0	total mass of arm, rotor, and platform
m	0.5	mass of arm
r	0.3	distance from **A/R** joint to arm center of mass
I	0.06	moment of inertia of arm about **A/R** joint
J	0.005	moment of inertia of rotor
D	0.5	moment of inertia of platform
k_1	64.0	translational spring constant of **P/B** connection
k_2	3600	rotational spring constant of **P/B** connection
k_3	8.0	rotational spring constant of **A/R** joint
b_1	1.6	translational friction coefficient of **P/B** connection
b_2	12.0	rotational friction coefficient of **P/B** connection
b_3	0.04	rotational friction coefficient of **A/R** joint
b_4	0.007	rotational friction coefficient of **R/P** connection

Table 7.1: Parameters for the ARP system, in *mks* units.

7.3.1 Truth model and design model

We assume that there is no motion of the ARP system in the vertical direction and model only the motion in the horizontal plane. There will be a total of five degrees of freedom in our truth model: three for the platform (translational position (x, y) and rotational position β), one for the rotor of the motor which drives the arm (rotational position α), and one for the arm itself (rotational position θ). All coordinates are defined with respect to the fixed base, and the three angles α, β, and θ are measured with respect to the x-axis as shown in Figure 7.3. We assume that the point (x, y) is the center of mass of both the platform and the rotor. The platform is subject to translational and rotational restoring forces proportional to its deviation from the rest position $(x, y, \beta) = (0, 0, 0)$. Also, the arm is subject to a restoring force proportional to its deviation $(\theta - \alpha)$ from alignment with the rotor. All motions are subject to viscous friction forces proportional to velocities. The control input is the torque u generated by the motor.

The control objective is to make the variable $\vartheta := \theta - \beta$, which is the angle of the arm relative to the platform, asymptotically track a given reference signal $\vartheta_r(t)$ while maintaining internal boundedness. We assume that this reference signal and its derivatives are available to the controller. The only other signals available to the controller are the angular positions and velocities of the arm and rotor relative to the platform, namely, $(\theta - \beta)$, $(\dot{\theta} - \dot{\beta})$, $(\alpha - \beta)$, and $(\dot{\alpha} - \dot{\beta})$.

We will use the Lagrangian formulation to derive a dynamic truth model of the ARP system. The total kinetic energy T of the three bodies (arm, rotor, and platform) is

$$\begin{aligned} T &= \tfrac{1}{2}(M-m)\big[\dot{x}^2 + \dot{y}^2\big] + \tfrac{1}{2}m\big[\dot{a}^2 + \dot{b}^2\big] \\ &\quad + \tfrac{1}{2}(I-mr^2)\dot{\theta}^2 + \tfrac{1}{2}J\dot{\alpha}^2 + \tfrac{1}{2}D\dot{\beta}^2 \end{aligned} \tag{7.85}$$

where (a, b) denotes the coordinates of the arm center of mass, that is,

$$\begin{bmatrix} a \\ b \end{bmatrix} = \begin{bmatrix} x + r\cos\theta \\ y + r\sin\theta \end{bmatrix} \tag{7.86}$$

Combining (7.85) with the derivative of (7.86), we obtain

$$T = \tfrac{1}{2}M\big[\dot{x}^2 + \dot{y}^2\big] + mr\dot{\theta}\left[-\dot{x}\sin\theta + \dot{y}\cos\theta\right]$$

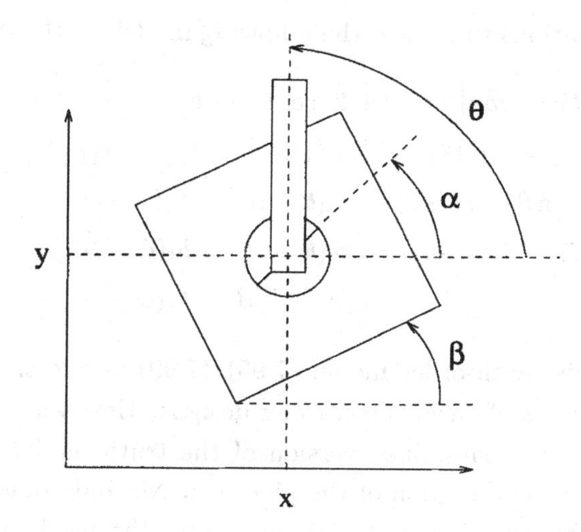

Figure 7.3: Top view of the ARP system. There are two translational coordinates (x, y) and three rotational coordinates (α, β, θ). All coordinates are defined with respect to the fixed base.

$$+\ \tfrac{1}{2}I\dot{\theta}^2\ +\ \tfrac{1}{2}J\dot{\alpha}^2\ +\ \tfrac{1}{2}D\dot{\beta}^2 \tag{7.87}$$

The potential energy V stored in the springs is

$$V\ =\ \tfrac{1}{2}k_1(x^2+y^2)\ +\ \tfrac{1}{2}k_2\beta^2\ +\ \tfrac{1}{2}k_3(\theta-\alpha)^2 \tag{7.88}$$

and Rayleigh's dissipation function $\mathcal{F}$ for viscous friction [47] is

$$\mathcal{F}\ =\ \tfrac{1}{2}b_1\left[\dot{x}^2+\dot{y}^2\right]\ +\ \tfrac{1}{2}b_2\dot{\beta}^2\ +\ \tfrac{1}{2}b_3(\dot{\theta}-\dot{\alpha})^2\ +\ \tfrac{1}{2}b_4(\dot{\alpha}-\dot{\beta})^2 \tag{7.89}$$

We calculate the required derivatives of the Lagrangian $L = T - V$:

$$\frac{d}{dt}\frac{\partial L}{\partial \dot{x}} - \frac{\partial L}{\partial x}\ =\ M\ddot{x}\ -\ mr\left[\ddot{\theta}\sin\theta+\dot{\theta}^2\cos\theta\right]\ +\ k_1x \tag{7.90}$$

$$\frac{d}{dt}\frac{\partial L}{\partial \dot{y}} - \frac{\partial L}{\partial y}\ =\ M\ddot{y}\ +\ mr\left[\ddot{\theta}\cos\theta-\dot{\theta}^2\sin\theta\right]\ +\ k_1y \tag{7.91}$$

$$\frac{d}{dt}\frac{\partial L}{\partial \dot{\theta}} - \frac{\partial L}{\partial \theta}\ =\ I\ddot{\theta}\ +\ mr[-\ddot{x}\sin\theta+\ddot{y}\cos\theta]\ +\ k_3(\theta-\alpha) \tag{7.92}$$

$$\frac{d}{dt}\frac{\partial L}{\partial \dot{\alpha}} - \frac{\partial L}{\partial \alpha}\ =\ J\ddot{\alpha}\ -\ k_3(\theta-\alpha) \tag{7.93}$$

$$\frac{d}{dt}\frac{\partial L}{\partial \dot{\beta}} - \frac{\partial L}{\partial \beta}\ =\ D\ddot{\beta}\ +\ k_2\beta \tag{7.94}$$

Lagrange's equations thus yield the following model for the ARP system:

$$\begin{aligned}
M\ddot{x} \;-\; mr\left[\ddot{\theta}\sin\theta + \dot{\theta}^2\cos\theta\right] \;+\; b_1\dot{x} \;+\; k_1 x &= 0 && (7.95)\\
M\ddot{y} \;+\; mr\left[\ddot{\theta}\cos\theta - \dot{\theta}^2\sin\theta\right] \;+\; b_1\dot{y} \;+\; k_1 y &= 0 && (7.96)\\
I\ddot{\theta} \;+\; mr[-\ddot{x}\sin\theta + \ddot{y}\cos\theta] \;+\; b_3(\dot{\theta} - \dot{\alpha}) \;+\; k_3(\theta - \alpha) &= 0 && (7.97)\\
J\ddot{\alpha} \;-\; b_3(\dot{\theta} - \dot{\alpha}) \;-\; k_3(\theta - \alpha) \;+\; b_4(\dot{\alpha} - \dot{\beta}) &= u && (7.98)\\
D\ddot{\beta} \;+\; b_2\dot{\beta} \;+\; k_2\beta \;-\; b_4(\dot{\alpha} - \dot{\beta}) &= -u && (7.99)
\end{aligned}$$

We will use this tenth-order model (7.95)–(7.99) as the truth model in all simulation tests of various controller designs. However, as our design model we will use a simplified version of the truth model in which we neglect the rotational motion of the platform. Not only does this reduce the order of the model, but it also eliminates the need for computing estimates of the unmeasured states β and $\dot{\beta}$ inside the dynamic controller. This simplification is justified because the rotational spring constant k_2 of the platform/base connection is large and the viscous friction coefficient b_4 of the rotor/platform connection is small. We therefore assume $\beta(t) \equiv 0$ for design and disregard the last equation (7.99).

Our motivation for neglecting the rotational dynamics of the platform is not just model order reduction. A more important reason for this simplification is that the presence of β causes the truth model to be non-minimum phase with respect to the output variable $\vartheta := \theta - \beta$ as was shown in [30]. Fortunately, this undesirable property is weak in that the unstable zeros introduced by these rotational dynamics are at a high frequency and can therefore be neglected during controller design.

To bring the design model into a useful state variable form, we make a partial change of coordinates $(x, \dot{x}, y, \dot{y}, \theta, \dot{\theta}) \leftrightarrow (\eta_1, \eta_2, \eta_3, \eta_4, \theta, \dot{\theta})$ as

$$\begin{bmatrix} \eta_1 \\ \eta_2 \\ \eta_3 \\ \eta_4 \end{bmatrix} = \begin{bmatrix} \sin\theta & 0 & -\cos\theta & 0 \\ 0 & \sin\theta & 0 & -\cos\theta \\ \cos\theta & 0 & \sin\theta & 0 \\ 0 & \cos\theta & 0 & \sin\theta \end{bmatrix} \begin{bmatrix} x \\ \dot{x} \\ y \\ \dot{y} \end{bmatrix} + \frac{mr}{M} \begin{bmatrix} 0 \\ -\dot{\theta} \\ 1 \\ 0 \end{bmatrix} \qquad (7.100)$$

with inverse transformation

$$\begin{bmatrix} x \\ \dot{x} \\ y \\ \dot{y} \end{bmatrix} = \begin{bmatrix} \sin\theta & 0 & \cos\theta & 0 \\ 0 & \sin\theta & 0 & \cos\theta \\ -\cos\theta & 0 & \sin\theta & 0 \\ 0 & -\cos\theta & 0 & \sin\theta \end{bmatrix} \begin{bmatrix} \eta_1 \\ \eta_2 + \frac{mr}{M}\dot{\theta} \\ \eta_3 - \frac{mr}{M} \\ \eta_4 \end{bmatrix} \tag{7.101}$$

Using (7.95) and (7.96), we differentiate both sides of (7.100) to obtain

$$\begin{bmatrix} \dot{\eta}_1 \\ \dot{\eta}_2 \\ \dot{\eta}_3 \\ \dot{\eta}_4 \end{bmatrix} = \begin{bmatrix} 0 & 1 & \dot{\theta} & 0 \\ -\frac{k_1}{M} & -\frac{b_1}{M} & 0 & \dot{\theta} \\ -\dot{\theta} & 0 & 0 & 1 \\ 0 & -\dot{\theta} & -\frac{k_1}{M} & -\frac{b_1}{M} \end{bmatrix} \begin{bmatrix} \eta_1 \\ \eta_2 \\ \eta_3 \\ \eta_4 \end{bmatrix} + \frac{mr}{M^2} \begin{bmatrix} 0 \\ -b_1\dot{\theta} \\ 0 \\ k_1 \end{bmatrix} \tag{7.102}$$

In these new coordinates, equation (7.97) becomes

$$\begin{aligned} \left[I - \frac{(mr)^2}{M}\right]\ddot{\theta} + \left[b_3 + b_1\left(\frac{mr}{M}\right)^2\right]\dot{\theta} + k_3\theta + \frac{mr}{M}(k_1\eta_1 + b_1\eta_2) & \\ = k_3\alpha + b_3\dot{\alpha} & \end{aligned} \tag{7.103}$$

while equation (7.98) remains unchanged. Thus the complete design model consists of the fourth-order nonlinear subsystem (7.102) plus the fourth-order linear subsystem

$$\begin{aligned} \begin{bmatrix} \dot{\theta} \\ \ddot{\theta} \\ \dot{\alpha} \\ \ddot{\alpha} \end{bmatrix} &= \begin{bmatrix} 0 & 1 & 0 & 0 \\ -a_1 & -a_2 & a_1 & a_3 \\ 0 & 0 & 0 & 1 \\ a_4 & a_5 & -a_4 & -a_5 - a_6 \end{bmatrix} \begin{bmatrix} \theta \\ \dot{\theta} \\ \alpha \\ \dot{\alpha} \end{bmatrix} \\ &\quad - \begin{bmatrix} 0 & 0 \\ p_1 & p_2 \\ 0 & 0 \\ 0 & 0 \end{bmatrix} \begin{bmatrix} \eta_1 \\ \eta_2 \end{bmatrix} + \begin{bmatrix} 0 \\ 0 \\ 0 \\ \frac{1}{J} \end{bmatrix} u \end{aligned} \tag{7.104}$$

where the positive constants a_i and p_j are given by

$$\begin{bmatrix} a_1 & a_2 & a_3 \\ a_4 & a_5 & a_6 \end{bmatrix} = \begin{bmatrix} \frac{k_3 M}{MI-(mr)^2} & \frac{b_3 M^2 + b_1 (mr)^2}{M[MI-(mr)^2]} & \frac{b_3 M}{MI-(mr)^2} \\ \frac{k_3}{J} & \frac{b_3}{J} & \frac{b_4}{J} \end{bmatrix} \tag{7.105}$$

$$\begin{bmatrix} p_1 & p_2 \end{bmatrix} = \frac{mr}{MI - (mr)^2} \begin{bmatrix} k_1 & b_1 \end{bmatrix} \tag{7.106}$$

In more compact notation with state variables $\eta := [\eta_1 \;\; \eta_2 \;\; \eta_3 \;\; \eta_4]^{\mathrm{T}}$ and $\chi := [\theta \;\; \dot{\theta} \;\; \alpha \;\; \dot{\alpha}]^{\mathrm{T}}$, the equations (7.102) and (7.104) can be written

$$\dot{\eta} = F(\dot{\theta})\,\eta + G(\dot{\theta}) \tag{7.107}$$

$$\dot{\chi} = A\chi + E\eta + Bu \tag{7.108}$$

where A, B, and E are constant matrices.

7.3.2 Full state feedback design

In this section we pretend that measurements of all states are available to the controller. An approach commonly used to solve nonlinear tracking problems with full state feedback is input/output linearization. In this approach we first derive a normal form in which we inspect the inverse dynamics for their stability properties [60].

We begin by selecting the output variable to be tracked. This variable was given to us as $\vartheta := \theta - \beta$, but because we have neglected β in our design model we choose θ as our output variable. We obtain the normal form for this output variable θ by computing its derivatives: from (7.104), (7.107), and (7.108) we have

$$\begin{aligned}\theta^{(3)} &= \begin{bmatrix}-a_1 & -a_2 & a_1 & a_3\end{bmatrix}\begin{bmatrix}A\chi + E\eta + Bu\end{bmatrix} \\ &\quad - \begin{bmatrix}p_1 & p_2 & 0 & 0\end{bmatrix}\begin{bmatrix}F(\dot{\theta})\,\eta + G(\dot{\theta})\end{bmatrix} \\ &= \begin{bmatrix}-a_1 & -a_2 & a_1 & a_3\end{bmatrix}\begin{bmatrix}A\chi + E\eta\end{bmatrix} \\ &\quad - \begin{bmatrix}p_1 & p_2 & 0 & 0\end{bmatrix}\begin{bmatrix}F(\dot{\theta})\,\eta + G(\dot{\theta})\end{bmatrix} + \frac{a_3}{J}u\end{aligned} \tag{7.109}$$

Because the control u appears for the first time in this expression for the third derivative of θ, we see that this system has relative degree three with respect to θ. To reduce (7.109) to a triple integrator, we employ the control transformation

$$\begin{aligned}u &= \frac{J}{a_3}\Big[\begin{bmatrix}a_1 & a_2 & -a_1 & -a_3\end{bmatrix}\begin{bmatrix}A\chi + E\eta\end{bmatrix} \\ &\quad + \begin{bmatrix}p_1 & p_2 & 0 & 0\end{bmatrix}\begin{bmatrix}F(\dot{\theta})\,\eta + G(\dot{\theta})\end{bmatrix} + v\Big]\end{aligned} \tag{7.110}$$

where v is a new control variable. This transformation yields $\theta^{(3)} = v$ from which it is straightforward to construct a control law for v such that θ asymptotically tracks the given reference signal.

This control transformation (7.110) also yields the following fifth-order inverse dynamics:

$$\dot{\eta} = F(\dot{\theta})\,\eta + G(\dot{\theta}) \tag{7.111}$$

$$a_3\dot{\alpha} = -a_1\alpha + a_1\theta + a_2\dot{\theta} + \ddot{\theta} + p_1\eta_1 + p_2\eta_2 \tag{7.112}$$

The zero dynamics are obtained by setting $\theta = \dot{\theta} = \ddot{\theta} = 0$. For this input/output linearization to succeed, we must show that η and α are bounded whenever θ, $\dot{\theta}$, and $\ddot{\theta}$ are bounded. Because k_1, b_1, and M are positive parameters, the matrix

$$H := \begin{bmatrix} 0 & 1 \\ -\frac{k_1}{M} & -\frac{b_1}{M} \end{bmatrix} \tag{7.113}$$

is Hurwitz. Thus there is a symmetric positive definite matrix $P \in I\!R^{2\times 2}$ such that the matrix $Q := -H^{\mathrm{T}}P - PH$ is positive definite. We see from (7.102) that

$$F(\dot{\theta}) = \begin{bmatrix} H & \dot{\theta}I_{2\times 2} \\ -\dot{\theta}I_{2\times 2} & H \end{bmatrix} \qquad G(\dot{\theta}) = \frac{mr}{M^2}\begin{bmatrix} 0 \\ -b_1\dot{\theta} \\ 0 \\ k_1 \end{bmatrix} \tag{7.114}$$

from which one can verify that

$$\left[F(\dot{\theta})\right]^{\mathrm{T}}\begin{bmatrix} P & 0 \\ 0 & P \end{bmatrix} + \begin{bmatrix} P & 0 \\ 0 & P \end{bmatrix}F(\dot{\theta}) = -\begin{bmatrix} Q & 0 \\ 0 & Q \end{bmatrix} \tag{7.115}$$

for every $\dot{\theta} \in I\!R$. It follows that

$$\frac{d}{dt}\left(\eta^{\mathrm{T}}\begin{bmatrix} P & 0 \\ 0 & P \end{bmatrix}\eta\right) = -\eta^{\mathrm{T}}\begin{bmatrix} Q & 0 \\ 0 & Q \end{bmatrix}\eta + 2\eta^{\mathrm{T}}\begin{bmatrix} P & 0 \\ 0 & P \end{bmatrix}G(\dot{\theta}) \tag{7.116}$$

If θ, $\dot{\theta}$, and $\ddot{\theta}$ are bounded, then we conclude from (7.116) and (7.112) that η and α are bounded.

A closer examination of the inverse dynamics (7.111)–(7.112) reveals that, in spite of its internal boundedness, this input/output linearizing control law is not practical. To see why, we note that the associated zero dynamics are block triangular with one 4×4 block and one 1×1 block.

The 1×1 block contributes a stable zero located at $-a_1/a_3 = -200$. The input/output linearizing design cancels this zero by placing a pole at the same location. However, a controller must employ high gain to place such a fast pole, and this is undesirable for two reasons. First, the closed-loop bandwidth may become too large, and second, the actuator is more likely to reach its magnitude and rate limits. As we will show later, this input/output linearization for the output θ failed to achieve acceptable performance in the presence of actuator limits (see Figure 7.6 on page 192).

When an input/output linearization design fails, the literature suggests two alternatives. The first is to ignore the presence of the troublesome zero and thus increase the relative degree by one [51, 52]. In our case this corresponds to dropping the term $\frac{a_3}{J}u$ from (7.109) and differentiating once more so that the control u appears in the expression for $\theta^{(4)}$. Such an approximate design was completed in [30], and the achieved performance was inferior to that obtained using the second alternative which we now pursue.

Instead of neglecting the stable zero located at $-a_1/a_3 = -200$, we follow [103] and redefine the output variable so that this zero disappears. The relative degree with respect to this new output variable will be four instead of three, and we can perform input/output linearization as before. We must then filter the given reference signal to obtain a reference for the new output variable. In contrast to the approach of neglecting the fast stable zero, this second approach is exact (for the design model), not approximate.

Our new output variable ζ is given by

$$\zeta \;:=\; \theta - \frac{a_3}{a_1 - a_2 a_3}\left[\dot{\theta} - a_3\alpha\right] \tag{7.117}$$

Because $a_1 \approx 140$ and $a_2 \approx a_3 \approx 0.7$, this new output variable ζ is close to the old output variable θ. We now show that the design model (7.107)–(7.108) has relative degree four with respect to the output variable ζ. We compute derivatives of ζ as follows:

$$\begin{aligned}\dot{\zeta} &= \dot{\theta} - \frac{a_3}{a_1 - a_2 a_3}\left[-a_1\theta - a_2\dot{\theta} + a_1\alpha - \begin{bmatrix}p_1 & p_2 & 0 & 0\end{bmatrix}\eta\right] \\ &= \frac{a_1}{a_1 - a_2 a_3}\dot{\theta} - \frac{a_3}{a_1 - a_2 a_3}\left[-a_1\theta + a_1\alpha - \begin{bmatrix}p_1 & p_2 & 0 & 0\end{bmatrix}\eta\right] \qquad (7.118)\end{aligned}$$

The second derivative is given by

$$(a_1 - a_2 a_3)\,\ddot{\zeta} = -a_1^2\theta - a_1(a_2 - a_3)\dot{\theta} + a_1^2\alpha - a_1\big[p_1 \;\; p_2 \;\; 0 \;\; 0\big]\eta + a_3\big[p_1 \;\; p_2 \;\; 0 \;\; 0\big]\big[F(\dot{\theta})\,\eta + G(\dot{\theta})\big] \tag{7.119}$$

The coefficients in the definition of ζ in (7.117) have been chosen so that the state variable $\dot{\alpha}$ does not appear in (7.119). As a result, we will need to take two more derivatives of (7.119) before the control u appears. Using the fact that the second derivatives of $F(\dot{\theta})$ and $G(\dot{\theta})$ with respect to $\dot{\theta}$ vanish, we compute as follows:

$$\begin{aligned}(a_1 - a_2 a_3)\,\zeta^{(4)} = {} & -a_1^2\ddot{\theta} - a_1(a_2 - a_3)\theta^{(3)} + a_1^2\ddot{\alpha} \\ & - a_1\big[p_1 \;\; p_2 \;\; 0 \;\; 0\big]\Big[\big[F(\dot{\theta})\big]^2\eta + F(\dot{\theta})\,G(\dot{\theta}) \\ & + \ddot{\theta}\Big(\frac{\partial F}{\partial\dot{\theta}}\eta + \frac{\partial G}{\partial\dot{\theta}}\Big)\Big] \\ & + a_3\big[p_1 \;\; p_2 \;\; 0 \;\; 0\big]\Big[\Big(\big[F(\dot{\theta})\big]^2 + 2\ddot{\theta}\frac{\partial F}{\partial\dot{\theta}}\Big)\big[F(\dot{\theta})\,\eta + G(\dot{\theta})\big] \\ & + \big[\theta^{(3)} I_{4\times4} + \ddot{\theta}F(\dot{\theta})\big]\Big(\frac{\partial F}{\partial\dot{\theta}}\eta + \frac{\partial G}{\partial\dot{\theta}}\Big)\Big]\end{aligned} \tag{7.120}$$

As can be seen from (7.104) and (7.109), the control u now appears through $\ddot{\alpha}$ and $\theta^{(3)}$. Extracting the explicit dependence on u, we obtain

$$\begin{aligned}(a_1 - a_2 a_3)\,\zeta^{(4)} = {} & -a_1^2\ddot{\theta} - a_1(a_2 - a_3)\Big[\theta^{(3)} - \frac{a_3}{J}u\Big] + a_1^2\Big[\ddot{\alpha} - \frac{1}{J}u\Big] \\ & - a_1\big[p_1 \;\; p_2 \;\; 0 \;\; 0\big]\Big[\big[F(\dot{\theta})\big]^2\eta + F(\dot{\theta})\,G(\dot{\theta}) \\ & + \ddot{\theta}\Big(\frac{\partial F}{\partial\dot{\theta}}\eta + \frac{\partial G}{\partial\dot{\theta}}\Big)\Big] \\ & + a_3\big[p_1 \;\; p_2 \;\; 0 \;\; 0\big]\Big[\Big(\big[F(\dot{\theta})\big]^2 + 2\ddot{\theta}\frac{\partial F}{\partial\dot{\theta}}\Big)\big[F(\dot{\theta})\,\eta + G(\dot{\theta})\big] \\ & + \Big[\Big(\theta^{(3)} - \frac{a_3}{J}u\Big)I_{4\times4} + \ddot{\theta}F(\dot{\theta})\Big]\Big(\frac{\partial F}{\partial\dot{\theta}}\eta + \frac{\partial G}{\partial\dot{\theta}}\Big)\Big] \\ & + \frac{1}{J}\Big[a_1(a_1 - a_2 a_3 + a_3^2) \\ & + a_3^2\big[p_1 \;\; p_2 \;\; 0 \;\; 0\big]\Big(\frac{\partial F}{\partial\dot{\theta}}\eta + \frac{\partial G}{\partial\dot{\theta}}\Big)\Big]u\end{aligned} \tag{7.121}$$

The coefficient in front of the control u in (7.121) is an affine function of η which vanishes only for large η. We conclude that the reduced model has relative degree four with respect to the new output variable ζ, at least over the expected operating range. The input/output linearizing control transformation is

$$
\begin{aligned}
u = & \frac{J}{a_1(a_1-a_2a_3+a_3^2)+a_3^2\begin{bmatrix}p_1 & p_2 & 0 & 0\end{bmatrix}\left(\frac{\partial F}{\partial \dot\theta}\eta+\frac{\partial G}{\partial \dot\theta}\right)}\Bigg[a_1^2\ddot\theta \\
& + a_1(a_2-a_3)\left[\theta^{(3)}-\frac{a_3}{J}u\right] - a_1^2\left[\ddot\alpha-\frac{1}{J}u\right] \\
& + a_1\begin{bmatrix}p_1 & p_2 & 0 & 0\end{bmatrix}\left[\left[F(\dot\theta)\right]^2\eta + F(\dot\theta)\,G(\dot\theta) + \ddot\theta\left(\frac{\partial F}{\partial \dot\theta}\eta+\frac{\partial G}{\partial \dot\theta}\right)\right] \\
& - a_3\begin{bmatrix}p_1 & p_2 & 0 & 0\end{bmatrix}\Bigg[\left(\left[F(\dot\theta)\right]^2+2\ddot\theta\frac{\partial F}{\partial \dot\theta}\right)\left[F(\dot\theta)\,\eta+G(\dot\theta)\right] \\
& + \left[\left(\theta^{(3)}-\frac{a_3}{J}u\right)I_{4\times4}+\ddot\theta F(\dot\theta)\right]\left(\frac{\partial F}{\partial \dot\theta}\eta+\frac{\partial G}{\partial \dot\theta}\right)\Bigg] + (a_1-a_2a_3)\,v\Bigg]
\end{aligned}
\tag{7.122}
$$

where v is a new control input. This control transformation is valid because after substituting for $\ddot\theta$, $\theta^{(3)}$, and $\ddot\alpha$, the control u disappears from the right-hand side of (7.122). We thus obtain the chain of integrators $\zeta^{(4)} = v$ from which it is straightforward to choose v such that ζ asymptotically tracks any given reference signal $\zeta_r(t)$:

$$
v = \zeta_r^{(4)} - c_0(\zeta-\zeta_r) - c_1(\dot\zeta-\dot\zeta_r) - c_2(\ddot\zeta-\ddot\zeta_r) - c_3(\zeta^{(3)}-\zeta_r^{(3)}) \tag{7.123}
$$

for appropriate positive design parameters c_i. The resulting fourth-order inverse dynamics are given by (7.111) which we have already shown to be bounded-input/bounded-state stable..

We have left to generate the reference signal $\zeta_r(t)$ from the given reference signal $\vartheta_r(t)$. This is an open-loop motion planning problem that is independent of the feedback stabilization problem solved by the input/output linearizing control law (7.122). If the parameter a_3 were zero, we would see from (7.117) that $\zeta = \theta$ which means $\zeta_r(t) = \vartheta_r(t)$ would be an appropriate choice for the new reference signal. Even though a_3 is nonzero, we still have the option of choosing $\zeta_r(t) = \vartheta_r(t)$, and we will call this the *unfiltered* version of the control (7.122)–(7.123).

Simulation results for $(c_0, c_1, c_2, c_3) = (10^4, 3500, 500, 35)$ show that this unfiltered version of (7.122)–(7.123) provides good tracking of reference signals $\vartheta_r(t) \equiv 1$ and $\vartheta_r(t) = 3\sin(5t)$ (Figures 7.4 and 7.5, respectively). Although actuator limits were not taken into account during

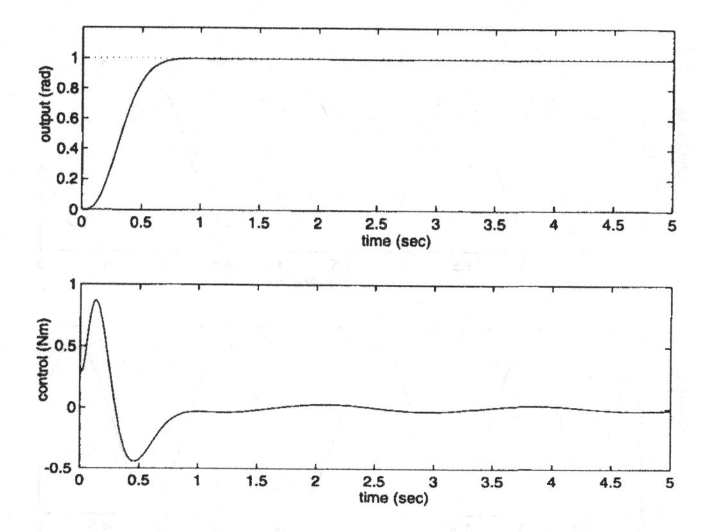

Figure 7.4: Simulation results for the control (7.122)–(7.123) (unfiltered) under control magnitude and rate saturation: tracking output ϑ (top plot) and control signal u (bottom plot).

the design, we performed the simulations under a control magnitude saturation of $\pm 5\, N\!\cdot\! m$ and a control rate saturation of $\pm 30\, N\!\cdot\! m/s$. These actuator limits did not affect the performance of the controller (7.122)–(7.123) for the two reference signals shown. In contrast, Figure 7.6 shows that the high gain input/output linearizing controller (7.110), with similarly placed closed-loop poles, cannot track the reference $\vartheta_r(t) = 3\sin(5t)$ in the presence of such actuator limits. Note that these figures show the true output variable $\vartheta = \theta - \beta$ for the simulated truth model.

Instead of simply assigning $\zeta_r(t) = \vartheta_r(t)$, another option is to generate the new reference signal ζ_r from the given one ϑ_r through a nonlinear filter. We solve for α in (7.117) and substitute into (7.118) to obtain

$$\dot{\zeta}_r = -\frac{a_1}{a_3}\zeta_r + \frac{a_1(a_3^2+a_1-a_2a_3)}{a_3(a_1-a_2a_3)}\vartheta_r + \frac{a_3}{a_1-a_2a_3}\begin{bmatrix} p_1 & p_2 & 0 & 0 \end{bmatrix}\eta_r \qquad (7.124)$$

where the subscript r has been added everywhere because we are now working with reference signals rather than state variables. If we define the reference signal

$$\sigma(t) := \overset{(4)}{\zeta_r}(t) + c_3\overset{(3)}{\zeta_r}(t) + c_2\ddot{\zeta}_r(t) + c_1\dot{\zeta}_r(t) + c_0\zeta_r(t) \qquad (7.125)$$

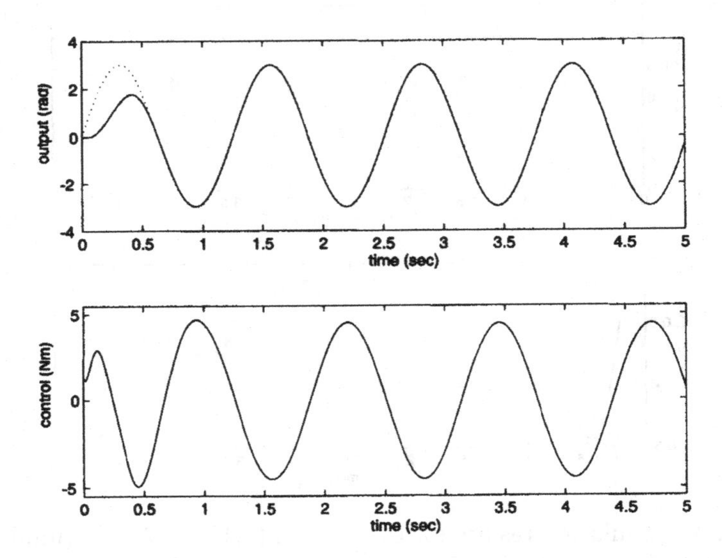

Figure 7.5: Simulation results for the control (7.122)–(7.123) (unfiltered) under control magnitude and rate saturation: tracking output ϑ (top plot) and control signal u (bottom plot).

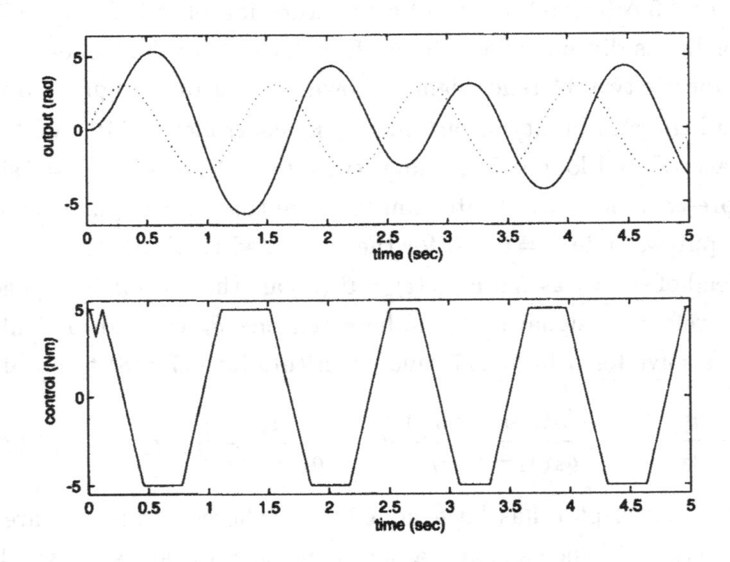

Figure 7.6: Simulation results for the high gain control (7.110) under control magnitude and rate saturation: tracking output ϑ (top plot) and control signal u (bottom plot).

which is the signal needed by the control law (7.123), then from (7.124) it satisfies

$$\begin{aligned}\dot{\sigma} &= -\frac{a_1}{a_3}\sigma + \frac{a_1(a_3^2+a_1-a_2a_3)}{a_3(a_1-a_2a_3)}\left[\vartheta_r^{(4)} + c_3\vartheta_r^{(3)} + c_2\ddot{\vartheta}_r + c_1\dot{\vartheta}_r + c_0\vartheta_r\right] \\ &+ \frac{a_3}{a_1-a_2a_3}\left[p_1 \;\; p_2 \;\; 0 \;\; 0\right]\left[\eta_r^{(4)} + c_3\eta_r^{(3)} + c_2\ddot{\eta}_r + c_1\dot{\eta}_r + c_0\eta_r\right] \quad (7.126)\end{aligned}$$

We generate the reference signal η_r from $\dot{\vartheta}_r$ through the stable fourth-order filter

$$\dot{\eta}_r = F(\dot{\vartheta}_r)\,\eta_r + G(\dot{\vartheta}_r) \quad (7.127)$$

Higher-order derivatives of η_r are given by the equations

$$\ddot{\eta}_r = F(\dot{\vartheta}_r)\,\dot{\eta}_r + \ddot{\vartheta}_r\left(\frac{\partial F}{\partial \dot{\theta}}\eta_r + \frac{\partial G}{\partial \dot{\theta}}\right) \quad (7.128)$$

$$\eta_r^{(3)} = F(\dot{\vartheta}_r)\,\ddot{\eta}_r + 2\ddot{\vartheta}_r\frac{\partial F}{\partial \dot{\theta}}\,\dot{\eta}_r + \vartheta_r^{(3)}\left(\frac{\partial F}{\partial \dot{\theta}}\eta_r + \frac{\partial G}{\partial \dot{\theta}}\right) \quad (7.129)$$

$$\eta_r^{(4)} = F(\dot{\vartheta}_r)\,\eta_r^{(3)} + 3\ddot{\vartheta}_r\frac{\partial F}{\partial \dot{\theta}}\,\ddot{\eta}_r + 3\vartheta_r^{(3)}\frac{\partial F}{\partial \dot{\theta}}\,\dot{\eta}_r + \vartheta_r^{(4)}\left(\frac{\partial F}{\partial \dot{\theta}}\eta_r + \frac{\partial G}{\partial \dot{\theta}}\right) \quad (7.130)$$

Our *filtered* version of the control law (7.122)–(7.123) thus includes the stable nonlinear filters (7.126) and (7.127) for the generation of the reference signal $\sigma(t)$. Note that the filter (7.126), which has a fast pole located at $-a_1/a_3 = -200$, is not in the feedback path and will not suffer the drawbacks of high gain. Of course, the effectiveness of this filter depends on the accuracy of the ARP system parameters listed in Table 7.1.

Finally, we obtain a third version of the control law (7.122)–(7.123) by approximating the nonlinear filter (7.126). Because a_1/a_3 is large, the third term in (7.126) is small compared to the first two terms, and we make the approximation

$$\begin{aligned}\dot{\sigma} &= -\frac{a_1}{a_3}\sigma \\ &+ \frac{a_1(a_3^2+a_1-a_2a_3)}{a_3(a_1-a_2a_3)}\left[\vartheta_r^{(4)} + c_3\vartheta_r^{(3)} + c_2\ddot{\vartheta}_r + c_1\dot{\vartheta}_r + c_0\vartheta_r\right] \quad (7.131)\end{aligned}$$

Thus our *approximated filtered* version of the control law (7.122)–(7.123) includes only the stable first-order *linear* filter (7.131) and is therefore much simpler than the fifth-order nonlinear filtered version.

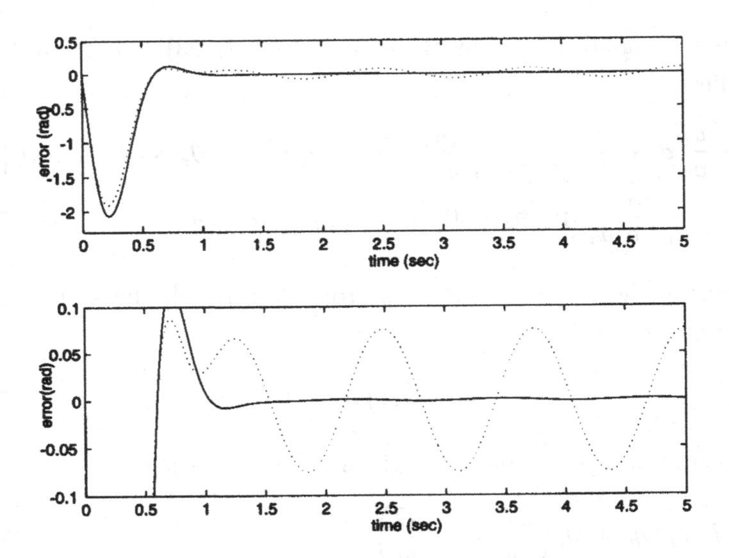

Figure 7.7: Comparison of the tracking error $\vartheta - \vartheta_r$ generated by the unfiltered (dotted) and approximated filtered (solid) versions of the control law (7.122)–(7.123). The top and bottom plots show the same signals in different scales.

The advantage of filtering the reference signal can be seen in Figure 7.7 which compares the tracking errors generated by the unfiltered and approximated filtered versions of the control law (7.122)–(7.123) for the reference signal $\vartheta_r(t) = 3\sin(5t)$. The two plots show the same signals in different scales: the second plot clearly indicates that the approximated filtered version leads to a much smaller steady-state tracking error.

7.3.3 Partial state feedback design

The controllers we designed in the preceding section cannot be implemented because the full state is not available for feedback; the only measured variables are $(\theta - \beta)$, $(\dot{\theta} - \dot{\beta})$, $(\alpha - \beta)$, and $(\dot{\alpha} - \dot{\beta})$. Because we have neglected the rotation of the platform in our design model, we will assume that these relative variables are close to the absolute variables θ, $\dot{\theta}$, α, and $\dot{\alpha}$. We will therefore design partial state feedback controllers in this section assuming measurements of the absolute variables are available, but we will replace these with measurements of the relative variables in the implementation and simulation.

Our task in this section is to construct controllers for the design model (7.107)–(7.108) which use measurements of the variable χ but *not* of the variable η. The simplest solution would be to use an open-loop observer for η of the form

$$\dot{\omega} \;=\; F(\dot{\theta})\,\omega + G(\dot{\theta}) \tag{7.132}$$

and then replace η with ω in the full state feedback control laws of the preceding section. From (7.115), the observation error $\eta-\omega$ would satisfy

$$\frac{d}{dt}\left([\eta-\omega]^{\mathrm{T}}\begin{bmatrix} P & 0 \\ 0 & P \end{bmatrix}[\eta-\omega]\right) \;=\; -\,[\eta-\omega]^{\mathrm{T}}\begin{bmatrix} Q & 0 \\ 0 & Q \end{bmatrix}[\eta-\omega] \tag{7.133}$$

and would thus converge to zero (provided solutions exist for all $t \geq 0$). Such a certainty equivalence scheme may therefore result in asymptotic tracking with internal stability (this is not true in general but may hold for this example). However, the performance of such a scheme would be limited by the convergence properties of the open-loop observer (7.132). Figure 7.8 shows the step response generated by the approximated filtered version of the control law (7.122)–(7.123) with this open-loop observer (7.132). The certainty equivalence combination is apparently stable, but there is no proof of stability and the settling time dictated by the open-loop dynamics is too long.

To guarantee closed-loop stability, we proceed to design a partial state feedback controller using the backstepping technique of Section 7.2. Because the backstepping controller uses a feedback observer, we expect an improvement in settling time over the open-loop observer design.

We begin by defining the error variable $z_1 := \zeta - \zeta_r$, where ζ and ζ_r are the output variable and reference signal of the preceding section. From (7.118) we have

$$\dot{z}_1 \;=\; \frac{1}{a_1 - a_2 a_3}\left[a_1\dot{\theta} + a_1 a_3 \theta - a_1 a_3 \alpha + a_3 \begin{bmatrix} p_1 & p_2 & 0 & 0 \end{bmatrix}\eta\right] \;-\; \dot{\zeta}_r \tag{7.134}$$

We will apply the backstepping construction of Lemma 7.2 a total of three times to obtain a dynamic controller. Defining the new variable

$$\begin{aligned} z_2 \;&:=\; s_1 z_1 + \frac{1}{a_1 - a_2 a_3}\Big[a_1\dot{\theta} + a_1 a_3 \theta \\ &\qquad - a_1 a_3 \alpha + a_3 \begin{bmatrix} p_1 & p_2 & 0 & 0 \end{bmatrix}\omega\Big] \;-\; \dot{\zeta}_r \end{aligned} \tag{7.135}$$

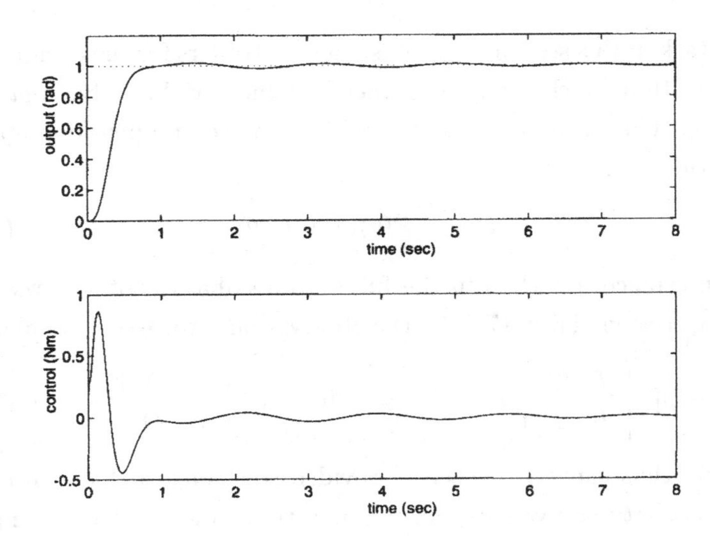

Figure 7.8: Step response with the approximated filtered version of the control law (7.122)–(7.123) with open-loop observer (7.132): tracking output ϑ (top plot) and control signal u (bottom plot).

we obtain

$$\dot{z}_1 = -s_1 z_1 + z_2 + \frac{a_3}{a_1 - a_2 a_3} \begin{bmatrix} p_1 & p_2 & 0 & 0 \end{bmatrix} [\eta - \omega] \tag{7.136}$$

where $s_1 > 0$ is a design parameter. From (7.135) and (7.136) we have

$$\begin{aligned} \dot{z}_2 = & -s_1^2 z_1 + s_1 z_2 + \frac{1}{a_1 - a_2 a_3} \Big[s_1 a_3 \begin{bmatrix} p_1 & p_2 & 0 & 0 \end{bmatrix} [\eta - \omega] - a_1^2(\theta - \alpha) \\ & - a_1(a_2 - a_3)\dot{\theta} - a_1 \begin{bmatrix} p_1 & p_2 & 0 & 0 \end{bmatrix} \eta \\ & + a_3 \begin{bmatrix} p_1 & p_2 & 0 & 0 \end{bmatrix} \dot{\omega} \Big] - \ddot{\zeta}_r \end{aligned} \tag{7.137}$$

We define the tuning function Ω_2 as follows:

$$\begin{aligned} \Omega_2 := & \; F(\dot{\theta})\,\omega + G(\dot{\theta}) \\ & + \frac{\gamma}{a_1 - a_2 a_3} \begin{bmatrix} P^{-1} \\ 0 \end{bmatrix} \begin{bmatrix} p_1 \\ p_2 \end{bmatrix} \left[a_3 z_1 + \frac{s_1 a_3 - a_1}{d_2} z_2 \right] \end{aligned} \tag{7.138}$$

where $\gamma > 0$ and $d_2 > 0$ are design parameters, and the matrix P is symmetric, positive definite, and such that the Lyapunov equation

$$H^{\mathrm{T}} P + P H = -Q \tag{7.139}$$

holds for some symmetric positive definite matrix $Q \in I\!R^{2\times 2}$ (here H from (7.113) is Hurwitz). This tuning function would have replaced the right-hand side of the open-loop observer (7.132) had the control u appeared in (7.137). The control has not yet appeared, however, so we continue the backstepping design and define

$$\begin{aligned} z_3 &:= (d_2 - s_1^2) z_1 + (s_1 + s_2) z_2 + \frac{1}{a_1 - a_2 a_3} \Big[-a_1^2(\theta - \alpha) - a_1(a_2 - a_3)\dot{\theta} \\ &\quad - a_1 \begin{bmatrix} p_1 & p_2 & 0 & 0 \end{bmatrix} \omega + a_3 \begin{bmatrix} p_1 & p_2 & 0 & 0 \end{bmatrix} \Omega_2 \Big] - \dot{\zeta}_r \qquad (7.140) \\ &= \lambda_1 z_1 + \lambda_2 z_2 + \frac{1}{a_1 - a_2 a_3} \Big[-a_1^2(\theta - \alpha) - a_1(a_2 - a_3)\dot{\theta} \\ &\quad - a_1 \begin{bmatrix} p_1 & p_2 & 0 & 0 \end{bmatrix} \omega \\ &\quad + a_3 \begin{bmatrix} p_1 & p_2 & 0 & 0 \end{bmatrix} \big[F(\dot{\theta})\,\omega + G(\dot{\theta}) \big] \Big] - \dot{\zeta}_r \qquad (7.141) \end{aligned}$$

where $s_2 > 0$ is a design parameter and λ_1 and λ_2 are the scalar constants

$$\lambda_1 := d_2 - s_1^2 + \frac{\gamma a_3^2}{(a_1 - a_2 a_3)^2} \begin{bmatrix} p_1 & p_2 \end{bmatrix} P^{-1} \begin{bmatrix} p_1 \\ p_2 \end{bmatrix} \qquad (7.142)$$

$$\lambda_2 := s_1 + s_2 + \frac{\gamma a_3 (s_1 a_3 - a_1)}{d_2 (a_1 - a_2 a_3)^2} \begin{bmatrix} p_1 & p_2 \end{bmatrix} P^{-1} \begin{bmatrix} p_1 \\ p_2 \end{bmatrix} \qquad (7.143)$$

Substituting (7.140) into (7.137) yields

$$\begin{aligned} \dot{z}_2 &= -d_2 z_1 - s_2 z_2 + z_3 + \frac{s_1 a_3 - a_1}{a_1 - a_2 a_3} \begin{bmatrix} p_1 & p_2 & 0 & 0 \end{bmatrix} [\eta - \omega] \\ &\quad + \frac{a_3}{a_1 - a_2 a_3} \begin{bmatrix} p_1 & p_2 & 0 & 0 \end{bmatrix} [\dot{\omega} - \Omega_2] \qquad (7.144) \end{aligned}$$

We next compute $\dot{z}_3$ from (7.141) as follows:

$$\begin{aligned} \dot{z}_3 &= \frac{1}{a_1 - a_2 a_3} \Bigg[\Big[-a_1(a_2 - a_3) + a_3 \begin{bmatrix} p_1 & p_2 & 0 & 0 \end{bmatrix} \left(\frac{\partial F}{\partial \dot{\theta}} \omega + \frac{\partial G}{\partial \dot{\theta}} \right) \Big] \cdot \\ &\quad \Big(-a_1(\theta - \alpha) - a_2 \dot{\theta} + a_3 \dot{\alpha} - \begin{bmatrix} p_1 & p_2 & 0 & 0 \end{bmatrix} \eta \Big) \\ &\quad - a_1^2(\dot{\theta} - \dot{\alpha}) + \begin{bmatrix} p_1 & p_2 & 0 & 0 \end{bmatrix} \big[-a_1 I_{4\times 4} + a_3 F(\dot{\theta}) \big] \dot{\omega} \Bigg] - \zeta_r^{(3)} \\ &\quad - (\lambda_1 s_1 + \lambda_2 d_2) z_1 + (\lambda_1 - \lambda_2 s_2) z_2 + \lambda_2 z_3 \\ &\quad + \frac{\lambda_1 a_3 + \lambda_2 (s_1 a_3 - a_1)}{a_1 - a_2 a_3} \begin{bmatrix} p_1 & p_2 & 0 & 0 \end{bmatrix} [\eta - \omega] \\ &\quad + \frac{\lambda_2 a_3}{a_1 - a_2 a_3} \begin{bmatrix} p_1 & p_2 & 0 & 0 \end{bmatrix} [\dot{\omega} - \Omega_2] \qquad (7.145) \end{aligned}$$

The next tuning function Ω_3 is

$$\begin{aligned}\Omega_3 \quad := \quad & \Omega_2 + \frac{\gamma}{d_2 d_3(a_1 - a_2 a_3)} \begin{bmatrix} P^{-1} \\ 0 \end{bmatrix} \begin{bmatrix} p_1 \\ p_2 \end{bmatrix} \Big[\lambda_1 a_3 + \lambda_2(s_1 a_3 - a_1) \\ & + a_1(a_2 - a_3) \ - \ a_3 \begin{bmatrix} p_1 & p_2 & 0 & 0 \end{bmatrix} \left(\frac{\partial F}{\partial \dot{\theta}} \omega + \frac{\partial G}{\partial \dot{\theta}} \right) \Big] \, z_3 \qquad (7.146)\end{aligned}$$

where $d_3 > 0$ is a design parameter. The control u did not appear in (7.145), so we continue the backstepping procedure and define

$$\begin{aligned} z_4 \ := \ & d_3 \left(1 + \frac{a_3}{a_1 - a_2 a_3} \begin{bmatrix} p_1 & p_2 & 0 & 0 \end{bmatrix} \frac{\Omega_3 - \Omega_2}{z_3} \right) z_2 \ + \ s_3 z_3 \\ & + \frac{1}{a_1 - a_2 a_3} \Bigg[\Big[-a_1(a_2 - a_3) + a_3 \begin{bmatrix} p_1 & p_2 & 0 & 0 \end{bmatrix} \left(\frac{\partial F}{\partial \dot{\theta}} \omega + \frac{\partial G}{\partial \dot{\theta}} \right) \Big] \cdot \\ & \Big(-a_1(\theta - \alpha) - a_2 \dot{\theta} + a_3 \dot{\alpha} - \begin{bmatrix} p_1 & p_2 & 0 & 0 \end{bmatrix} \omega \Big) \\ & - \ a_1^2(\dot{\theta} - \dot{\alpha}) + \begin{bmatrix} p_1 & p_2 & 0 & 0 \end{bmatrix} \Big[-a_1 I_{4\times 4} + a_3 F(\dot{\theta}) \Big] \Omega_3 \Bigg] \ - \ \zeta_r^{(3)} \\ & - \ (\lambda_1 s_1 + \lambda_2 d_2) z_1 \ + \ (\lambda_1 - \lambda_2 s_2) z_2 \ + \ \lambda_2 z_3 \\ & + \frac{\lambda_2 a_3}{a_1 - a_2 a_3} \begin{bmatrix} p_1 & p_2 & 0 & 0 \end{bmatrix} [\Omega_3 - \Omega_2] \qquad (7.147)\end{aligned}$$

where $s_3 > 0$ is a design parameter. Note that the division by z_3 in the top line of (7.147) is well-defined because from (7.146) we see that $\Omega_3 - \Omega_2$ contains z_3 as a factor. Substituting (7.147) into (7.145), we obtain

$$\begin{aligned} \dot{z}_3 \ = \ & -d_3 \left(1 + \frac{a_3}{a_1 - a_2 a_3} \begin{bmatrix} p_1 & p_2 & 0 & 0 \end{bmatrix} \frac{\Omega_3 - \Omega_2}{z_3} \right) z_2 \ - \ s_3 z_3 \ + \ z_4 \\ & + \frac{1}{a_1 - a_2 a_3} \Big[\lambda_1 a_3 + \lambda_2(s_1 a_3 - a_1) + a_1(a_2 - a_3) \qquad (7.148) \\ & - a_3 \begin{bmatrix} p_1 & p_2 & 0 & 0 \end{bmatrix} \left(\frac{\partial F}{\partial \dot{\theta}} \omega + \frac{\partial G}{\partial \dot{\theta}} \right) \Big] \begin{bmatrix} p_1 & p_2 & 0 & 0 \end{bmatrix} [\eta - \omega] \\ & + \frac{1}{a_1 - a_2 a_3} \begin{bmatrix} p_1 & p_2 & 0 & 0 \end{bmatrix} \Big[(\lambda_2 a_3 - a_1) I_{4\times 4} + a_3 F(\dot{\theta}) \Big] [\dot{\omega} - \Omega_3] \end{aligned}$$

Finally, the control u will appear through $\ddot{\alpha}$ when we compute the derivative of (7.147):

$$\begin{aligned} \dot{z}_4 \ = \ & -\frac{\gamma a_3^2}{d_2(a_1 - a_2 a_3)^2} \begin{bmatrix} p_1 & p_2 \end{bmatrix} P^{-1} \begin{bmatrix} p_1 \\ p_2 \end{bmatrix} \begin{bmatrix} p_1 & p_2 & 0 & 0 \end{bmatrix} \frac{\partial F}{\partial \dot{\theta}} \dot{\omega} \, z_2 \\ & + d_3 \left(1 + \frac{a_3}{a_1 - a_2 a_3} \begin{bmatrix} p_1 & p_2 & 0 & 0 \end{bmatrix} \frac{\Omega_3 - \Omega_2}{z_3} \right) \dot{z}_2 \ + \ s_3 \dot{z}_3 \end{aligned}$$

$$
\begin{aligned}
&+ \frac{1}{a_1 - a_2 a_3} \bigg[a_3 \big[p_1 \;\; p_2 \;\; 0 \;\; 0 \big] \frac{\partial F}{\partial \dot{\theta}} \dot{\omega} \Big(-a_1(\theta - \alpha) - a_2 \dot{\theta} \\
&+ a_3 \dot{\alpha} - \big[p_1 \;\; p_2 \;\; 0 \;\; 0 \big] \omega \Big) \\
&+ \bigg[-a_1(a_2 - a_3) + a_3 \big[p_1 \;\; p_2 \;\; 0 \;\; 0 \big] \Big(\frac{\partial F}{\partial \dot{\theta}} \omega + \frac{\partial G}{\partial \dot{\theta}} \Big) \bigg] \cdot \\
&\Big(-a_1(\dot{\theta} - \dot{\alpha}) - a_2 \ddot{\theta} + a_3 \ddot{\alpha} - \big[p_1 \;\; p_2 \;\; 0 \;\; 0 \big] \dot{\omega} \Big) \\
&- a_1^2 (\ddot{\theta} - \ddot{\alpha}) + a_3 \ddot{\theta} \big[p_1 \;\; p_2 \;\; 0 \;\; 0 \big] \frac{\partial F}{\partial \dot{\theta}} \Omega_3 \\
&+ \big[p_1 \;\; p_2 \;\; 0 \;\; 0 \big] \big[-a_1 I_{4\times 4} + a_3 F(\dot{\theta}) \big] \dot{\Omega}_3 \bigg] \\
&- \zeta_r^{(4)} - (\lambda_1 s_1 + \lambda_2 d_2) \dot{z}_1 + (\lambda_1 - \lambda_2 s_2) \dot{z}_2 + \lambda_2 \dot{z}_3 \\
&+ \frac{\lambda_2 a_3}{a_1 - a_2 a_3} \big[p_1 \;\; p_2 \;\; 0 \;\; 0 \big] [\dot{\Omega}_3 - \dot{\Omega}_2]
\end{aligned} \tag{7.149}
$$

Algebraic manipulations of this expression lead to our dynamic backstepping controller. We define our final tuning function Ω_4 by

$$
\Omega_4 \;\; := \;\; \Omega_3 + \frac{\gamma}{d_2 d_3 d_4} \begin{bmatrix} P^{-1} \\ 0 \end{bmatrix} \begin{bmatrix} p_1 \\ p_2 \end{bmatrix} \cdot \Upsilon \cdot z_4 \tag{7.150}
$$

where $d_4 > 0$ is a design parameter and the scalar function Υ is the coefficient of $[p_1 \;\; p_2 \;\; 0 \;\; 0]\eta$ in (7.149). This function Υ can be calculated explicitly by expanding (7.149), and we omit the resulting expression. The dynamic part of our backstepping controller is given by the observer

$$
\begin{aligned}
\dot{\omega} &= \Omega_4 \\
&= F(\dot{\theta})\,\omega + G(\dot{\theta}) + \gamma \big[\Lambda_1 z_1 + \Lambda_2 z_2 + \Lambda_3 z_3 + \Lambda_4 z_4 \big]
\end{aligned} \tag{7.151}
$$

where the Λ_i are nonlinear functions of measured variables. Thus our backstepping dynamics consist of the open-loop observer (7.132) modified by nonlinear correction terms. The positive design parameter γ represents a type of observer gain.

Our final task is to choose the control law itself. This construction, outlined in the proof of Lemma 7.2, is based on the expressions for $\dot{z}_4$ and Ω_4 in (7.149) and (7.150). We omit the resulting expression for the control law, noting only that its construction introduces a final design parameter $s_4 > 0$.

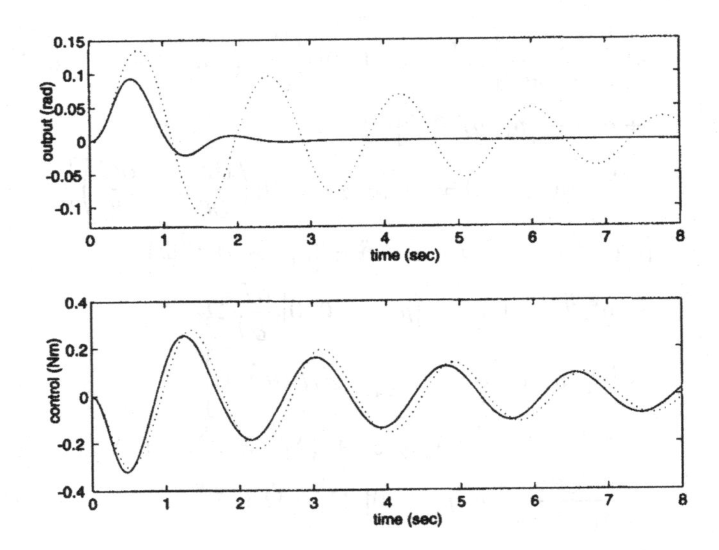

Figure 7.9: Signals generated by the approximated filtered version of the control law (7.122)–(7.123) with open-loop observer (7.132) (dotted) and by the backstepping controller (solid): tracking output ϑ (top plot) and control signal u (bottom plot).

There are several constant design parameters in this dynamic backstepping controller. These include the eight constants s_1, s_2, s_3, s_4, d_2, d_3, d_4, and γ, as well as the symmetric positive definite matrix Q in the Lyapunov equation (7.139). For simulation purposes, we first chose $\gamma = 1$ and $Q = I_{2\times 2}$ and then used numerical optimization with an *ad hoc* cost to obtain the values listed in Table 7.2.

Q	$I_{2\times 2}$	γ	1.0
s_1	10.6858	d_2	45.7867
s_2	0.0334	d_3	47.9097
s_3	17.8677	d_4	24.1476
s_4	6.413		

Table 7.2: Design parameter values for the backstepping controller.

We expect the backstepping controller to result in improved settling time compared to the approximated filtered version of the control law (7.122)–(7.123) with open-loop observer (7.132). Figure 7.9 shows simulation results for regulation, that is, for the reference signal $\vartheta_r(t) \equiv 0$: the backstepping controller generates a much shorter settling time with the same control effort. The step response for the backstepping controller is shown in Figure 7.10. As compared to the step response of Figure 7.8, this step response has a larger transient but a much shorter settling time. This can be seen more clearly in Figure 7.11 which displays the tracking errors associated with the two step responses. The two plots show the same signals in different scales: the first plot illustrates that the backstepping controller generates a larger transient, while the second plot illustrates that the backstepping controller generates a shorter settling time. It is feasible that the transient behavior of the backstepping controller can be improved through different choices of design parameters; see also [86, 85] for a discussion of methods for improving the transient performance of backstepping designs.

7.4 Summary

We introduced a class of extended strict feedback systems for which the design of globally convergent dynamic partial state feedback tracking controllers is feasible. We then presented a recursive backstepping method for constructing such controllers. These results extend the applicability of the tuning function method of [84] beyond the adaptive control problem.

We constructed various nonlinear controllers for a mechanical ARP system. Full state feedback controllers were developed based on input/output linearization techniques. These controllers were then used as the basis for constructing partial state feedback controllers via the backstepping method.

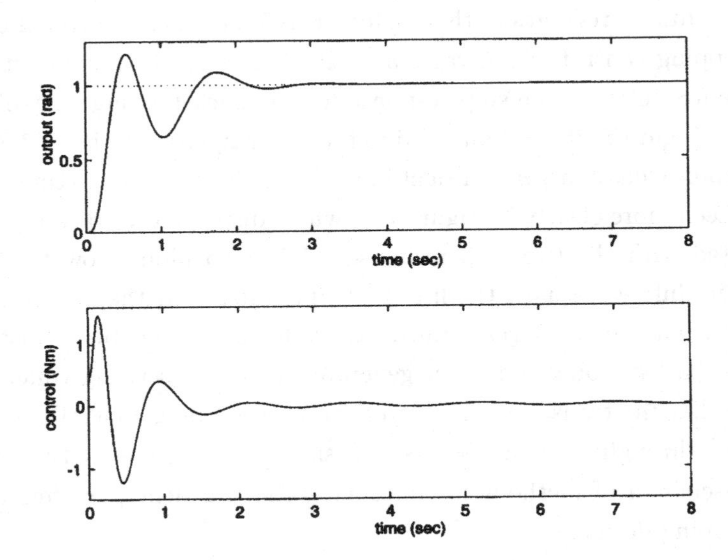

Figure 7.10: Step response with the backstepping controller: tracking output ϑ (top plot) and control signal u (bottom plot).

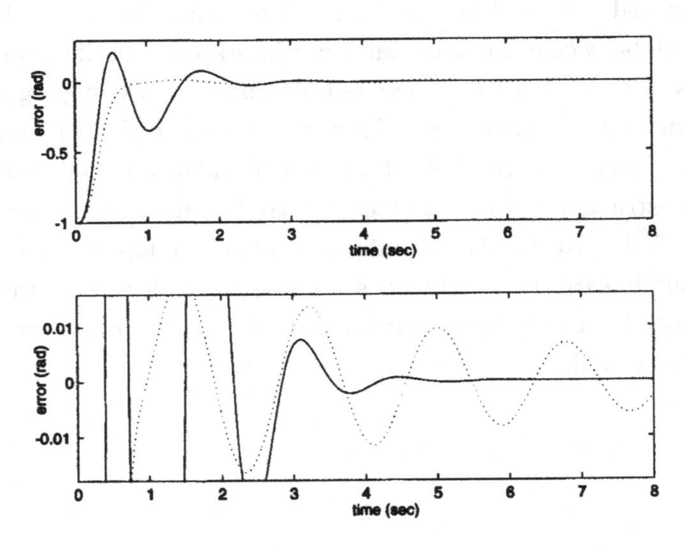

Figure 7.11: Step response tracking error $\vartheta - \vartheta_r$ generated by the approximated filtered version of the control law (7.122)–(7.123) with open-loop observer (7.132) (dotted) and by the backstepping controller (solid). The two plots show the same signals in different scales.

Chapter 8

Robust Nonlinear PI Control

Thus far we have represented system uncertainty by a disturbance input w allowed to have an arbitrarily fast time variation. Its only constraint was the pointwise condition $w \in W$ where W was some known set possibly depending on the state x and control u. We now address a more specific situation in which our system contains some uncertain nonlinearity $\phi(x)$. Suppose that all we know about ϕ is a set-valued map $\Phi(x)$ such that $\phi(x) \in \Phi(x)$ for all $x \in \mathcal{X}$. We could assign $w := \phi(x)$ and $W(x) := \Phi(x)$ and proceed as in Chapters 3–6, but we would be throwing away a crucial piece of information about the uncertainty ϕ, namely, that $\phi(x)$ *does not explicitly depend on time t*. Our goal in this chapter is to illustrate how we can take advantage of this additional information to design less conservative robust controllers.

We will consider nonlinear systems having both types of uncertainties, namely, disturbances w and uncertain functions $\phi(x)$. The robust backstepping techniques of Chapter 5 will be used to accommodate the disturbances w, while the dynamic backstepping techniques of Chapter 7 will be used to accommodate the uncertain nonlinearities $\phi(x)$. Our design objective will be set-point tracking (the tracking of constant reference signals). We will interpret the dynamic part of our controller as a type of robust integral action, hence the title of this chapter.

For the special case in which the uncertain nonlinearity $\phi(x)$ is simply an unknown constant parameter $\phi(x) \equiv \theta$, our results represent a robust version of the tuning function adaptive control design in [84, 85]. In this special case, we are no longer restricted to tracking constant reference signals. Related robust adaptive control designs are presented in [115, 157], but our results differ from these: whereas we guarantee internal boundedness and achieve the convergence of the tracking error to zero when the disturbance w is zero, the design in [157] does not guarantee the boundedness of internal controller signals, while the design in [115] does not guarantee convergence of the tracking error to zero when $w = 0$.

8.1 Problem formulation

We consider the nonlinear system

$$\dot{x} = F(x) + G(x)\,u + P(x)\,\phi(x) + Q(x)\,w \tag{8.1}$$

$$y = H(x) \tag{8.2}$$

where $x \in I\!R^n$ is the state, $u \in I\!R^m$ is the control input, $y \in I\!R^m$ is the variable to be tracked, $\phi : I\!R^n \to I\!R^p$ is an uncertain nonlinearity, $w \in I\!R^q$ is an unmeasured exogenous disturbance input, and F, G, H, P, and Q are known smooth functions of appropriate dimensions. In this chapter we assume that the full state x is available for feedback. In Chapters 5–7, the control input u was a scalar variable, but the results were easily extendible to multi-input systems. To illustrate such multi-input backstepping, we now allow the control input u to be a vector rather than a scalar.

8.1.1 Class of systems

We characterize the class of systems to be considered by a series of assumptions about the general system (8.1)–(8.2). Our first assumption is that the nominal system

$$\dot{x} = F(x) + G(x)\,u \tag{8.3}$$

$$y = H(x) \tag{8.4}$$

is globally input/output linearizable with no zero dynamics, namely,

N1: the system (8.3)–(8.4) has a uniform (constant) vector relative degree $\{r_1, \ldots, r_m\}$ at every point in $I\!R^n$, and furthermore there are no zero dynamics, that is, $r_1 + \ldots + r_m = n$.

This assumption can be relaxed, especially the requirement that there be no zero dynamics, but we will not pursue this issue here. As a consequence of this assumption, one can build a global diffeomorphism $\xi := \Xi(x)$ and a smooth state feedback $u = \alpha(x) + \beta(x)\,v$ (where $v \in I\!R^m$ is a new control input) such that the system (8.1)–(8.2) becomes

$$\dot{\xi}_1 = \begin{bmatrix} \xi_2 \\ v_1 \end{bmatrix} + P_1(\xi)\,\phi(\xi) + Q_1(\xi)\,w \tag{8.5}$$

$$\dot{\xi}_2 = \begin{bmatrix} \xi_3 \\ v_2 \end{bmatrix} + P_2(\xi)\,\phi(\xi) + Q_2(\xi)\,w$$

$$\vdots$$

$$\dot{\xi}_r = v_r + P_r(\xi)\,\phi(\xi) + Q_r(\xi)\,w$$

$$y = \xi_1 \tag{8.6}$$

for suitably defined smooth functions P_i and Q_i, where we have allowed the abuse of notation $\phi(\xi) := \phi(\Xi^{-1}(\xi))$. Here the new state variable ξ and the new control variable v have been partitioned as $\xi = [\xi_1^{\mathrm{T}} \dots \xi_r^{\mathrm{T}}]^{\mathrm{T}}$ and $v = [v_1^{\mathrm{T}} \dots v_r^{\mathrm{T}}]^{\mathrm{T}}$, where $r := \max\{r_i\}$ and $\dim(\xi_i) = \mathrm{card}\big\{j : r_j \geq i\big\}$. Note that any of the components v_i (except the last component v_r) may be vacuous, that is, of dimension zero. We next assume that the functions P_i and Q_i satisfy a strict feedback condition similar to the one introduced in Chapter 5:

N2: for each $i \in \{1, \dots, r-1\}$, the functions P_i and Q_i are independent of ξ_j for $j \in \{i+1, \dots, r\}$, that is, $P_i(\xi) = P_i(\xi_1, \dots, \xi_i)$ and $Q_i(\xi) = Q_i(\xi_1, \dots, \xi_i)$.

Our final assumption complements N2 and is a structural condition on the uncertain nonlinearity ϕ:

N3: for $\ell := \min\big\{j : P_j \not\equiv 0\big\}$, if $\ell < r$ then the uncertainty ϕ is independent of ξ_j for all $j \in \{\ell+1, \dots, r\}$, that is, $\phi(\xi) = \phi(\xi_1, \dots, \xi_\ell)$.

Assumption N3 is most restrictive when $\ell = 1$ and $r > 1$, in which case the uncertain nonlinearity ϕ must be a function of the output y alone. For simplicity, we will consider only the case $\ell = 1$ and assume $\phi = \phi(y)$ for the remainder of this chapter. The less restrictive cases where $\ell > 1$ are straightforward extensions of the case $\ell = 1$.

We assume knowledge of two bounds on the uncertain nonlinearity $\phi(y)$ at each point $y \in I\!R^m$: one bound on ϕ itself and one bound on the local Lipschitz constant of ϕ near y. To be precise, we know a C^∞ function $\rho : I\!R^m \times I\!R^m \to I\!R_+$ and an usc set-valued map $\Phi : I\!R^m \rightsquigarrow I\!R^p$ with nonempty compact convex values such that

U1: $\phi(y) \in \Phi(y)$ for all $y \in I\!R^m$, and
U2: $|\phi(y_1) - \phi(y_2)| \le \rho(y_1, y_2)\, |y_1 - y_2|$ for all $y_1, y_2 \in I\!R^m$.

The set-valued map Φ in U1 describes a tube Graph(Φ) $\subset I\!R^m \times I\!R^p$ inside of which the uncertainty ϕ is known to lie. Also, one can show that the knowledge of the function ρ which satisfies U2 is equivalent to the knowledge of a bound on the local Lipschitz constant of ϕ near every point in $I\!R^m$.

There is a special characterization of the uncertainty ϕ which falls into our framework. Suppose we have a parameterization of the uncertainty ϕ, namely, suppose $\phi(y) = \psi(y, \theta)$ for some *known* locally Lipschitz function ψ and some unknown constant parameter θ belonging to a known compact set Θ. In this case, we can use our knowledge of ψ to construct the bounds Φ and ρ. In particular, we can take $\Phi(y) := \mathrm{co}\, \psi(y, \Theta)$ and choose ρ to satisfy

$$\rho(y_1, y_2) \;\ge\; \max\Big\{ \big\|\partial_y \psi(y, \theta)\big\| \;:\; y \in [y_1, y_2],\; \theta \in \Theta \Big\} \qquad (8.7)$$

where $\partial_y \psi$ denotes the Clarke partial generalized Jacobian [20]. This special characterization of ϕ includes the robust adaptive control setting of [115, 157] where $\phi(y) \equiv \theta$, in which case we take $\Phi \equiv \mathrm{co}\, \Theta$ and $\rho \equiv 0$.

8.1.2 Design objective

We wish to construct a dynamic state feedback controller for the system (8.1)–(8.2) which provides robust tracking of arbitrary constant set-points $y_s \in I\!R^m$. Our controller will be of the form

$$u = \mu(x, \omega, y_s) \qquad (8.8)$$

$$\dot{\omega} = \Omega(x, \omega, y_s) + d \qquad (8.9)$$

for smooth functions μ and Ω, where $\omega \in I\!R^p$ is the internal state of the controller and $y_s \in I\!R^m$ is the desired output set-point. We have also included an additional unmeasured exogenous disturbance input $d \in I\!R^p$ whose role will be interpreted below. Note that the order p of the controller is the same as the dimension of the uncertain vector ϕ. Because several uncertainties appearing in the same state equation can be combined into a single uncertain function, the dimension of the vector ϕ, and thus the dimension of the controller, can always be bounded from above by the dimension n of the system state. In contrast, adaptive controllers based on parameterizations of ϕ will be of the same order as the number of parameters, which can be higher than the dimension of the state.

If we let $\overline{w} := [w^{\mathrm{T}}\ d^{\mathrm{T}}]^{\mathrm{T}} \in I\!R^{q \times p}$ denote the combined exogenous disturbance input, then our design objective is to construct smooth functions μ and Ω such that system (8.1)–(8.2) with controller (8.8)–(8.9) has the following properties: for every pair $x(0)$, $\omega(0)$ of initial conditions, every bounded disturbance $\overline{w}(t) \in L_\infty$, every set-point $y_s \in I\!R^m$, and every uncertainty ϕ satisfying U1–U2, we have

Q1: *(global boundedness)* the solutions $x(t)$ and $\omega(t)$ exist and are bounded for all $t \geq 0$,

Q2: *(asymptotic tracking)* if $\overline{w}(t) \equiv 0$, then $y(t) \to y_s$ and $\omega(t) \to \Phi(y_s)$ as $t \to \infty$,

Q3: *(finite L_∞-gain)* the tracking error $[y(t) - y_s]$ satisfies

$$\|y - y_s\|_\infty \ \leq \ \lambda \|\overline{w}\|_\infty + \lambda_0 \tag{8.10}$$

where λ and λ_0 are constants independent of $\overline{w}$, with λ also independent of $x(0)$ and $\omega(0)$, and,

Q4: *(finite L_2-gain)* if $d(t) \equiv 0$ and $w(t) \in L_2$, then $[y(t) - y_s] \in L_2$ and

$$\|y - y_s\|_2 \ \leq \ \gamma \|w\|_2 + \gamma_0 \tag{8.11}$$

where γ and γ_0 are constants independent of w, with γ also independent of $x(0)$ and $\omega(0)$.

Note that in the asymptotic tracking property Q2, we require not only the convergence of the tracking error to zero, but also the convergence of the controller state ω to the compact convex set $\Phi(y_s)$ where the uncertainty $\phi(y_s)$ is known to lie. Also, the finite-gain properties Q3 and Q4 are with respect to the tracking error only; we do not require a finite gain between the disturbance and the other signals in the system.

The exogenous disturbance d in the controller dynamics (8.9) seems artificial because these dynamics are generated inside the controller and should not be subject to such a disturbance. Nevertheless, we include this disturbance for two reasons. First, robustness with respect to d will guarantee robustness with respect to small errors in the measurement of the state x. As a result, the controller state ω will not exhibit unboundedness. Second, robustness with respect to d will guarantee robustness with respect to time variations in the uncertainty ϕ or in the set-point y_s. This second issue will be made clear at the end of Section 8.2.3.

8.2 Controller design

Our main result, stated in Section 8.2.1, is that the design objectives Q1–Q4 can be met for systems satisfying assumptions N1–N3. We prove this result in Sections 8.2.2–8.2.6 by constructing the desired controller together with an associated Lyapunov function. This controller design is similar to the dynamic feedback design we presented in Chapter 7.

8.2.1 Main result

Theorem 8.1 *If the system* (8.1)–(8.2) *satisfies assumptions* N1–N3, *then there exist* C^∞ *functions* μ *and* Ω *such that the controller* (8.8)–(8.9) *meets the design objective characterized by* Q1–Q4. *Furthermore, the constants* λ *and* γ *in* Q3 *and* Q4 *can be reduced arbitrarily using high control gain.*

As a result of assumption N1, it suffices to design a controller for the transformed system (8.5)–(8.6), that is,

$$v = \mu_r(\xi, \omega, y_s) \tag{8.12}$$

$$\dot{\omega} = \Omega_r(\xi, \omega, y_s) + d \tag{8.13}$$

for smooth functions μ_r and Ω_r. The controller design will be presented in Sections 8.2.3–8.2.5, and the proof of Theorem 8.1 will be completed in Section 8.2.6.

8.2.2 Technical lemma

The proof of the following lemma is adapted from the proof of Whitney's theorem [15, p. 24].

Lemma 8.2 *Let $K : I\!R^m \rightsquigarrow I\!R^p$ be usc with nonempty compact convex values. Then for any C^∞ function $M : I\!R^m \to I\!R^{p \times p}$ with symmetric positive definite values, there exist a positive function $\delta : I\!R^m \to I\!R_+$ and a C^∞ function $L : I\!R^p \times I\!R^m \to I\!R^p$ such that*

(i) $L(x,y) = 0$ *for all* $(x,y) \in \mathrm{Graph}(K^{-1})$

(ii) $(k-x)^{\mathrm{T}} L(x,y) > 0$ *whenever* $(x,y) \in I\!R^p \times I\!R^m \backslash \mathrm{Graph}(K^{-1})$ *and* $k \in K(y)$

(iii) $(k-x)^{\mathrm{T}} L(x,y) \geq (k-x)^{\mathrm{T}} M(y)\,(k-x) - \delta(y)$ *whenever* $(x,y) \in I\!R^p \times I\!R^m$ *and* $k \in K(y)$

Proof: Because K is usc with nonempty closed values, the set

$$G := I\!R^p \times I\!R^m \backslash \mathrm{Graph}(K^{-1})$$

is an open subset of $I\!R^p \times I\!R^m$. Fix $z = (\overline{x}, \overline{y}) \in G$. Then $\overline{x} \in I\!R^p \backslash K(\overline{y})$, and because $K(\overline{y})$ is closed, there exists $\varepsilon > 0$ such that $\overline{x} \in I\!R^p \backslash F$ where $F := K(\overline{y}) + \varepsilon B$ and B denotes the closed unit ball in $I\!R^p$. Because F is closed and convex, there exists a neighborhood $U_p \subset I\!R^p \backslash F$ of $\overline{x}$ and a unit vector $L_z \in I\!R^p$ such that $(k-x)^{\mathrm{T}} L_z > 0$ for all $x \in U_p$ and all $k \in F$. Also, because K is usc, there exists a neighborhood $U_m \subset I\!R^m$ of $\overline{y}$ such that $K(y) \subset F$ for all $y \in U_m$. It follows that there exists an open ball $N_z \subset U_p \times U_m \subset G$ containing $(\overline{x}, \overline{y})$ such that $(k-x)^{\mathrm{T}} L_z > 0$ whenever $(x,y) \in N_z$ and $k \in K(y)$. Now the open cover $\{N_z\}$ of G has a countable subcover $\{N_{z_i}\}$, and we define $N_i := N_{z_i}$ and $L_i := L_{z_i}$. For each $i \geq 1$, let $\psi_i : I\!R^p \times I\!R^m \to I\!R_+$ be a C^∞ function such that $\psi_i(x,y) > 0$ for $(x,y) \in N_i$ and $\psi_i(x,y) = 0$ for $(x,y) \in I\!R^p \times I\!R^m \backslash N_i$. Because each ψ_i has compact support, for every $i \geq 1$ there exists $\varepsilon_i > 0$ such that the function $\varepsilon_i \psi_i$ and all of its partial derivatives of order less than i are bounded in magnitude by 2^{-i}. We can therefore define a C^0 function $S : I\!R^p \times I\!R^m \to I\!R^p$ by the uniformly convergent series

$$S(x,y) \;:=\; \sum_{i=1}^{\infty} \varepsilon_i \, \psi_i(x,y)\, L_i \tag{8.14}$$

All of the term-by-term partial derivatives of the right-hand side of (8.14) are also uniformly convergent, and it follows that S is C^∞. Now for each $i \geq 1$, if $(x,y) \in N_i$ then $\psi_i(x,y) > 0$ and $(k-x)^{\mathrm{T}}L_i > 0$ for all $k \in K(y)$, and otherwise $\psi_i(x,y) = 0$. Thus $S(x,y) = 0$ for all $(x,y) \in \mathrm{Graph}(K^{-1})$ and $(k-x)^{\mathrm{T}}S(x,y) > 0$ whenever $(x,y) \in G$ and $k \in K(y)$.

Let $\kappa : I\!R^m \to I\!R^p$ be any C^∞ function, and let $c : I\!R^m \to I\!R_+$ be any C^∞ function which satisfies

$$c(y) \;\geq\; \max_{k \in K(y)} \, [\kappa(y) - k]^{\mathrm{T}} M(y)\, [\kappa(y) - k] \tag{8.15}$$

for all $y \in I\!R^m$. Let $\alpha : R \to [0,1]$ be C^∞ and such that $\alpha(r) = 0$ for $r \leq 0$ and $\alpha(r) = 1$ for $r \geq 1$. We define a C^∞ function $\beta : I\!R^p \times I\!R^m \to [0,1]$ by

$$\beta(x,y) \;:=\; \alpha\Big([\kappa(y) - x]^{\mathrm{T}} M(y)\, [\kappa(y) - x] \;-\; c(y)\Big) \tag{8.16}$$

Finally, we define $L : I\!R^p \times I\!R^m \to I\!R^p$ by

$$L(x,y) \;:=\; 2\beta(x,y)M(y)\,[\kappa(y) - x] + \big[1 - \beta(x,y)\big]\, S(x,y) \tag{8.17}$$

Clearly L is C^∞, and we have left to verify conditions (i)–(iii). We fix $(x,y) \in I\!R^p \times I\!R^m$ and let $k \in K(y)$. Suppose $(x,y) \in \mathrm{Graph}(K^{-1})$; then from (8.15) and (8.16) we have $\beta(x,y) = 0$ which implies $L(x,y) = 0$. Next suppose $(x,y) \in G$. Then $(k-x)^{\mathrm{T}}S(x,y) > 0$, and from (8.15),

$$\begin{aligned} 2(k-x)^{\mathrm{T}} M(y)\,[\kappa(y) - x] &= 2\,[\kappa(y) - x]^{\mathrm{T}} M(y)\,[\kappa(y) - x] \\ &\quad + 2\,[k - \kappa(y)]^{\mathrm{T}} M(y)\,[\kappa(y) - x] \\ &\geq [\kappa(y) - x]^{\mathrm{T}} M(y)\,[\kappa(y) - x] \,- \\ &\quad [k - \kappa(y)]^{\mathrm{T}} M(y)\,[k - \kappa(y)] \\ &\geq [\kappa(y) - x]^{\mathrm{T}} M(y)\,[\kappa(y) - x] \;-\; c(y) \end{aligned} \tag{8.18}$$

If $\beta(x,y) > 0$, then the right-hand side of (8.18) is strictly positive and thus $(k-x)^{\mathrm{T}}L(x,y) > 0$. Otherwise $(k-x)^{\mathrm{T}}L(x,y) = (k-x)^{\mathrm{T}}S(x,y) > 0$. We conclude that L satisfies condition (ii). We now show that L satisfies condition (iii) with the function δ given by

$$\begin{aligned} \delta(y) \;:=\; c(y) \;-\; \inf\Big\{&[1 - \beta(x,y)]\,(k-x)^{\mathrm{T}}\big[S(x,y) - 2M(y)\,[\kappa(y) - x]\big] \\ &: x \in I\!R^p,\; k \in K(y)\Big\} \end{aligned} \tag{8.19}$$

Note that $0 \leq \delta(y) < \infty$ because $\beta(\cdot, y) = 1$ except on a bounded subset of $\mathbb{R}^p$. From (8.17) we have

$$\begin{aligned}
(k-x)^{\mathrm{T}} L(x,y) &= 2(k-x)^{\mathrm{T}} M(y)\,[\kappa(y) - x] \\
&\quad + [1 - \beta(x,y)]\,(k-x)^{\mathrm{T}}\big[S(x,y) - 2M(y)\,[\kappa(y) - x]\big] \\
&\geq 2(k-x)^{\mathrm{T}} M(y)\,(k-x) \\
&\quad + 2(k-x)^{\mathrm{T}} M(y)\,[\kappa(y) - k] + c(y) - \delta(y) \\
&\geq (k-x)^{\mathrm{T}} M(y)\,(k-x) \\
&\quad - [\kappa(y) - k]^{\mathrm{T}} M(y)\,[\kappa(y) - k] + c(y) - \delta(y) \\
&\geq (k-x)^{\mathrm{T}} M(y)\,(k-x) - \delta(y)
\end{aligned} \tag{8.20}$$

where (8.20) follows from (8.15). This completes the proof. The function $\kappa(y)$ in this proof was arbitrary and can be considered as a design parameter. By choosing $\kappa(y)$ to take values close to the "centers" of the sets $K(y)$, we can reduce the magnitude of the function $\delta(y)$. ■

8.2.3 Controller design for $r = 1$

When $r = 1$, the system (8.5)–(8.6) can be written

$$\dot{\xi}_1 = v + P_1(\xi_1)\,\phi(y) + Q_1(\xi_1)\,w \tag{8.21}$$

$$y = \xi_1 \tag{8.22}$$

If we consider a controller of the form (8.12)–(8.13), then the closed-loop system becomes

$$\dot{\xi}_1 = \mu_1(\xi_1, \omega, y_s) + P_1(\xi_1)\,\phi(y) + Q_1(\xi_1)\,w \tag{8.23}$$

$$\dot{\omega} = \Omega_1(\xi_1, \omega, y_s) + d$$

$$y = \xi_1 \tag{8.24}$$

We wish to construct smooth functions μ_1 and Ω_1 such that Q1–Q4 are satisfied for this system. We define the tracking error $z_1 := (y - y_s) = (\xi_1 - y_s)$ and consider the Lyapunov function

$$V_1(\xi_1, \omega) := z_1^{\mathrm{T}} C_1 z_1 + \big[\phi(y_s) - \omega\big]^{\mathrm{T}} E\,\big[\phi(y_s) - \omega\big] \tag{8.25}$$

where C_1 and E are any symmetric positive definite matrices. These matrices are design parameters allowed to depend smoothly on the set-point, that is, $C_1 = C_1(y_s)$ and $E = E(y_s)$. Such dependence on y_s will be allowed for all design parameters we encounter from now on, and we will make no further mention of it. Calculating the derivative $\dot{V}_1$ along solutions of (8.23)–(8.24), we obtain

$$\begin{aligned} \dot{V}_1 &= 2z_1^{\mathrm{T}}C_1\big[\mu_1 + P_1\phi(y) + Q_1 w\big] \\ &\quad + 2\big[\phi(y_s) - \omega\big]^{\mathrm{T}} E\big[-\Omega_1 - d\big] \qquad (8.26) \\ &= 2z_1^{\mathrm{T}}C_1\big[\mu_1 + P_1\phi(y_s) + P_1[\phi(y) - \phi(y_s)] + Q_1 w\big] \\ &\quad + 2\big[\phi(y_s) - \omega\big]^{\mathrm{T}} E\big[-\Omega_1 - d\big] \\ &= 2z_1^{\mathrm{T}}C_1\big[\mu_1 + P_1\omega + P_1[\phi(y) - \phi(y_s)] + Q_1 w\big] \\ &\quad + 2\big[\phi(y_s) - \omega\big]^{\mathrm{T}}\big[P_1^{\mathrm{T}}C_1 z_1 - E\Omega_1 - Ed\big] \qquad (8.27) \end{aligned}$$

Let us now bound some of the uncertain terms in (8.27). It follows from U2 that there exists a smooth function $N_1 : I\!R^m \times I\!R^m \to I\!R^{m\times m}$ such that

$$2\, z_1^{\mathrm{T}}C_1P_1(\xi_1)\big[\phi(y) - \phi(y_s)\big] \;\leq\; 2\, z_1^{\mathrm{T}}C_1N_1(\xi_1, y_s)\, z_1 \qquad (8.28)$$

for all ξ_1 and y_s and all uncertainties ϕ satisfying U1–U2. Indeed, a conservative choice for N_1 would be $N_1(\xi_1, y_s) = \rho_1(\xi_1, y_s)I$, where ρ_1 is any smooth function satisfying

$$\rho_1(\xi_1, y_s) \;\geq\; \frac{\rho(\xi_1, y_s)}{\lambda_{min}(C_1)}\,\big|C_1P_1(\xi_1)\big| \qquad (8.29)$$

where ρ is from U2 and $\lambda_{min}(C_1)$ denotes the minimum eigenvalue of C_1. Next, from Young's inequality we have the following:

$$2z_1^{\mathrm{T}}C_1Q_1(\xi_1)\, w \;\leq\; \frac{1}{\varepsilon}\, z_1^{\mathrm{T}}C_1Q_1(\xi_1)\big[Q_1(\xi_1)\big]^{\mathrm{T}}C_1z_1 + \varepsilon|w|^2 \qquad (8.30)$$

$$-2\big[\phi(y_s) - \omega\big]^{\mathrm{T}}Ed \;\leq\; \frac{1}{\varepsilon}\big[\phi(y_s) - \omega\big]^{\mathrm{T}}E^2\big[\phi(y_s) - \omega\big] + \varepsilon|d|^2 \qquad (8.31)$$

where $\varepsilon > 0$ is a design parameter. Let us now choose

$$\mu_1(\xi_1, \omega, y_s) \;=\; -\Big[C_1^{-1}D_1 + \frac{1}{2\varepsilon}Q_1Q_1^{\mathrm{T}}C_1 + N_1\Big]z_1 \;-\; P_1\omega \qquad (8.32)$$

$$\Omega_1(\xi_1, \omega, y_s) \;=\; E^{-1}L(\omega, y_s) \;+\; E^{-1}P_1^{\mathrm{T}}C_1z_1 \qquad (8.33)$$

where D_1 is a symmetric positive definite design parameter and the smooth function $L : I\!R^p \times I\!R^m \to I\!R^p$ is to be chosen below. Substituting (8.28), (8.30), (8.31), (8.32), and (8.33) into (8.27), we obtain

$$\dot{V}_1 \;\leq\; -2z_1^{\mathrm{T}}D_1z_1 \;-\; 2\big[\phi(y_s) - \omega\big]^{\mathrm{T}}L(\omega, y_s) \;+$$

$$\frac{1}{\varepsilon}\left[\phi(y_s)-\omega\right]^{\mathrm{T}}E^2\left[\phi(y_s)-\omega\right] + \varepsilon|\overline{w}|^2 \tag{8.34}$$

where $\overline{w} := [w^{\mathrm{T}}\ d^{\mathrm{T}}]^{\mathrm{T}}$. Also, when $d = 0$ we have

$$\dot{V}_1 \ \le\ -2z_1{}^{\mathrm{T}}D_1z_1 - 2\left[\phi(y_s)-\omega\right]^{\mathrm{T}}L(\omega,y_s) + \varepsilon|w|^2 \tag{8.35}$$

as can be seen from (8.27) prior to using the inequality (8.31). We have left to choose the function L. Suppose we were to choose $L \equiv 0$; then our controller for the system (8.21)–(8.22) would be

$$v \ =\ -K_1(\xi_1,y_s)\,z_1 - P_1(\xi_1)\,\omega \tag{8.36}$$

$$\dot{\omega} \ =\ K_2(\xi_1,y_s)\,z_1 + d \tag{8.37}$$

for functions K_1 and K_2 defined from (8.32)–(8.33). If K_1, K_2, and P_1 were independent of ξ_1, then this controller would represent a *linear* PI controller with gains possibly scheduled by the set-point y_s. We can therefore interpret the controller (8.36)–(8.37) as a *nonlinear* PI controller.

It follows from (8.35) and standard Lyapunov analysis [75, Theorem 4.8] that the choice $L \equiv 0$ would guarantee property Q1 in the absence of the disturbance $\overline{w}$. This choice would also guarantee the convergence of the tracking error z_1 to zero, but there would be no guarantee on the convergence of ω to the set $\Phi(y_s)$ as is required by property Q2. Furthermore, small nonzero disturbances $\overline{w}(t)$ could cause some closed-loop signals to be unbounded. Thus the choice $L \equiv 0$ is not sufficient to guarantee the robustness we desire, and we need to modify the nonlinear PI controller (8.36)–(8.37). Recall that the function L is not allowed to depend on ϕ because it represents part of the controller dynamics (8.33). From Lemma 8.2 there exists a C^∞ function L, depending on the set-valued map Φ, which has the following properties for every uncertainty ϕ satisfying U1–U2:

L1: $L(\omega,y_s) = 0$ for all $\omega \in \Phi(y_s)$,

L2: $-2[\phi(y_s)-\omega]^{\mathrm{T}}L(\omega,y_s) < 0$ for all $\omega \notin \Phi(y_s)$, and,

L3: $-2[\phi(y_s)-\omega]^{\mathrm{T}}L(\omega,y_s) \le -[\phi(y_s)-\omega]^{\mathrm{T}}\left[M + \frac{1}{\varepsilon}E^2\right][\phi(y_s)-\omega] + \delta$ for all $\omega \in I\!R^p$,

where M is a symmetric positive definite design parameter and $\delta \ge 0$ is a constant which depends on M, E, ε, y_s, and Φ. Such a function L is easy to construct when the set $\Phi(y_s)$ has a convenient shape (like a box or a ball) and depends smoothly on y_s. Note that when ω is inside the set $\Phi(y_s)$, our controller (8.32)–(8.33) reduces to the nonlinear PI controller (8.36)–(8.37).

The desired properties L1–L3 of the function L are deduced directly from the Lyapunov derivatives (8.34)–(8.35). These properties have been identified by other authors as desirable [156], and they can be used to motivate a variant of the σ-modification of [59]. As noted in [157], the function L should be C^{r-2} rather than merely continuous so that it can be used in the recursive design outlined below for $r > 2$. For this reason, Lemma 8.2 is significant because it proves the existence of a C^∞ function L with these properties. With the choice for L described by L1–L3, it follows from (8.34) and L3 that

$$\dot{V}_1 \;\leq\; -2z_1^{\mathrm{T}}D_1z_1 \;-\; \big[\phi(y_s)-\omega\big]^{\mathrm{T}}M\big[\phi(y_s)-\omega\big] \;+\;\delta\;+\;\varepsilon|\overline{w}|^2 \tag{8.38}$$

Also, it follows from (8.35), L1, and L2 that

$$\overline{w}=0 \;\Longrightarrow\; \dot{V}_1 \leq 0 \tag{8.39}$$

$$\overline{w}=0 \text{ and } (z_1,\omega)\notin\{0\}\times\Phi(y_s) \;\Longrightarrow\; \dot{V}_1 < 0 \tag{8.40}$$

Finally, using (8.38), (8.39), (8.40), and standard Lyapunov arguments, we conclude that properties Q1–Q4 are satisfied. We postpone the details until Section 8.2.6.

Let us briefly describe one consequence of the robustness we have achieved with respect to the disturbance d. Suppose the uncertain nonlinearity ϕ is actually changing with time, that is, $\phi=\phi(y,t)$, and assume that U1–U2 are satisfied uniformly in t. Suppose furthermore that the partial derivative $\partial\phi/\partial t$ exists almost everywhere. Then in the above computation of $\dot{V}_1$, instead of (8.26) we would achieve

$$\begin{aligned}\dot{V}_1 \;&=\; 2z_1^{\mathrm{T}}C_1\big[\mu_1+P_1\phi(y,t)+Q_1w\big]\\ &\quad+2\big[\phi(y_s,t)-\omega\big]^{\mathrm{T}}E\left[\frac{\partial\phi}{\partial t}(y_s,t)-\Omega_1-d\right]\end{aligned} \tag{8.41}$$

Because the disturbance d appears in this expression matched with $\partial\phi/\partial t$, the above controller design which achieves robustness with respect to d also achieves robustness with respect to this time variation in ϕ. Similar arguments apply in the situation in which the set-point y_s changes with time, that is, $y_s = y_s(t)$.

8.2.4 Backstepping construction

The design for $r > 1$ is based on a recursive backstepping procedure in which the design for $r = i$ is used to generate the design for $r = i+1$. The backstepping construction, which is similar to the one in Section 7.2.4, is the same at each recursion step.

We consider a system of the form

$$\begin{aligned} \dot{\xi}_a &= B\xi_b + A(\xi_a, \omega, y_s) + P_a(\xi_a)\,\phi(y) + Q_a(\xi_a)\,w \qquad (8.42) \\ \dot{\xi}_b &= u + P_b(\xi_a, \xi_b)\,\phi(y) + Q_b(\xi_a, \xi_b)\,w \\ y &= J\xi_a \qquad (8.43) \end{aligned}$$

where B and J are given matrices and A, P_a, P_b, Q_a, and Q_b are given smooth functions. We wish to design a dynamic controller

$$u = \mu(\xi_a, \xi_b, \omega, y_s) \qquad (8.44)$$

$$\dot{\omega} = \Omega(\xi_a, \xi_b, \omega, y_s) + d \qquad (8.45)$$

such that the derivative of a certain Lyapunov function is negative.

We assume knowledge of a smooth conceptual controller

$$u = \mu_c(\xi_a, \omega, y_s) \qquad (8.46)$$

$$\dot{\omega} = \Omega_c(\xi_a, \omega, y_s) + d \qquad (8.47)$$

for the reduced-order ξ_a-subsystem

$$\dot{\xi}_a = Bu + A(\xi_a, \omega, y_s) + P_a(\xi_a)\,\phi(y) + Q_a(\xi_a)\,w \qquad (8.48)$$

$$y = J\xi_a \qquad (8.49)$$

together with an associated conceptual Lyapunov function

$$V_c(\xi_a, \omega) = U(\xi_a, \omega) + \left[\phi(y_s) - \omega\right]^{\mathrm{T}} E \left[\phi(y_s) - \omega\right] \qquad (8.50)$$

where U is a C^∞ function and E is a symmetric positive definite matrix. This conceptual controller is such that the conceptual Lyapunov derivative denoted by

$$\dot{V}_c \Big|_{\substack{\xi_b = \mu_c \\ \dot{\omega} = \Omega_c + d}} \qquad (8.51)$$

is negative. We will use this conceptual controller (8.46)–(8.47) for the ξ_a-subsystem (8.48)–(8.49) to construct an actual controller (8.44)–(8.45) for the complete system (8.42)–(8.43).

We first note that the actual derivative of V_c (along solutions to the actual closed-loop system) can be written as

$$\dot{V}_c = \dot{V}_c\Big|_{\substack{\xi_b = \mu_c \\ \dot{\omega} = \Omega_c + d}} + \frac{\partial U}{\partial \xi_a} B\left[\xi_b - \mu_c\right] + \frac{\partial U}{\partial \omega}\left[\Omega - \Omega_c\right] + 2\left[\phi(y_s) - \omega\right]^{\mathrm{T}} E\left[\Omega_c - \Omega\right] \tag{8.52}$$

We define the error variable

$$z_b := \xi_b - \mu_c(\xi_a, \omega, y_s) \tag{8.53}$$

Calculating $\dot{z}_b$ from (8.53) and (8.42)–(8.43), we obtain

$$\begin{aligned} \dot{z}_b &= \mu + P_b\phi(y) + Q_b w \\ &\quad - \frac{\partial \mu_c}{\partial \xi_a}\left[B\xi_b + A + P_a\phi(y) + Q_a w\right] - \frac{\partial \mu_c}{\partial \omega}\left[\Omega + d\right] \\ &= \mu + \left[P_b - \frac{\partial \mu_c}{\partial \xi_a} P_a\right]\phi(y) + \left[\left(Q_b - \frac{\partial \mu_c}{\partial \xi_a} Q_a\right) \;\middle|\; -\frac{\partial \mu_c}{\partial \omega}\right]\overline{w} \\ &\quad - \frac{\partial \mu_c}{\partial \xi_a}\left[B\xi_b + A\right] - \frac{\partial \mu_c}{\partial \omega}\Omega \end{aligned} \tag{8.54}$$

where $\overline{w} := [w^{\mathrm{T}}\ d^{\mathrm{T}}]^{\mathrm{T}}$. Now let C_b be a symmetric positive definite matrix and consider the Lyapunov function

$$V(\xi_a, \xi_b, \omega) = V_c(\xi_a, \omega) + z_b^{\mathrm{T}} C_b z_b \tag{8.55}$$

We compute the derivative $\dot{V}$ from (8.52) and (8.54) as follows:

$$\begin{aligned} \dot{V} &= \dot{V}_c\Big|_{\substack{\xi_b = \mu_c \\ \dot{\omega} = \Omega_c + d}} + \frac{\partial U}{\partial \xi_a} B z_b + \frac{\partial U}{\partial \omega}\left[\Omega - \Omega_c\right] \\ &\quad + 2\left[\phi(y_s) - \omega\right]^{\mathrm{T}} E\left[\Omega_c - \Omega\right] + 2 z_b^{\mathrm{T}} C_b \dot{z}_b \\ &= \dot{V}_c\Big|_{\substack{\xi_b = \mu_c \\ \dot{\omega} = \Omega_c + d}} + \frac{\partial U}{\partial \omega}\left[\Omega - \Omega_c\right] + 2\left[\phi(y_s) - \omega\right]^{\mathrm{T}} E\left[\Omega_c - \Omega\right] \\ &\quad + 2 z_b^{\mathrm{T}} C_b\left[\mu + A_e + P_e\phi(y) + Q_e\overline{w} - \frac{\partial \mu_c}{\partial \omega}\Omega\right] \end{aligned} \tag{8.56}$$

where

$$P_e := \left[P_b - \frac{\partial \mu_c}{\partial \xi_a} P_a \right] \tag{8.57}$$

$$Q_e := \left[\left(Q_b - \frac{\partial \mu_c}{\partial \xi_a} Q_a \right) \;\middle|\; -\frac{\partial \mu_c}{\partial \omega} \right] \tag{8.58}$$

$$A_e := \frac{1}{2} C_b^{-1} B^{\mathrm{T}} \left[\frac{\partial U}{\partial \xi_a} \right]^{\mathrm{T}} - \frac{\partial \mu_c}{\partial \xi_a} \left[B\xi_b + A \right] \tag{8.59}$$

are smooth functions which do not depend on ϕ or $\overline{w}$. Let us now bound some of the uncertain terms in (8.56). It follows from U2 that given a symmetric positive definite matrix D_1 and a constant $r > 0$, there exists a smooth function $N_b(\xi_a, \xi_b, \omega, y_s)$ such that

$$2\, z_b^{\mathrm{T}} C_b P_e \left[\phi(y) - \phi(y_s) \right] \;\le\; 2\, z_b^{\mathrm{T}} C_b N_b z_b \,+\, \frac{1}{r} z_1^{\mathrm{T}} D_1 z_1 \tag{8.60}$$

where $z_1 := (y - y_s) = (J\xi_a - y_s)$. Also, from Young's inequality,

$$2 z_b^{\mathrm{T}} C_b Q_e \overline{w} \;\le\; \frac{1}{\varepsilon}\, z_b^{\mathrm{T}} C_b Q_e Q_e^{\mathrm{T}} C_b z_b \,+\, \varepsilon |\overline{w}|^2 \tag{8.61}$$

for any $\varepsilon > 0$. Substituting (8.60) and (8.61) into (8.56), we obtain

$$\begin{aligned} \dot{V} \;\le\; & \dot{V}_c\Big|_{\substack{\xi_b = \mu_c \\ \dot{\omega} = \Omega_c + d}} + \frac{\partial U}{\partial \omega} \left[\Omega - \Omega_c \right] + 2\left[\phi(y_s) - \omega \right]^{\mathrm{T}} E \left[\Omega_c - \Omega \right] \\ & + 2 z_b^{\mathrm{T}} C_b \left[N_b z_b + \frac{1}{2\varepsilon} Q_e Q_e^{\mathrm{T}} C_b z_b + \mu + A_e + P_e \phi(y_s) - \frac{\partial \mu_c}{\partial \omega} \Omega \right] \\ & + \frac{1}{r} z_1^{\mathrm{T}} D_1 z_1 \,+\, \varepsilon |\overline{w}|^2 \end{aligned} \tag{8.62}$$

Let us now choose the function $\Omega(\xi_a, \xi_b, \omega, y_s)$ to be

$$\Omega \;=\; \Omega_c + E^{-1} P_e^{\mathrm{T}} C_b z_b \tag{8.63}$$

which is smooth and does not depend on ϕ or $\overline{w}$. Substituting (8.63) into (8.62), we obtain

$$\begin{aligned} \dot{V} \;\le\; & \dot{V}_c\Big|_{\substack{\xi_b = \mu_c \\ \dot{\omega} = \Omega_c + d}} + \frac{\partial U}{\partial \omega} E^{-1} P_e^{\mathrm{T}} C_b z_b \,+\, \frac{1}{r} z_1^{\mathrm{T}} D_1 z_1 \,+\, \varepsilon |\overline{w}|^2 \\ & + 2 z_b^{\mathrm{T}} C_b \Big[N_b z_b + \frac{1}{2\varepsilon} Q_e Q_e^{\mathrm{T}} C_b z_b + \mu + A_e + P_e \omega \\ & \qquad - \frac{\partial \mu_c}{\partial \omega} \left(\Omega_c + E^{-1} P_e^{\mathrm{T}} C_b z_b \right) \Big] \end{aligned} \tag{8.64}$$

Finally, we choose the function $\mu(\xi_a, \xi_b, \omega, y_s)$ to be

$$\begin{aligned}\mu \;=\; & -\left[\frac{1}{2}C_b^{-1}D_b + \frac{1}{2\varepsilon}Q_eQ_e^{\mathrm{T}}C_b + N_b\right]z_b - A_e - P_e\omega \\ & + \frac{\partial\mu_c}{\partial\omega}\left(\Omega_c + E^{-1}P_e^{\mathrm{T}}C_bz_b\right) - \frac{1}{2}P_eE^{-1}\left[\frac{\partial U}{\partial\omega}\right]^{\mathrm{T}} \end{aligned} \tag{8.65}$$

where D_b is a design parameter. This function μ is smooth and does not depend on ϕ or $\overline{w}$, and upon substitution into (8.64) yields

$$\dot{V}_b \;\leq\; \dot{V}_c\Big|_{\substack{\xi_b=\mu_c \\ \dot{\omega}=\Omega_c+d}} - z_b^{\mathrm{T}}D_bz_b + \frac{1}{r}z_1^{\mathrm{T}}D_1z_1 + \varepsilon|\overline{w}|^2 \tag{8.66}$$

To summarize, we have found smooth functions μ and Ω such that derivative of V along solutions of the complete system (8.42)–(8.43) is related to the conceptual derivative of V_c along solutions of the ξ_a-subsystem (8.48)–(8.49) through this inequality (8.66). This construction of μ and Ω will be the basis of the controller design for $r > 1$.

8.2.5 Controller design for $r \geq 2$

In Section 8.2.3 we constructed smooth functions μ_1 and Ω_1 such that the closed-loop system

$$\dot{\xi}_1 \;=\; \mu_1(\xi_1, \omega, y_s) + P_1(\xi_1)\,\phi(y) + Q_1(\xi_1)\,w \tag{8.67}$$

$$\begin{aligned}\dot{\omega} \;&=\; \Omega_1(\xi_1, \omega, y_s) + d \\ y \;&=\; \xi_1\end{aligned} \tag{8.68}$$

satisfies Q1–Q4. Specifically, the derivative of the Lyapunov function

$$V_1(\xi_1, \omega) \;:=\; z_1^{\mathrm{T}}C_1z_1 + \left[\phi(y_s) - \omega\right]^{\mathrm{T}}E\left[\phi(y_s) - \omega\right] \tag{8.69}$$

along solutions to (8.67)–(8.68) satisfies

$$\dot{V}_1 \;\leq\; -2z_1^{\mathrm{T}}D_1z_1 - \left[\phi(y_s) - \omega\right]^{\mathrm{T}}M\left[\phi(y_s) - \omega\right] + \delta + \varepsilon|\overline{w}|^2 \tag{8.70}$$

and, when $d = 0$,

$$\dot{V}_1 \;\leq\; -2z_1^{\mathrm{T}}D_1z_1 - 2\left[\phi(y_s) - \omega\right]^{\mathrm{T}}L(\omega, y_s) + \varepsilon|w|^2 \tag{8.71}$$

Let us now suppose $r = 2$ and consider the task of constructing a controller for the system

$$\dot{\xi}_1 = \begin{bmatrix} \xi_2 \\ v_1 \end{bmatrix} + P_1(\xi_1)\,\phi(y) + Q_1(\xi_1)\,w \tag{8.72}$$

$$\begin{aligned} \dot{\xi}_2 &= v_2 + P_2(\xi_1,\xi_2)\,\phi(y) + Q_2(\xi_1,\xi_2)\,w \\ y &= \xi_1 \end{aligned} \tag{8.73}$$

We use the backstepping construction of Section 8.2.4 and base our design on the knowledge of the functions μ_1 and Ω_1 above. If the first component v_1 of the control variable is not vacuous, we assign it to be $v_1 = [0\;\; I]\mu_1$. We have left to construct smooth functions μ_2 and Ω_2 such that the closed-loop system

$$\dot{\xi}_1 = \begin{bmatrix} I \\ 0 \end{bmatrix}\xi_2 + \begin{bmatrix} 0 & 0 \\ 0 & I \end{bmatrix}\mu_1(\xi_1,\omega,y_s) + P_1(\xi_1)\,\phi(y) + Q_1(\xi_1)\,w \tag{8.74}$$

$$\begin{aligned} \dot{\xi}_2 &= \mu_2(\xi_1,\xi_2,\omega,y_s) + P_2(\xi_1,\xi_2)\,\phi(y) + Q_2(\xi_1,\xi_2)\,w \\ \dot{\omega} &= \Omega_2(\xi_1,\xi_2,\omega,y_s) + d \\ y &= \xi_1 \end{aligned} \tag{8.75}$$

has properties Q1–Q4. This system (8.74)–(8.75) is in the form (8.42)–(8.43) with $\xi_a = \xi_1$, $\xi_b = \xi_2$, $\mu = \mu_2$, $\Omega = \Omega_2$, etc. Moreover, with $V_c = V_1$, $\mu_c = [I\;\; 0]\mu_1$, and $\Omega_c = \Omega_1$, it follows from (8.70) and (8.71) that the conceptual derivative of V_1 satisfies

$$\dot{V}_1\bigg|_{\substack{\xi_2 = [I\;\;0]\mu_1 \\ \dot{\omega} = \Omega_1 + d}} \leq -2z_1^{\mathrm{T}}D_1z_1 - \big[\phi(y_s) - \omega\big]^{\mathrm{T}}M\,\big[\phi(y_s) - \omega\big] + \delta + \varepsilon|\overline{w}|^2 \tag{8.76}$$

and, when $d = 0$,

$$\dot{V}_1\bigg|_{\substack{\xi_2 = [I\;\;0]\mu_1 \\ \dot{\omega} = \Omega_1 + d}} \leq -2z_1^{\mathrm{T}}D_1z_1 - 2\big[\phi(y_s) - \omega\big]^{\mathrm{T}}L(\omega,y_s) + \varepsilon|w|^2 \tag{8.77}$$

We define the new variable $z_2 := \xi_2 - [I\;\; 0]\mu_1(\xi_1,\omega,y_s)$ which corresponds to z_b defined in (8.53), and we consider the Lyapunov function

$$V_2(\xi_1,\xi_2,\omega) = V_1(\xi_1,\omega) + z_2^{\mathrm{T}}C_2z_2 \tag{8.78}$$

which corresponds to V in (8.55), where C_2 is a symmetric positive definite design parameter. We then follow the construction of Section 8.2.4 to obtain smooth functions $\mu_2\,(=\mu)$ and $\Omega_2\,(=\Omega)$ such that the derivative of V_2 along solutions to (8.74)–(8.75) satisfies, from (8.66),

$$\dot{V}_2 \;\le\; \dot{V}_1\Big|_{\substack{\xi_2=[I\;\;0]\mu_1 \\ \dot{\omega}=\Omega_1+d}} - z_2^{\mathrm{T}} D_2 z_2 + \frac{1}{r} z_1^{\mathrm{T}} D_1 z_1 + \varepsilon|\overline{w}|^2 \tag{8.79}$$

where D_2 is a symmetric positive definite design parameter. It follows from (8.76)–(8.77) that

$$\begin{aligned}\dot{V}_2 \;\le\; & -\left[2-\frac{1}{r}\right] z_1^{\mathrm{T}} D_1 z_1 - z_2^{\mathrm{T}} D_2 z_2 \\ & - \left[\phi(y_s)-\omega\right]^{\mathrm{T}} M \left[\phi(y_s)-\omega\right] + \delta + 2\varepsilon|\overline{w}|^2\end{aligned} \tag{8.80}$$

and, when $d=0$,

$$\begin{aligned}\dot{V}_2 \;\le\; & -\left[2-\frac{1}{r}\right] z_1^{\mathrm{T}} D_1 z_1 - z_2^{\mathrm{T}} D_2 z_2 \\ & - 2\left[\phi(y_s)-\omega\right]^{\mathrm{T}} L(\omega,y_s) + 2\varepsilon|w|^2\end{aligned} \tag{8.81}$$

It follows (see Section 8.2.6 below) that properties Q1–Q4 are satisfied for the $r=2$ system (8.72)–(8.73).

Before we move on to the designs for $r\ge 3$, let us take a closer look at the controller dynamics we obtain when $r=2$. From (8.33), (8.57), and (8.63) we see that these dynamics are

$$\dot{\omega} = E^{-1}L(\omega,y_s) + E^{-1}P_1^{\mathrm{T}}C_1 z_1 + E^{-1}\left[P_2 - \frac{\partial\mu_1}{\partial\xi_1}P_1\right]^{\mathrm{T}} C_2 z_2 + d \tag{8.82}$$

Recall that the term $E^{-1}P_1^{\mathrm{T}}C_1 z_1$ represents nonlinear integral action on the tracking error z_1, and that L represents a modification of this integral action when $\omega\notin\Phi(y_s)$. The third term in (8.82) represents integral action on the new error variable z_2. As can be seen from (8.63), such integral action is added at each step of the recursive design.

The designs for $r\ge 3$ can be obtained in the same manner as above through the recursive application of the backstepping construction of Section 8.2.4. For example, for $r=3$ we would assign $\xi_a = [\xi_1^{\mathrm{T}}\;\;\xi_2^{\mathrm{T}}]^{\mathrm{T}}$,

$\xi_b = \xi_3$, $\mu_c = \mu_2$, $\Omega_c = \Omega_2$, $V_c = V_2$, etc. We would define a new variable $z_3 := \xi_3 - [I \;\; 0]\mu_2(\xi_1, \xi_2, \omega, y_s)$ and consider the Lyapunov function

$$V_3(\xi_1, \xi_2, \xi_3, \omega) = V_2(\xi_1, \xi_2, \omega) + z_3^{\mathrm{T}} C_3 z_3 \tag{8.83}$$

which corresponds to V in (8.55), where C_3 is a symmetric positive definite design parameter. We would then obtain smooth functions μ_3 $(= \mu)$ and Ω_3 $(= \Omega)$ such that

$$\begin{aligned} \dot{V}_3 \leq & -\left[2 - \frac{2}{r}\right] z_1^{\mathrm{T}} D_1 z_1 - z_2^{\mathrm{T}} D_2 z_2 - z_3^{\mathrm{T}} D_3 z_3 \\ & - \left[\phi(y_s) - \omega\right]^{\mathrm{T}} M \left[\phi(y_s) - \omega\right] + \delta + 3\varepsilon|\overline{w}|^2 \end{aligned} \tag{8.84}$$

and, when $d = 0$,

$$\begin{aligned} \dot{V}_3 \leq & -\left[2 - \frac{2}{r}\right] z_1^{\mathrm{T}} D_1 z_1 - z_2^{\mathrm{T}} D_2 z_2 - z_3^{\mathrm{T}} D_3 z_3 \\ & - 2\left[\phi(y_s) - \omega\right]^{\mathrm{T}} L(\omega, y_s) + 3\varepsilon|w|^2 \end{aligned} \tag{8.85}$$

where D_3 is a symmetric positive definite design parameter. At the final step of the general recursive design for $r > 3$, we will have a complete Lyapunov function

$$V(\xi, \omega) = z^{\mathrm{T}} C z + \left[\phi(y_s) - \omega\right]^{\mathrm{T}} E \left[\phi(y_s) - \omega\right] \tag{8.86}$$

where $C = \mathrm{diag}\{C_1, \ldots, C_r\}$ and $z := [z_1^{\mathrm{T}} \;\; \cdots \;\; z_r^{\mathrm{T}}]^{\mathrm{T}}$. The variable z consists of the tracking error $z_1 := (y - y_s) = (\xi_1 - y_s)$ together with the auxiliary error variables

$$z_{i+1} := \xi_{i+1} - [I \;\; 0]\mu_i(\xi_1, \ldots, \xi_i, \omega, y_s) \tag{8.87}$$

The derivative $\dot{V}$ along closed-loop trajectories will satisfy

$$\dot{V} \leq -z^{\mathrm{T}} D z - \left[\phi(y_s) - \omega\right]^{\mathrm{T}} M \left[\phi(y_s) - \omega\right] + \delta + r\varepsilon|\overline{w}|^2 \tag{8.88}$$

and, when $d = 0$,

$$\dot{V} \leq -z^{\mathrm{T}} D z - 2\left[\phi(y_s) - \omega\right]^{\mathrm{T}} L(\omega, y_s) + r\varepsilon|w|^2 \tag{8.89}$$

where $D = \mathrm{diag}\{D_1, \ldots, D_r\}$. This completes the controller design.

8.2.6 Proof of the main result

In the previous sections, we have constructed a controller (8.12)–(8.13) for the system (8.5)–(8.6) together with a Lyapunov function (8.86) whose derivative satisfies (8.88)–(8.89). From standard Lyapunov analysis we conclude that $z(t)$ and $\omega(t)$ exist and are bounded for all $t \geq 0$. It then follows from the definition of each variable z_i that $\xi(t)$ is bounded, and we conclude that Q1 is satisfied. From L1, L2, and (8.89) we see that

$$\overline{w} = 0 \implies \dot{V} \leq 0 \tag{8.90}$$

$$\overline{w} = 0 \text{ and } (z,\omega) \notin \{0\} \times \Phi(y_s) \implies \dot{V} < 0 \tag{8.91}$$

It follows from [75, Theorem 4.8] that $z(t) \to 0$ and $\omega(t) \to \Phi(y_s)$ as $t \to \infty$, and because z_1 is the tracking error we see that Q2 is satisfied. Next, from (8.86) and (8.88) we see that there exists $\sigma > 0$, depending only on the design parameters C, D, E, and M, such that

$$\dot{V} \leq -\sigma V + \delta + r\varepsilon|\overline{w}|^2 \tag{8.92}$$

If we define $V(t) := V(\xi(t), \omega(t))$, then (8.92) implies that $\sigma V(t) \leq \max\left\{\sigma V(0),\, \delta + r\varepsilon\|\overline{w}\|_\infty^2\right\}$ for all $t \geq 0$. Now $z_1{}^{\mathrm{T}} C_1 z_1 \leq V$ from (8.86), and it follows that

$$|z_1(t)|^2 \leq \frac{V(0)}{\lambda_{min}(C_1)} + \frac{\delta}{\sigma\lambda_{min}(C_1)} + \frac{r\varepsilon}{\sigma\lambda_{min}(C_1)}\|\overline{w}\|_\infty^2 \tag{8.93}$$

for all $t \geq 0$, where $\lambda_{min}(C_1)$ denotes the minimum eigenvalue of C_1. We conclude that

$$\|y - y_s\|_\infty \leq \sqrt{\frac{V(0)}{\lambda_{min}(C_1)} + \frac{\delta}{\sigma\lambda_{min}(C_1)}} + \sqrt{\frac{r\varepsilon}{\sigma\lambda_{min}(C_1)}}\|\overline{w}\|_\infty \tag{8.94}$$

which means Q3 is satisfied. Finally, suppose $d(t) \equiv 0$ and $w(t) \in L_2$. From L1, L2, and (8.89) we have $\dot{V} \leq -z_1{}^{\mathrm{T}} D_1 z_1 + r\varepsilon|w|^2$, which means

$$\lambda_{min}(D_1)\int_0^t |z_1(\tau)|^2\, d\tau \leq V(0) - V(t) + r\varepsilon\int_0^t |w(\tau)|^2\, d\tau \tag{8.95}$$

for all $t \geq 0$. Taking limits as $t \to \infty$, we obtain

$$\|y - y_s\|_2 \leq \sqrt{\frac{V(0)}{\lambda_{min}(D_1)}} + \sqrt{\frac{r\varepsilon}{\lambda_{min}(D_1)}}\|w\|_2 \tag{8.96}$$

which means Q4 is satisfied. Note that the coefficients of the disturbance norms in (8.94) and (8.96) can be reduced arbitrarily by the choice of the design parameter ε. This completes the proof of Theorem 8.1.

8.3 Design example

We consider the task of controlling the speed of a fan driven by a DC motor. The system model is

$$J\dot{\nu} = \kappa_1 I - \tau_L - \tau_D(\nu) \tag{8.97}$$
$$L\dot{I} = u - \kappa_2\nu - RI$$
$$y = \nu \tag{8.98}$$

where ν is the fan speed, I is the armature current, u is the armature voltage (our control input), J, κ_1, κ_2, L, and R are known positive constants, τ_L is an uncertain constant load torque, and $\tau_D(\nu)$ is an uncertain, speed-dependent drag torque. We define $\xi_1 = \nu$ and $\xi_2 = (\kappa_1/J)\,I$ and apply a preliminary linear feedback $u = \kappa_2\nu + RI + (JL/\kappa_1)\,v$ to obtain

$$\dot{\xi}_1 = \xi_2 - \frac{1}{J}\big[\tau_L + \tau_D(y)\big] \tag{8.99}$$
$$\dot{\xi}_2 = v$$
$$y = \xi_1 \tag{8.100}$$

where v is a new control variable. We assume that $|\tau_L| \leq J\tau$ for some known constant τ. We also assume that the drag torque $\tau_D(y)$ is monotone increasing with $\tau_D(0) = 0$ (more drag at higher speeds, no drag at rest) and exhibits growth in y somewhere between linear and cubic growth. To summarize,

$$\frac{1}{J}\tau_L \in \big[-\tau,\, \tau\big] \tag{8.101}$$
$$\frac{1}{J}\,y\,\tau_D(y) \in \big[ay^2,\; ay^2 + by^4\big] \tag{8.102}$$

for known positive constants τ, a, and b. If we define the uncertain nonlinearity

$$\phi(y) := -\frac{1}{J}\big[\tau_L + \tau_D(y)\big] \tag{8.103}$$

then we can use (8.101)–(8.102) to construct a set-valued map $\Phi(y)$ which satisfies property U1:

$$\begin{aligned}\Phi(y) &= \Big[-\tau - ay - \max\{0, by^3\},\; \tau - ay - \min\{0, by^3\}\Big] \\ &:= \big[\psi_1(y),\; \psi_2(y)\big]\end{aligned} \tag{8.104}$$

We also assume knowledge of the function ρ in U2, which is equivalent to the knowledge of a bound on the local Lipschitz constant of $\tau_D(y)$ near every point y.

The first step in our design is to construct a conceptual controller for the ξ_1-subsystem: we seek smooth functions $\mu_1(\xi_1, \omega, y_s)$ and $\Omega_1(\xi_1, \omega, y_s)$ such that the conceptual closed-loop system

$$\dot{\xi}_1 = \mu_1(\xi_1, \omega, y_s) + \phi(y) \tag{8.105}$$

$$\dot{\omega} = \Omega_1(\xi_1, \omega, y_s) + d$$

$$y = \xi_1 \tag{8.106}$$

has properties Q1–Q4, where $\omega \in I\!R$ is the internal state of our controller. Note that our controller is of dynamic order one because ϕ is a scalar function. We consider the Lyapunov function V_1 from (8.25), namely,

$$V_1(\xi_1, \omega) = C_1 z_1^2 + E\big[\phi(y_s) - \omega\big]^2 \tag{8.107}$$

where $z_1 := (y - y_s)$ denotes the tracking error. Computing the conceptual Lyapunov derivative $\dot{V}_1$ along solutions of (8.105)–(8.106), we obtain

$$\dot{V}_1\Bigg|_{\substack{\xi_2 = \mu_1 \\ \dot{\omega} = \Omega_1 + d}} = 2C_1 z_1\Big(\mu_1 + \omega + \big[\phi(y) - \phi(y_s)\big]\Big) + 2\big[\phi(y_s) - \omega\big]\big[C_1 z_1 - E\Omega_1 - Ed\big] \tag{8.108}$$

Recall that, because of physical considerations, the function $-\phi$ is monotone increasing; as a result, we have $z_1\big[\phi(y) - \phi(y_s)\big] \leq 0$ which means

$$\dot{V}_1\Bigg|_{\substack{\xi_2 = \mu_1 \\ \dot{\omega} = \Omega_1 + d}} \leq 2C_1 z_1\big[\mu_1 + \omega\big] + 2\big[\phi(y_s) - \omega\big]\big[C_1 z_1 - E\Omega_1 - Ed\big] \tag{8.109}$$

We can therefore choose

$$\mu_1(\xi_1, \omega, y_s) = -\frac{1}{C_1}\sigma(z_1) - \omega \tag{8.110}$$

$$\Omega_1(\xi_1, \omega, y_s) = \frac{1}{E}L(\omega, y_s) + \frac{C_1}{E}z_1 \tag{8.111}$$

to obtain

$$\dot{V}_1\Big|_{\substack{\xi_2 = \mu_1 \\ \dot{\omega} = \Omega_1 + d}} \leq -2z_1\sigma(z_1) - 2\Big[\phi(y_s) - \omega\Big]\Big[L(\omega, y_s) + Ed\Big] \quad (8.112)$$

where the functions σ and L are yet to be chosen. In the design of Section 8.2.3, we chose the linear function $\sigma(z_1) = D_1 z_1$ when we constructed μ_1 in (8.32). For this example, however, we will choose σ to be a saturation function so that its derivative $d\sigma/dz_1$ has compact support; namely, we will choose a smooth function σ such that

$$\begin{aligned} z_1\sigma(z_1) &= \sigma_1|z_1| \quad \text{when } |z_1| \geq (\sigma_1/\sigma_2) \\ z_1\sigma(z_1) &\geq \sigma_2 z_1^2 \quad \text{when } |z_1| \leq (\sigma_1/\sigma_2) \end{aligned} \quad (8.113)$$

for design parameters $\sigma_1 > 0$ and $\sigma_2 > 0$ (possibly depending on y_s). We choose L to satisfy L1–L3:

$$L(\omega, y_s) = \begin{cases} -\alpha\Big|\omega - \psi_2(y_s)\Big| & \text{when } \psi_2(y_s) \leq \omega \\ 0 & \text{when } \psi_1(y_s) \leq \omega \leq \psi_2(y_s) \\ \alpha\Big|\omega - \psi_1(y_s)\Big| & \text{when } \omega \leq \psi_1(y_s) \end{cases} \quad (8.114)$$

for some design parameter $\alpha > 0$, where the functions ψ_i come from the definition of Φ in (8.104). This choice for L is C^0 rather than C^∞ which is sufficient for this example, but in general for $r > 2$ one would choose a C^{r-2} function L by replacing the absolute value in (8.114) by something smoother.

In the second and final step of the design, we construct a controller for the actual system (8.99)–(8.100), namely, we construct smooth functions $\mu_2(\xi_1, \xi_2, \omega, y_s)$ and $\Omega_2(\xi_1, \xi_2, \omega, y_s)$ such that the system

$$\begin{aligned} \dot{\xi}_1 &= \xi_2 + \phi(y) && (8.115) \\ \dot{\xi}_2 &= \mu_2(\xi_1, \xi_2, \omega, y_s) \\ \dot{\omega} &= \Omega_2(\xi_1, \xi_2, \omega, y_s) + d \\ y &= \xi_1 && (8.116) \end{aligned}$$

has properties Q1–Q4. We define $z_2 := \xi_2 - \mu_1(\xi_1, \omega, y_s)$ and consider the Lyapunov function

$$V_2(\xi_1, \xi_2, \omega) = V_1(\xi_1, \omega) + C_2 z_2^2 \quad (8.117)$$

Computing the derivative $\dot{V}_2$ along solutions of (8.115)–(8.116), we obtain

$$\dot{V}_2 \leq \dot{V}_1\Big|_{\substack{\xi_2=\mu_1 \\ \dot{\omega}=\Omega_1+d}} + 2C_1 z_1 z_2 + 2E\big[\phi(y_s)-\omega\big]\big[\Omega_1-\Omega_2\big] + 2C_2 z_2 \dot{z}_2$$

$$\leq \dot{V}_1\Big|_{\substack{\xi_2=\mu_1 \\ \dot{\omega}=\Omega_1+d}} + 2E\big[\phi(y_s)-\omega\big]\big[\Omega_1-\Omega_2\big]$$

$$+ 2C_2 z_2\left[\frac{C_1}{C_2} z_1 + \mu_2 + \frac{1}{C_1}\frac{d\sigma}{dz_1}\big[\xi_2+\phi(y)\big] + \Omega_2 + d\right] \quad (8.118)$$

We choose

$$\Omega_2(\xi_1,\xi_2,\omega,y_s) = \Omega_1(\xi_1,\omega,y_s) + \frac{C_2}{C_1 E}\frac{d\sigma}{dz_1} z_2 \quad (8.119)$$

which upon substitution into (8.118) yields

$$\dot{V}_2 \leq \dot{V}_1\Big|_{\substack{\xi_2=\mu_1 \\ \dot{\omega}=\Omega_1+d}} + 2C_2 z_2\Bigg[\frac{C_1}{C_2} z_1 + \mu_2$$

$$+ \frac{1}{C_1}\frac{d\sigma}{dz_1}\big[\xi_2+\phi(y)-\phi(y_s)+\omega\big]$$

$$+ \Omega_1 + \frac{C_2}{C_1 E}\frac{d\sigma}{dz_1} z_2 + d\Bigg] \quad (8.120)$$

We have left to choose the function μ_2. If we choose

$$\mu_2(\xi_1,\xi_2,\omega,y_s) = -\left[\frac{D_2}{2C_2} + \frac{C_2}{2\varepsilon} + N_2\right] z_2 - \frac{C_1}{C_2} z_1$$

$$- \frac{1}{C_1}\frac{d\sigma}{dz_1}\big[\xi_2+\omega\big] - \Omega_1 - \frac{C_2}{C_1 E}\frac{d\sigma}{dz_1} z_2 \quad (8.121)$$

for positive design parameters D_2, ε, and N_2, then (8.120) becomes

$$\dot{V}_2 \leq -2z_1\sigma(z_1) - 2\big[\phi(y_s)-\omega\big]\big[L(\omega,y_s)+Ed\big] - D_2 z_2^2$$

$$- 2C_2 N_2 z_2^2 + \varepsilon d^2 + 2z_2\frac{C_2}{C_1}\frac{d\sigma}{dz_1}\big[\phi(y)-\phi(y_s)\big] \quad (8.122)$$

where we have substituted the inequality (8.112).

When $|z_1| \geq (\sigma_1/\sigma_2)$ we have $d\sigma/dz_1 = 0$ by our choice of the saturation function σ, and it follows from (8.113) and (8.122) that

$$\dot{V}_2 \leq -2\sigma_1|z_1| - 2\big[\phi(y_s)-\omega\big]\big[L(\omega,y_s)+Ed\big] - D_2 z_2^2 + \varepsilon d^2 \quad (8.123)$$

for $|z_1| \geq (\sigma_1/\sigma_2)$. On the other hand, when $|z_1| \leq (\sigma_1/\sigma_2)$, from (8.113) and (8.122) we have

$$\begin{aligned} \dot{V}_2 \;\leq\; & -2\sigma_2 z_1^2 - 2\big[\phi(y_s) - \omega\big]\big[L(\omega, y_s) + Ed\big] \\ & - D_2 z_2^2 - 2C_2 N_2 z_2^2 + \varepsilon d^2 + 2k(y_s)|z_1 z_2| \end{aligned} \tag{8.124}$$

where $k(y_s)$ is a constant which is calculated from the maximum value of $\rho(y, y_s)$ on the set $|z_1| \leq (\sigma_1/\sigma_2)$ (recall that ρ is the function in property U2). It follows that choosing $N_2 = (k^2/2C_2\sigma_2)$ yields

$$\dot{V}_2 \;\leq\; -\sigma_2 z_1^2 - 2\big[\phi(y_s) - \omega\big]\big[L(\omega, y_s) + Ed\big] - D_2 z_2^2 + \varepsilon d^2 \tag{8.125}$$

for $|z_1| \leq (\sigma_1/\sigma_2)$. Thus (8.123) and (8.125) prove that the controller

$$v \;=\; \mu_2(\xi_1, \xi_2, \omega, y_s) \tag{8.126}$$

$$\dot{\omega} \;=\; \Omega_2(\xi_1, \xi_2, \omega, y_s) + d \tag{8.127}$$

for the system (8.99)–(8.100) meets the design objective described by properties Q1–Q4. In particular, because we have included the artificial disturbance d in (8.127), this design is robust to time variations in the uncertain nonlinearity $\tau_D(\nu)$. One can check from (8.119) and (8.121) that this controller (8.126)–(8.127), although nonlinear, satisfies a global Lipschitz condition in ξ_1, ξ_2, and ω. This linear growth was achieved by choosing σ in (8.110) to be a saturation function rather than a linear function. Also, no parameterization of the uncertain nonlinear drag torque $\tau_D(\nu)$ was needed for this design. If such a parameterization were available, an adaptive controller designed using the methods of [72, 84, 85, 115, 157] would be of higher order than our controller (8.126)–(8.127).

8.4 Summary

We have combined the backstepping techniques of Chapters 5 and 7 to obtain a design procedure for systems having uncertain nonlinearities $\phi(x)$ as well as exogenous disturbances w. Controller dynamics supply the integral action required for the asymptotic tracking of constant set-points. The closed-loop system is robust to exogenous disturbances w, small state measurement disturbances, time variations in the uncertain nonlinearity $\phi(x)$, and time variations in the set-point. For the special case in which the uncertainty $\phi(x)$ is simply an unknown constant parameter $\phi(x) \equiv \theta$, our design becomes a robust version of the tuning function adaptive control design in [84, 85].

We illustrated our design method using the example of fan speed control. The drag torque was represented by an unknown nonlinear function of the fan speed, and our controller included robust integral action to account for this unknown nonlinearity.

Appendix: Local $\mathcal{K}$-continuity in metric spaces

In this appendix we describe some new notions of continuity in metric spaces which are related to Lipschitz and Hölder continuity. The results we present include key inequalities used in Chapter 6 for the control of nonlinear systems subject to measurement disturbances. We will introduce *local $\mathcal{K}$-continuity*, a property of mappings between metric spaces which is weaker than local Lipschitz or local Hölder continuity but stronger than continuity. We will show that, unlike local Lipschitz or local Hölder continuity, local $\mathcal{K}$-continuity is a uniform property.[1] Moreover, the most elementary uniform property, namely, uniform continuity, is stronger than local $\mathcal{K}$-continuity. We also introduce a metric property called *$C\mathcal{K}$-continuity* and show that on proper metric spaces,[2] the notions of $C\mathcal{K}$-continuity, local $\mathcal{K}$-continuity, and continuity are all equivalent. The concepts discussed in this appendix are related to the concept of the *modulus of continuity* of a function of a real variable [145].

Terminology in this appendix is from the real analysis textbook [123]. Throughout, we let X, Z, and W denote metric spaces, and we let $\mathbb{R}$ and $\mathbb{R}_+$ denote the sets of real and nonnegative real numbers, respectively. Recall from Lyapunov stability theory [49, 77] that a class $\mathcal{K}$

[1] Uniform properties are those preserved under uniform homeomorphisms. Likewise, topological and metric properties are those preserved under homeomorphisms and isometries, respectively.

[2] A metric space is said to be *proper* when every closed and bounded subset is compact, *locally compact* when every point is contained in an open set with compact closure, and *σ-compact* when it is the countable union of compact sets. A proper metric space is both locally compact and σ-compact.

function is a continuous, strictly increasing function $\gamma : \mathbb{R}_+ \to \mathbb{R}_+$ such that $\gamma(0) = 0$. Also, a class $\mathcal{K}_\infty$ function is a class $\mathcal{K}$ function which is onto $\mathbb{R}_+$ (and which therefore has a class $\mathcal{K}_\infty$ inverse). Given two class $\mathcal{K}$ functions γ_1 and γ_2, we write the order relation $\gamma_1 = o(\gamma_2)$ when $\limsup_{r \downarrow 0} \gamma_1(r)/\gamma_2(r) = 0$.

One of the main consequences of the results in this appendix is that for any continuous function ϕ from a proper metric space X to a metric space Z, there exist a continuous function $\rho : X \to \mathbb{R}_+$ and a class $\mathcal{K}$ function γ such that

$$d(\phi(x), \phi(y)) \;\leq\; \rho(y) \cdot \gamma(d(x, y)) \tag{A.1}$$

for all $x, y \in X$. Furthermore, ρ and γ can be taken to be locally Lipschitz whenever ϕ is locally Lipschitz. We use inequality (A.1) in Chapter 6 to obtain global bounds on the effects of measurement errors in nonlinear control systems. For example, suppose x is the true state of a control system and y is the measured state. Then the distance between the true and measured values of any continuous nonlinearity ϕ can be bounded from above *globally* by a function which depends only on the measured quantity y and the measurement error $d(x, y)$.

A.1 $\mathcal{K}$-continuity

We use class $\mathcal{K}$ functions to generalize two standard notions of continuity in metric spaces, namely, Lipschitz and Hölder continuity. Recall that a function $\phi : X \to Z$ is said to be Lipschitz continuous on a set $E \subset X$ when there exists $M \in \mathbb{R}_+$ such that

$$d(\phi(x), \phi(y)) \;\leq\; M \cdot d(x, y) \tag{A.2}$$

for all $x, y \in E$. The constant M is called a Lipschitz constant of ϕ on E. Lipschitz continuity can be generalized by replacing the distance function $d(x, y)$ in the right-hand side of (A.2) by some power of the distance function: a function $\phi : X \to Z$ is said to be Hölder continuous on $E \subset X$ when there exist $c > 0$ and $M \in \mathbb{R}_+$ such that

$$d(\phi(x), \phi(y)) \;\leq\; M\,[d(x, y)]^c \tag{A.3}$$

for all $x, y \in E$. We obtain a further generalization of these notions by replacing the power of $d(x,y)$ in (A.3) by a class $\mathcal{K}$ function of $d(x,y)$:

Definition A.1 *Let $\gamma \in \mathcal{K}$ and let $E \subset X$. A function $\phi : X \to Z$ is* **γ-continuous on** *E when there exists $M \in I\!R_+$ such that for all $x, y \in E$,*

$$d(\phi(x), \phi(y)) \;\leq\; M \cdot \gamma(d(x,y)) \tag{A.4}$$

The number M is called a **γ-constant of** *ϕ* **on** *E. We say that a function ϕ is* **$\mathcal{K}$-continuous on** *E when there exists $\gamma \in \mathcal{K}$ such that ϕ is γ-continuous on E.*

We can characterize the class of $\mathcal{K}$-continuous functions as follows:

Lemma A.2 *$\phi : X \to Z$ is $\mathcal{K}$-continuous on $E \subset X$ if and only if*

(i) *ϕ is uniformly continuous on E, and*

(ii) *for every $\varepsilon > 0$ there exists $\delta > 0$ such that for all $x, y \in E$ satisfying $d(x,y) \leq \varepsilon$ we have $d(\phi(x), \phi(y)) \leq \delta$.*

Proof: First suppose there exist $\gamma \in \mathcal{K}$ and a constant $M \in I\!R_+$ such that for all $x, y \in E$ we have $d(\phi(x), \phi(y)) \leq M \cdot \gamma(d(x,y))$. Fix $\varepsilon > 0$ and let $\delta > 0$ be such that $M \cdot \gamma(\delta) \leq \varepsilon$; then $d(x,y) \leq \delta$ implies $d(\phi(x), \phi(y)) \leq M \cdot \gamma(d(x,y)) \leq \varepsilon$ and thus ϕ is uniformly continuous on E. Next, fix $\varepsilon > 0$ and let $\delta = M \cdot \gamma(\varepsilon)$; then $d(x,y) \leq \varepsilon$ implies $d(\phi(x), \phi(y)) \leq M \cdot \gamma(d(x,y)) \leq \delta$ and thus ϕ satisfies (ii). Now suppose ϕ satisfies (i) and (ii). For each $r \in I\!R_+$ we define

$$\mu(r) := \sup\Bigl\{ d(\phi(x), \phi(y)) \;:\; x, y \in E,\; d(x,y) \leq r \Bigr\}$$

It follows from (ii) that $\mu : I\!R_+ \to I\!R_+$ has finite values. Also, μ is increasing with $\mu(0) = 0$, and by the uniform continuity of ϕ we have $\mu(r) \to 0$ as $r \to 0$. It follows that there exists $\gamma \in \mathcal{K}$ such that $\mu(r) \leq \gamma(r)$ for all $r \geq 0$. Let $x, y \in E$; then $d(\phi(x), \phi(y)) \leq \mu(d(x,y)) \leq \gamma(d(x,y))$, which means ϕ is γ-continuous on E. ∎

Note that property (ii) in Lemma A.2 looks like the definition of uniform continuity, except that the roles of ε and δ are reversed. Every bounded function satisfies property (ii). The following example shows that there are uniformly continuous functions which do *not* satisfy property (ii) (even when the domain space is proper), and furthermore that $\mathcal{K}$-continuity is not a uniform property:

Example A.3 Let X^* denote the metric space generated by the following metric for the set of real numbers:

$$d^*(x,y) \quad := \quad \frac{|x-y|}{1+|x-y|} + |x^{\frac{1}{3}} - y^{\frac{1}{3}}| \tag{A.5}$$

We first show that X^* is uniformly equivalent to $I\!R$. Fix $\varepsilon > 0$. By the uniform continuity of the cube root, there exists $\delta_0 > 0$ such that $|x-y| \le \delta_0$ implies $|x^{\frac{1}{3}} - y^{\frac{1}{3}}| \le \frac{1}{2}\varepsilon$. Let $\delta = \min\{\delta_0, \frac{1}{2}\varepsilon/(1+\varepsilon)\}$. First suppose $|x-y| \le \delta$; then $|x-y|/(1+|x-y|) \le |x-y| \le \frac{1}{2}\varepsilon$ and also $|x^{\frac{1}{3}} - y^{\frac{1}{3}}| \le \frac{1}{2}\varepsilon$, and it follows that $d^*(x,y) \le \varepsilon$. Next suppose $d^*(x,y) \le \delta$; then $|x-y|/(1+|x-y|) \le \varepsilon/(1+\varepsilon)$ which means $|x-y| \le \varepsilon$. Therefore X^* and $I\!R$ are uniformly equivalent. We next show that X^* is proper. It is sufficient to show that the set $\{x \in X^* : d^*(x,0) \le c\}$ is compact for every $c \ge 0$. Because X^* is equivalent to $I\!R$, we need only show that the set $\{x \in I\!R : d^*(x,0) \le c\}$ is bounded (in $I\!R$) for every $c \ge 0$, but this is clearly true because $d^*(x,0) \le c$ implies $|x| \le c^3$. Finally, we show that the identity map from X^* to $I\!R$ (which is uniformly continuous on X^* by the uniform equivalence of X^* and $I\!R$) is *not* $\mathcal{K}$-continuous on X^*. Fix $\delta > 0$, let $x > 0$ be such that $\delta^3 \le x^2$, and let $y = x + 2\delta$. From the mean value theorem we have $|x^{\frac{1}{3}} - y^{\frac{1}{3}}| \le \frac{1}{3}x^{-\frac{2}{3}}(2\delta) \le 1$ which implies $d^*(x,y) \le 2$. Thus for every $\delta > 0$ we have found $x, y \in X^*$ such that $d^*(x,y) \le 2$ but $|x-y| > \delta$, and it follows from Lemma A.2 that the identity map from X^* to $I\!R$ is not $\mathcal{K}$-continuous on X^*. ■

In some cases, uniform continuity alone implies $\mathcal{K}$-continuity. For example, if E is totally bounded,[3] then every uniformly continuous function on E is bounded on E and thus $\mathcal{K}$-continuous on E by Lemma A.2. We next show that uniform continuity implies $\mathcal{K}$-continuity on convex subsets of normed linear spaces:

Lemma A.4 *Let E be a convex subset of a normed linear space V. Then a function $\phi : V \to Z$ is uniformly continuous on E if and only if it is $\mathcal{K}$-continuous on E.*

Proof: Suppose ϕ is uniformly continuous; then there exists $\delta > 0$ such that for all $\xi, \zeta \in E$ satisfying $\|\xi - \zeta\| \le \delta$ we have $d(\phi(\xi), \phi(\zeta)) \le 1$. Fix

[3] A subset of a metric space is said to be *totally bounded* when for each $\varepsilon > 0$ the subset can be covered by a finite number of balls of radius ε.

$\varepsilon > 0$ and let $N \geq 1$ be an integer such that $\varepsilon \leq N\delta$. Let $x, y \in E$ be such that $\|x-y\| \leq \varepsilon$, and for $i = 0, \ldots, N$ define $\xi_i = x + \frac{i}{N}(y-x)$. Note that $\xi_0 = x$ and $\xi_N = y$, and that $\|\xi_i - \xi_{i-1}\| \leq \delta$ for $i = 1, \ldots, N$. By the convexity of E we have $\xi_i \in E$ for $i = 0, \ldots, N$ from which it follows that $d(\phi(x), \phi(y)) \leq \sum_{i=1}^{N} d(\phi(\xi_i), \phi(\xi_{i-1})) \leq N$. Thus for every $\varepsilon > 0$ we have found $N \geq 1$ such that for all $x, y \in E$ satisfying $\|x - y\| \leq \varepsilon$ we have $d(\phi(x), \phi(y)) \leq N$, which means ϕ satisfies property (ii) of Lemma A.2. We conclude that ϕ is $\mathcal{K}$-continuous on E. The converse is immediate from Lemma A.2. ■

We end this section with a corollary to Lemma A.2:

Corollary A.5 *If $\phi : X \to Z$ is continuous, then ϕ is $\mathcal{K}$-continuous on every compact set $F \subset X$. The converse is true if X is locally compact.*

Proof: Suppose ϕ is continuous and let $F \subset X$ be compact; then ϕ is uniformly continuous and bounded on F, and the result follows from Lemma A.2. Conversely, suppose X is locally compact and fix $x \in X$; then there exists a compact neighborhood U of x. By assumption there exists $\gamma \in \mathcal{K}$ such that ϕ is γ-continuous on U, which in particular implies that ϕ is continuous at x. ■

A.2 Local $\mathcal{K}$-continuity

The properties of Lipschitz and Hölder continuity can be made weaker through localization, and the same is true of $\mathcal{K}$-continuity:

Definition A.6 *Let $\gamma \in \mathcal{K}$. A function $\phi : X \to Z$ is* **locally γ-continuous** *when every $x \in X$ has a neighborhood U such that ϕ is γ-continuous on U. We say that ϕ is* **locally $\mathcal{K}$-continuous** *when there exists $\gamma \in \mathcal{K}$ such that ϕ is locally γ-continuous.*

In contrast to local Lipschitz or Hölder continuity, local $\mathcal{K}$-continuity is a uniform property, that is, it is preserved under uniform homeomorphisms. Before we prove this fact, we give an example of a homeomorphism which is not locally $\mathcal{K}$-continuous and thus establish that local $\mathcal{K}$-continuity is not a topological property:

Example A.7 Let K denote the metric space obtained by assigning the discrete metric to the set of class $\mathcal{K}_\infty$ functions, and let $X = K \times \mathbb{R}_+$ (with the usual product metric). Define $\phi : X \to X$ by the equation $\phi(\gamma, r) := (\gamma, \gamma(r))$. We first show that ϕ is a homeomorphism. Note that ϕ is bijective and its inverse is given by $\phi^{-1}(\gamma, r) = (\gamma, \gamma^{-1}(r))$. Fix $(\gamma_0, r_0) \in X$ and let $\{(\gamma_i, r_i)\}_{i=1}^{\infty}$ be a sequence in X which converges to (γ_0, r_0). Then the sequence $\{\gamma_i\}_{i=1}^{\infty}$ converges to γ_0 in K, which means that $\gamma_i = \gamma_0$ for i sufficiently large. Thus for i large we have $\phi(\gamma_i, r_i) = \phi(\gamma_0, r_i) = (\gamma_0, \gamma_0(r_i))$ which by the continuity of γ_0 converges to $(\gamma_0, \gamma_0(r_0)) = \phi(\gamma_0, r_0)$. Therefore ϕ is continuous at (γ_0, r_0) (and thus on X because (γ_0, r_0) was arbitrary). A similar argument shows that ϕ^{-1} is continuous. We next show that ϕ is not locally $\mathcal{K}$-continuous. Let $\gamma \in \mathcal{K}$; then $\sqrt{\gamma} \in K$. Let $U \subset X$ be a neighborhood of $(\sqrt{\gamma}, 0) \in X$; then U contains a set of the form $U_0 = \{\sqrt{\gamma}\} \times [0, \varepsilon) \subset X$ for some $\varepsilon > 0$. If ϕ were γ-continuous on U_0, then there would exist $M \in \mathbb{R}_+$ such that $\sqrt{\gamma(r)} \leq M \cdot \gamma(r)$ for all $r \in [0, \varepsilon)$, which is impossible. Therefore ϕ is not γ-continuous on U, which means ϕ is not locally γ-continuous. Because γ was arbitrary, we conclude that ϕ is not locally $\mathcal{K}$-continuous. ∎

The next two lemmas show that local $\mathcal{K}$-continuity is a uniform property:

Lemma A.8 *If $\phi : X \to Z$ is uniformly continuous, then ϕ is locally $\mathcal{K}$-continuous.*

Proof: As in the proof of Lemma A.2 above, we define $\mu : \mathbb{R}_+ \to \mathbb{R}_+ \cup \{\infty\}$ by $\mu(r) := \sup\big\{d(\phi(x), \phi(y)) : x, y \in X,\ d(x, y) \leq r\big\}$; note that μ may take the value ∞. Nevertheless, μ is increasing with $\mu(0) = 0$, and by the uniform continuity of ϕ we have $\mu(r) \to 0$ as $r \to 0$. It follows that there exist $r_0 > 0$ and $\gamma \in \mathcal{K}$ such that $\mu(r) \leq \gamma(r)$ for all $r \in [0, 2r_0]$. Fix $x \in X$ and let $U = \{y \in X : d(x, y) < r_0\}$; then U is a neighborhood of x. For any two points $\xi, \zeta \in U$ we have $d(\xi, \zeta) \leq 2r_0$ which means $d(\phi(\xi), \phi(\zeta)) \leq \mu(d(\xi, \zeta)) \leq \gamma(d(\xi, \zeta))$, and it follows that ϕ is γ-continuous on U. The point x was arbitrary, and so ϕ is locally γ-continuous. ∎

Lemma A.9 *If $\phi : X \to Z$ and $\psi : Z \to W$ are locally $\mathcal{K}$-continuous, then the composition $\psi \circ \phi : X \to W$ is locally $\mathcal{K}$-continuous.*

Proof: Let $\gamma_1, \gamma_2 \in \mathcal{K}$ be such that ϕ and ψ are locally γ_1- and γ_2-continuous, respectively. Fix $x \in X$ and let $U_1 \subset X$ be a neighborhood of x such that ϕ has a γ_1-constant $M_1 > 0$ on U_1. Let $U_2 \subset Z$ be a neighborhood of $\phi(x)$ such that ψ has a γ_2-constant $M_2 \geq 0$ on U_2. Now ϕ is continuous and so $\phi^{-1}(U_2) \subset X$ is a neighborhood of x. It follows that $U = U_1 \cap \phi^{-1}(U_2) \cap \{y \in X : \gamma_1(2d(x,y)) < M_1^{-2}\}$ is a neighborhood of x. Let $\xi, \zeta \in U$; then $\gamma_1(d(\xi,\zeta)) \leq \gamma_1(d(\xi,x) + d(\zeta,x)) \leq \gamma_1(2\max\{d(\xi,x), d(\zeta,x)\}) < M_1^{-2}$ and it follows that $M_1 \cdot \gamma_1(d(\xi,\zeta)) \leq \sqrt{\gamma_1(d(\xi,\zeta))}$. If we define a class $\mathcal{K}$ function $\gamma_3 = \gamma_2 \circ \sqrt{\gamma_1}$, then we have $d(\psi\circ\phi(\xi), \psi\circ\phi(\zeta)) \leq M_2 \cdot \gamma_2(d(\phi(\xi),\phi(\zeta))) \leq M_2 \cdot \gamma_2(M_1 \cdot \gamma_1(d(\xi,\zeta))) \leq M_2 \cdot \gamma_3(d(\xi,\zeta))$. The point x was arbitrary, and we conclude that $\psi\circ\phi$ is locally γ_3-continuous. ∎

We next show that if X is locally compact and σ-compact, then every continuous function on X is also locally $\mathcal{K}$-continuous. We begin by proving that every countable collection of class $\mathcal{K}$ functions has an upper bound in $\mathcal{K}$ with respect to the order relation $o(\cdot)$. This observation is itself of independent interest.

Lemma A.10 *For every sequence of class $\mathcal{K}$ functions $\{\gamma_i\}_{i=1}^{\infty}$ there exists $\gamma \in \mathcal{K}$ such that $\gamma_i = o(\gamma)$ for all $i \geq 1$.*

Proof: Suppose the statement is false; then there exists a sequence of class $\mathcal{K}$ functions $\{\gamma_i\}_{i=1}^{\infty}$ such that for every $\gamma \in \mathcal{K}$ there exists $i \geq 1$ such that $\limsup_{r\downarrow 0} \gamma_i(r)/\gamma(r) > 0$. Given $\gamma \in \mathcal{K}$, let $j \geq 1$ be such that $\limsup_{r\downarrow 0} \gamma_j(r)/\sqrt{\gamma(r)} > 0$; it then follows that $\limsup_{r\downarrow 0} \gamma_j(r)/\gamma(r) = \infty$. In other words, for every $\gamma \in \mathcal{K}$ there exists $j \geq 1$ such that $\limsup_{r\downarrow 0} \gamma_j(r)/\gamma(r) = \infty$. We now construct a continuous function $\phi : [0,1] \to [0,1]$ as follows. We set $\phi(0) = 0$, and for $r \in (0,1]$ we set $\phi(r) = \min\{r,\ \gamma_{k+1}(r - \frac{1}{k+1}),\ \gamma_k(\frac{1}{k} - r)\}$ where $k \geq 1$ is the unique integer satisfying $r \in (\frac{1}{k+1}, \frac{1}{k}]$. Clearly ϕ is continuous on $(0,1]$, and because $0 \leq \phi(r) \leq r$ we see that ϕ is also continuous at $r = 0$. Thus from Corollary A.5 there exists $\gamma \in \mathcal{K}$ such that ϕ is γ-continuous on $[0,1]$. From above we know there exists $j \geq 1$ such that $\limsup_{r\downarrow 0} \gamma_j(r)/\gamma(r) = \infty$. Now from the construction of ϕ there exists $\varepsilon > 0$ such that $\phi(r) = \gamma_j(r - \frac{1}{j})$ for all $r \in [\frac{1}{j}, \frac{1}{j} + \varepsilon]$. It then follows from the γ-continuity of ϕ that there exists $M \in I\!R_+$ such that $\gamma_j(r - \frac{1}{j}) \leq M \cdot \gamma(r - \frac{1}{j})$

for $0 \leq r - \frac{1}{j} \leq \varepsilon$. This implies $\limsup_{r\downarrow 0} \gamma_j(r)/\gamma(r) \leq M$, which is a contradiction. ■

Lemma A.11 *Let X be locally compact and σ-compact. If $\phi : X \to Z$ is continuous, then ϕ is locally $\mathcal{K}$-continuous.*

Proof: Because X is locally compact and σ-compact, there exists an exhaustion of X, that is, there exists a sequence $\{U_i\}_{i=1}^{\infty}$ of open subsets of X such that $\overline{U}_i$ is a compact subset of U_{i+1} for all $i \geq 1$ and furthermore $X = \bigcup_{i=1}^{\infty} U_i$. It follows from Corollary A.5 that for each $i \geq 1$ there exists $\gamma_i \in \mathcal{K}$ such that ϕ is γ_i-continuous on U_i. It then follows from Lemma A.10 that there exists $\gamma \in \mathcal{K}$ such that $\gamma_i = o(\gamma)$ for all $i \geq 1$. We now show that ϕ is locally γ-continuous. Fix $x \in X$ and let $i \geq 1$ be such that $x \in U_i$. Now $\overline{U}_i$ is compact which means $\operatorname{diam}(U_i) < \infty$, and so we may define $L = \sup\bigl\{\gamma_i(r)/\gamma(r) \,:\, 0 < r \leq \operatorname{diam}(U_i)\bigr\}$. Because $\gamma_i = o(\gamma)$ we have $0 < L < \infty$. Let $M \in I\!R_+$ be a γ_i-constant of ϕ on U_i; then for any $\xi, \zeta \in U_i$ we have $d(\phi(\xi), \phi(\zeta)) \leq M \cdot \gamma_i(d(\xi,\zeta)) \leq ML \cdot \gamma(d(\xi,\zeta))$ which means ML is a γ-constant of ϕ on U_i. Thus x has a neighborhood U_i such that ϕ is γ-continuous on U_i. The choice for x was arbitrary, and therefore ϕ is locally γ-continuous. ■

We end this section by generalizing the fact that locally Lipschitz functions are Lipschitz on compact sets:

Lemma A.12 *Let $\gamma \in \mathcal{K}$. If $\phi : X \to Z$ is locally γ-continuous, then ϕ is γ-continuous on every compact set $F \subset X$. The converse is true if X is locally compact.*

Proof: Let ϕ be locally γ-continuous, and suppose there exists a compact set $F \subset X$ such that ϕ is not γ-continuous on F. Then for all $i \geq 1$ there exists $(x_i, y_i) \in F \times F$ such that $d(\phi(x_i), \phi(y_i)) > i \cdot \gamma(d(x_i, y_i))$. Now $F \times F$ is compact, so the sequence $\{(x_i, y_i)\}_{i=1}^{\infty}$ in $F \times F$ has a subsequence $\{(x_{i_j}, y_{i_j})\}_{j=1}^{\infty}$ converging to some $(x_0, y_0) \in F \times F$. Because ϕ is continuous and F is compact we have $\operatorname{diam}(\phi(F)) < \infty$, and thus

$$\begin{aligned}\gamma(d(x_0, y_0)) &= \lim_{j\to\infty} \gamma(d(x_{i_j}, y_{i_j})) \leq \lim_{j\to\infty} d(\phi(x_{i_j}), \phi(y_{i_j}))/i_j \\ &\leq \lim_{j\to\infty} \operatorname{diam}(\phi(F))/i_j = 0\end{aligned}$$

It follows that $d(x_0, y_0) = 0$. Now by the local γ-continuity of ϕ there exists a neighborhood U of x_0 such that ϕ is γ-continuous on U. Therefore there exists $M \in I\!R_+$ such that for j sufficiently large we have $d(\phi(x_{i_j}), \phi(y_{i_j})) \leq M \cdot \gamma(d(x_{i_j}, y_{i_j}))$, a contradiction.

Conversely, suppose X is locally compact and ϕ is γ-continuous on every compact set $F \subset X$. By the local compactness of X, every $x \in X$ has a compact neighborhood U, and by assumption ϕ is γ-continuous on U. We conclude that ϕ is locally γ-continuous. ∎

A.3 $\mathcal{CK}$-continuity

We have obtained one generalization of $\mathcal{K}$-continuity through localization. In this section we explore a different generalization by replacing the constant M in Definition A.1 with a continuous function on X:

Definition A.13 *Let $\rho : X \to I\!R_+$ be continuous, and let $\gamma \in \mathcal{K}$. A function $\phi : X \to Z$ is* **$\rho\gamma$-continuous** *when for all $x, y \in X$ we have*

$$d(\phi(x), \phi(y)) \leq \rho(x) \cdot \gamma(d(x, y)) \tag{A.6}$$

We say that ϕ is **$\mathcal{CK}$-continuous** *when there exist a continuous function $\rho : X \to I\!R_+$ and $\gamma \in \mathcal{K}$ such that ϕ is $\rho\gamma$-continuous.*

$\mathcal{CK}$-continuity is not a uniform property: the metric

$$d^*(x, y) := |x - y|/(1 + |x - y|)$$

for the set of real numbers is uniformly equivalent to the usual metric, but the identity map from the first corresponding space to the second is not $\mathcal{CK}$-continuous. Clearly, if a function ϕ is $\rho\gamma$-continuous, then it is also locally γ-continuous. We next prove a partial converse for the case where X is proper:

Lemma A.14 *Let X be proper, and suppose $\phi : X \to Z$ is locally γ_0-continuous for $\gamma_0 \in \mathcal{K}$. Then for every $\varepsilon > 0$ there exist a continuous function $\rho : X \to I\!R_+$ and $\gamma \in \mathcal{K}$ such that ϕ is $\rho\gamma$-continuous and $\gamma(r) = \gamma_0(r)$ for all $r \in [0, \varepsilon]$.*

Proof: Fix $x_0 \in X$, and for $r \geq 0$ let $B(r) = \{x \in X,\ d(x, x_0) \leq r\}$ denote the closed ball of radius r centered at x_0. Now X is proper which means $B(r)$ is compact for all $r \geq 0$, and it follows from Lemma A.12 that for every $r \geq 0$ there exists $M(r) \geq 0$ such that $M(r)$ is a γ_0-constant of ϕ on $B(r)$. We define $\mu : I\!R_+ \to I\!R_+$ by the equation $\mu(r) := 1 + \inf\{M(2s) : r < s\}$. This function μ is increasing and can therefore be bounded from above by a continuous function $L : I\!R_+ \to I\!R_+$. It follows from these definitions that for every $r \geq 0$ there exists $s > r$ such that $M(2s) \leq \mu(r) \leq L(r)$ which means $L(r)$ is a γ_0-constant of ϕ on $B(2s)$. Because $B(2r) \subset B(2s)$ when $r < s$, we see that $L(r)$ is a γ_0-constant of ϕ on $B(2r)$ for all $r \geq 0$. We next fix $\varepsilon > 0$ and construct functions ρ and γ as follows. Let $\psi : I\!R_+ \to I\!R_+$ be continuous and increasing and such that $\psi(r) = 1$ for $r \in [0, \varepsilon]$ and also $\psi(r) \geq L(r)$ for $r \geq \varepsilon + \delta$ for some $\delta > 0$. Then the function γ defined by the equation $\gamma(r) := \psi(r) \cdot \gamma_0(r)$ is of class $\mathcal{K}$, and furthermore there exists $c \geq 0$ such that $L(r) \leq c \cdot \psi(r)$ for all $r \geq 0$. Note that $\gamma(r) = \gamma_0(r)$ for all $r \in [0, \varepsilon]$ and also $\gamma(r) \geq \gamma_0(r)$ for all $r \geq 0$. We next define ρ by the equation $\rho(x) := \max\{L(d(x, x_0)),\ c\}$; clearly ρ is continuous on X. We have left to show that (A.6) holds for all $x, y \in X$. Suppose first that $d(x, y) \leq d(x, x_0)$; it follows that both x and y belong to $B(2d(x, x_0))$ and therefore $d(\phi(x), \phi(y)) \leq L(d(x, x_0)) \cdot \gamma_0(d(x, y)) \leq \rho(x) \cdot \gamma(d(x, y))$ as desired. Next suppose $d(x, x_0) \leq d(x, y)$; it follows that both x and y belong to $B(2d(x, y))$ and therefore $d(\phi(x), \phi(y)) \leq L(d(x, y)) \cdot \gamma_0(d(x, y)) \leq c \cdot \gamma(d(x, y)) \leq \rho(x) \cdot \gamma(d(x, y))$ as desired. ■

Because a proper metric space is both locally compact and σ-compact, we have the following simple consequence of Lemmas A.11 and A.14:

Corollary A.15 *If X is proper, then $C\mathcal{K}$-continuity, local $\mathcal{K}$-continuity, and continuity are all equivalent properties of a function $\phi : X \to Z$.*

This corollary is illustrated graphically in Figure A.1 which summarizes the relationships between the various continuity properties discussed in this appendix. We end with a corollary of Lemma A.14 which provides a characterization of locally Lipschitz functions:

Corollary A.16 *If X is proper, then $\phi : X \to Z$ is locally Lipschitz if and only if there exist a locally Lipschitz function $\rho : X \to I\!R_+$ and a locally Lipschitz function $\gamma \in \mathcal{K}$ such that ϕ is $\rho\gamma$-continuous.*

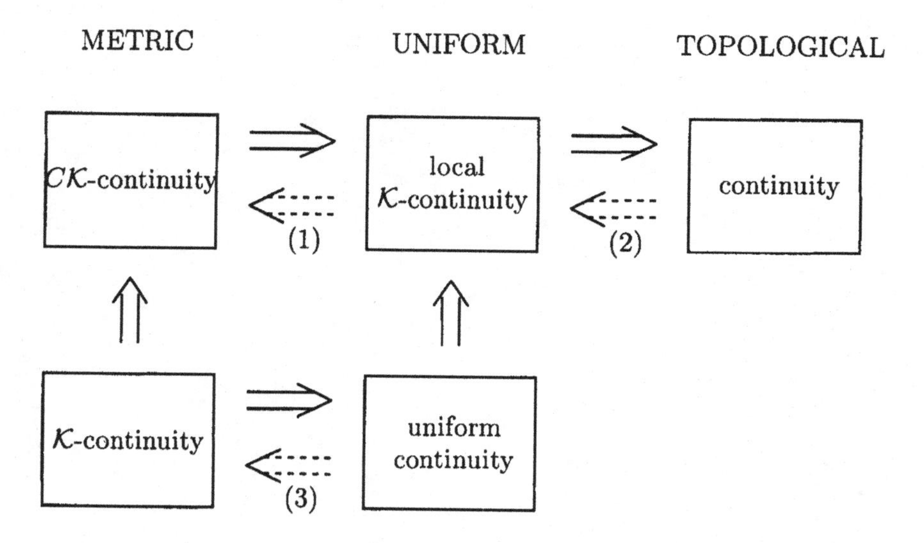

Figure A.1: Relationships between various metric, uniform, and topological continuity properties of a function $\phi : X \to Z$. The numbered implications with dotted arrows are true under the following (respectively numbered) conditions on the metric space X:

(1) X is proper (Lemma A.14)
(2) X is locally compact and σ-compact (Lemma A.11)
(3) X is a convex subset of a normed linear space (Lemma A.4)

Bibliography

[1] E. G. AL'BREKHT, *On the optimal stabilization of nonlinear systems*, J. Appl. Math. Mech., 25 (1962), pp. 1254–1266.

[2] B. D. O. ANDERSON AND J. B. MOORE, *Linear Optimal Control*, Prentice-Hall, Englewood Cliffs, New Jersey, 1971.

[3] Z. ARTSTEIN, *Stabilization with relaxed controls*, Nonlinear Anal., 7 (1983), pp. 1163–1173.

[4] M. ATHANS AND P. L. FALB, *Optimal Control*, McGraw-Hill Book Company, New York, 1966.

[5] J.-P. AUBIN, *Viability Theory*, Birkhäuser, Boston, 1991.

[6] J.-P. AUBIN AND A. CELLINA, *Differential Inclusions*, Springer-Verlag, Berlin, 1984.

[7] J.-P. AUBIN AND H. FRANKOWSKA, *Set-Valued Analysis*, Birkhäuser, Boston, 1990.

[8] T. BAŞAR AND P. BERNHARD, *H_∞-Optimal Control and Related Minimax Design Problems*, Birkhäuser, Boston, second ed., 1995.

[9] T. BAŞAR AND G. J. OLSDER, *Dynamic Noncooperative Game Theory*, Academic Press, London, 1982.

[10] J. A. BALL, J. W. HELTON, AND M. L. WALKER, *H_∞ control for nonlinear systems with output feedback*, IEEE Trans. Automat. Contr., 38 (1993), pp. 546–559.

[11] B. R. BARMISH, M. J. CORLESS, AND G. LEITMANN, *A new class of stabilizing controllers for uncertain dynamical systems*, SIAM J. Contr. Optimiz., 21 (1983), pp. 246–255.

[12] B. R. BARMISH AND G. LEITMANN, *On ultimate boundedness control of uncertain systems in the absence of matching conditions*, IEEE Trans. Automat. Contr., 27 (1982), pp. 153–158.

[13] S. Battilotti, *Sufficient conditions for global robust stabilization via measurement feedback for some classes of nonlinear systems*, in Proceedings of the 33rd IEEE Conference on Decision and Control, Lake Buena Vista, Florida, Dec. 1994, pp. 808–813.

[14] H. W. Bode, *Network Analysis and Feedback Amplifier Design*, Van Nostrand and Company, New York, 1945.

[15] T. Bröcker, *Differentiable Germs and Catastrophes*, Cambridge University Press, Cambridge, England, 1975.

[16] A. E. Bryson, Jr. and Y.-C. Ho, *Applied Optimal Control*, Hemisphere Publishing Corporation, New York, 1975.

[17] C. I. Byrnes and A. Isidori, *New results and examples in nonlinear feedback stabilization*, Syst. Contr. Lett., 12 (1989), pp. 437–442.

[18] Y. H. Chen, *A new matching condition for robust control design*, in Proceedings of the 1993 American Control Conference, San Francisco, California, June 1993, pp. 122–126.

[19] Y. H. Chen and G. Leitmann, *Robustness of uncertain systems in the absence of matching assumptions*, Int. J. Contr., 45 (1987), pp. 1527–1542.

[20] F. H. Clarke, *Optimization and Nonsmooth Analysis*, Society for Industrial and Applied Mathematics, Philadelphia, 1990.

[21] ——, *Qualitative properties of trajectories of control systems: A survey*, J. Dynam. Contr. Syst. To appear.

[22] M. J. Corless, *Robust stability analysis and controller design with quadratic Lyapunov functions*, in Variable Structure and Lyapunov Control, A. Zinober, ed., Springer-Verlag, Berlin, 1993.

[23] M. J. Corless and G. Leitmann, *Continuous state feedback guaranteeing uniform ultimate boundedness for uncertain dynamic systems*, IEEE Trans. Automat. Contr., 26 (1981), pp. 1139–1144.

[24] J.-M. Coron and L. Praly, *Adding an integrator for the stabilization problem*, Syst. Contr. Lett., 17 (1991), pp. 89–104.

[25] C. A. Desoer and M. Vidyasagar, *Feedback Systems: Input-Output Properties*, Academic Press, New York, 1975.

[26] J. FENG AND M. C. SMITH, *When is a controller optimal in the sense of H_∞ loop-shaping?*, IEEE Trans. Automat. Contr., 40 (1995), pp. 2026–2039.

[27] A. F. FILIPPOV, *On certain questions in the theory of optimal control*, SIAM J. Control, 1 (1962), pp. 76–84.

[28] R. A. FREEMAN, *A backstepping design of robust globally stabilizing feedback compensators for nonlinear systems*, Master's thesis, University of Illinois at Urbana-Champaign, 1992.

[29] ——, *Global internal stabilizability does not imply global external stabilizability for small sensor disturbances*, IEEE Trans. Automat. Contr., 40 (1995), pp. 2119–2122.

[30] ——, *Robust Control of Nonlinear Systems*, PhD thesis, University of California, Santa Barbara, California, Dec. 1995.

[31] R. A. FREEMAN AND P. V. KOKOTOVIĆ, *Backstepping design of robust controllers for a class of nonlinear systems*, in Proceedings of the IFAC Nonlinear Control Systems Design Symposium, Bordeaux, France, June 1992, pp. 307–312.

[32] ——, *Design and comparison of globally stabilizing controllers for an uncertain nonlinear system*, in Systems, Models, and Feedback: Theory and Applications, A. Isidori and T. J. Tarn, eds., Birkhäuser, Boston, 1992, pp. 249–264.

[33] ——, *Design of robust nonlinear feedback controls: Concepts, tools, and procedures*, in Proceedings of the 31st Allerton Conference on Communication, Control, and Computing, Monticello, Illinois, Sept. 1993, pp. 791–800.

[34] ——, *Design of 'softer' robust nonlinear control laws*, Automatica, 29 (1993), pp. 1425–1437.

[35] ——, *Global robustness of nonlinear systems to state measurement disturbances*, in Proceedings of the 32nd IEEE Conference on Decision and Control, San Antonio, Texas, Dec. 1993, pp. 1507–1512.

[36] ——, *Optimal nonlinear controllers for feedback linearizable systems*, in Proceedings of the Workshop on Robust Control via Variable Structure & Lyapunov Techniques, Benevento, Italy, Sept. 1994, pp. 286–293.

[37] ——, *Robust control Lyapunov functions: The measurement feedback case*, in Proceedings of the 33rd IEEE Conference on Decision and Control, Lake Buena Vista, Florida, Dec. 1994, pp. 3533–3538.

[38] ——, *Tools and procedures for robust control of nonlinear systems*, in Proceedings of the 33rd IEEE Conference on Decision and Control, Lake Buena Vista, Florida, Dec. 1994, pp. 3458–3463.

[39] ——, *Backstepping design with nonsmooth nonlinearities*, in Proceedings of the IFAC Nonlinear Control Systems Design Symposium, Tahoe City, California, June 1995, pp. 483–488.

[40] ——, *Optimal nonlinear controllers for feedback linearizable systems*, in Proceedings of the 1995 American Control Conference, Seattle, Washington, June 1995, pp. 2722–2726.

[41] ——, *Robust integral control for a class of uncertain nonlinear systems*, in Proceedings of the 34th IEEE Conference on Decision and Control, New Orleans, Louisiana, Dec. 1995, pp. 2245–2250.

[42] ——, *Tracking controllers for systems linear in the unmeasured states*, in Proceedings of the IFAC Nonlinear Control Systems Design Symposium, Tahoe City, California, June 1995, pp. 620–625.

[43] ——, *Inverse optimality in robust stabilization*, SIAM J. Contr. Optimiz., 34 (1996). To appear.

[44] ——, *Lyapunov design*, in The Control Handbook, W. S. Levine, ed., CRC Press, 1996, pp. 932–940.

[45] ——, *Tracking controllers for systems linear in the unmeasured states*, Automatica, (1996). To appear.

[46] S. T. GLAD, *Robustness of nonlinear state feedback—A survey*, Automatica, 23 (1987), pp. 425–435.

[47] H. GOLDSTEIN, *Classical Mechanics*, Addison-Wesley, Reading, Massachusetts, 1950.

[48] S. GUTMAN, *Uncertain dynamical systems—Lyapunov min-max approach*, IEEE Trans. Automat. Contr., 24 (1979), pp. 437–443.

[49] W. HAHN, *Stability of Motion*, Springer-Verlag, New York, 1967.

[50] C. A. HARVEY AND G. STEIN, *Quadratic weights for asymptotic regulator properties*, IEEE Trans. Automat. Contr., 23 (1978), pp. 378–387.

[51] J. HAUSER, S. SASTRY, AND P. V. KOKOTOVIĆ, *Nonlinear control via approximate input-output linearization: The ball and beam example*, IEEE Trans. Automat. Contr., 37 (1992), pp. 392–398.

[52] J. HAUSER, S. SASTRY, AND G. MEYER, *Nonlinear control design for slightly nonminimum phase systems: Application to V/STOL aircraft*, Automatica, 28 (1992), pp. 665–679.

[53] H. HERMES, *Asymptotically stabilizing feedback controls*, J. Differential Equations, 92 (1991), pp. 76–89.

[54] ——, *Asymptotically stabilizing feedback controls and the nonlinear regulator problem*, SIAM J. Contr. Optimiz., 29 (1991), pp. 185–196.

[55] D. J. HILL, *Dissipative nonlinear systems: Basic properties and stability analysis*, in Proceedings of the 31st IEEE Conference on Decision and Control, Tucson, Arizona, Dec. 1992, pp. 3259–3264.

[56] D. J. HILL AND P. J. MOYLAN, *The stability of nonlinear dissipative systems*, IEEE Trans. Automat. Contr., 21 (1976), pp. 708–711.

[57] ——, *Connections between finite-gain and asymptotic stability*, IEEE Trans. Automat. Contr., 25 (1980), pp. 931–936.

[58] J. HUANG AND W. J. RUGH, *Stabilization on zero-error manifolds and the nonlinear servomechanism problem*, in Proceedings of the 29th IEEE Conference on Decision and Control, Honolulu, Hawaii, Dec. 1990, pp. 1262–1267.

[59] P. A. IOANNOU AND A. DATTA, *Robust adaptive control: A unified approach*, Proceedings of the IEEE, 79 (1991), pp. 1736–1768.

[60] A. ISIDORI, *Nonlinear Control Systems*, Springer-Verlag, Berlin, second ed., 1989.

[61] A. ISIDORI AND A. ASTOLFI, *Disturbance attenuation and H_∞-control via measurement feedback in nonlinear systems*, IEEE Trans. Automat. Contr., 37 (1992), pp. 1283–1293.

[62] A. ISIDORI AND C. I. BYRNES, *Output regulation of nonlinear systems*, IEEE Trans. Automat. Contr., 35 (1990), pp. 131–140.

[63] D. H. JACOBSON, *Extensions of Linear-Quadratic Contol, Optimization and Matrix Theory*, Academic Press, London, 1977.

[64] D. H. Jacobson, D. H. Martin, M. Pachter, and T. Geveci, *Extensions of Linear-Quadratic Contol Theory*, vol. 27 of Lecture Notes in Control and Information Sciences, Springer-Verlag, Berlin, 1980.

[65] M. R. James and J. S. Baras, *Robust H_∞ output feedback control for nonlinear systems*, IEEE Trans. Automat. Contr., 40 (1995), pp. 1007–1017.

[66] Z.-P. Jiang, I. Mareels, and Y. Wang, *A Lyapunov formulation of nonlinear small gain theorem for interconnected systems*, in Proceedings of the IFAC Nonlinear Control Systems Design Symposium, Tahoe City, California, June 1995, pp. 666–671.

[67] Z.-P. Jiang, A. R. Teel, and L. Praly, *Small-gain theorem for ISS systems and applications*, Math. Control Signals Systems, 7 (1994), pp. 95–120.

[68] V. Jurdjevic and J. P. Quinn, *Controllability and stability*, J. Differential Equations, 28 (1978), pp. 381–389.

[69] R. E. Kalman, *Contributions to the theory of optimal control*, Bol. Soc. Matem. Mex., (1960), pp. 102–119.

[70] ——, *When is a linear control system optimal?*, Trans. ASME Ser. D: J. Basic Eng., 86 (1964), pp. 1–10.

[71] R. E. Kalman and J. E. Bertram, *Control system analysis and design via the "second method" of Lyapunov, I: Continuous-time systems*, J. Basic Engineering, 32 (1960), pp. 317–393.

[72] I. Kanellakopoulos, P. V. Kokotović, and A. S. Morse, *Systematic design of adaptive controllers for feedback linearizable systems*, IEEE Trans. Automat. Contr., 36 (1991), pp. 1241–1253.

[73] ——, *A toolkit for nonlinear feedback design*, Syst. Contr. Lett., 18 (1992), pp. 83–92.

[74] J. L. Kelley, *General Topology*, D. Van Nostrand Company, Princeton, New Jersey, 1955.

[75] H. K. Khalil, *Nonlinear Systems*, Macmillan, New York, 1992.

[76] ——, *Robust servomechanism output feedback controllers for a class of feedback linearizable systems*, in Proceedings of the IFAC 12th World Congress, vol. 8, Sydney, Australia, July 1993, pp. 35–38.

[77] ——, *Nonlinear Systems*, Prentice Hall, Upper Saddle River, New Jersey, second ed., 1996.

[78] P. P. KHARGONEKAR, I. R. PETERSEN, AND K. ZHOU, *Robust stabilization of uncertain linear systems: Quadratic stabilizability and H_∞ control theory*, IEEE Trans. Automat. Contr., 35 (1990), pp. 356–361.

[79] M. KISIELEWICZ, *Differential Inclusions and Optimal Control*, PWN—Polish Scientific Publishers, Warszawa, Poland, 1991.

[80] P. V. KOKOTOVIĆ, *The joy of feedback: Nonlinear and adaptive*, IEEE Control Systems Magazine, 12 (1992), pp. 7–17.

[81] P. V. KOKOTOVIĆ AND H. J. SUSSMANN, *A positive real condition for global stabilization of nonlinear systems*, Syst. Contr. Lett., 13 (1989), pp. 125–133.

[82] A. A. KRASOVSKY, *A new solution to the problem of a control system analytical design*, Automatica, 7 (1971), pp. 45–50.

[83] A. J. KRENER, *Necessary and sufficient conditions for nonlinear worst case (H_∞) control and estimation*, J. Math. Syst. Estim. Contr. To appear.

[84] M. KRSTIĆ, I. KANELLAKOPOULOS, AND P. V. KOKOTOVIĆ, *Adaptive nonlinear control without overparameterization*, Syst. Contr. Lett., 19 (1992), pp. 177–185.

[85] ——, *Nonlinear and Adaptive Control Design*, John Wiley & Sons, New York, 1995.

[86] M. KRSTIĆ, P. V. KOKOTOVIĆ, AND I. KANELLAKOPOULOS, *Transient performance improvement with a new class of adaptive controllers*, Syst. Contr. Lett., 21 (1993), pp. 451–461.

[87] J. KURZWEIL, *On the inversion of Liapunov's second theorem on stability of motion*, Ann. Math. Soc. Transl. Ser. 2, 24 (1956), pp. 19–77.

[88] E. B. LEE AND L. MARKUS, *Foundations of Optimal Control Theory*, John Wiley & Sons, New York, 1967.

[89] G. LEITMANN, *Guaranteed ultimate boundedness for a class of uncertain linear dynamical systems*, IEEE Trans. Automat. Contr., 23 (1978), pp. 1109–1110.

[90] ——, *Guaranteed asymptotic stability for some linear systems with bounded uncertainties*, ASME J. Dynam. Syst. Meas. Contr., 101 (1979), pp. 212–216.

[91] K. E. Lenz, P. P. Khargonekar, and J. C. Doyle, *When is a controller H_∞-optimal?*, Math. Control Signals Systems, 1 (1988), pp. 107–122.

[92] Y. Lin, *Lyapunov Function Techniques for Stabilization*, PhD thesis, Rutgers, The State University of New Jersey, New Brunswick, New Jersey, 1992.

[93] Y. Lin and E. D. Sontag, *Further universal formulas for Lyapunov approaches to nonlinear stabilization*, in Proc. Conf. Inform. Sci. and Systems, John Hopkins University, Mar. 1991, pp. 541–546.

[94] ——, *A universal formula for stabilization with bounded controls*, Syst. Contr. Lett., 16 (1991), pp. 393–397.

[95] Y. Lin, E. D. Sontag, and Y. Wang, *Recent results on Lyapunov-theoretic techniques for nonlinear stability*, in Proceedings of the 1994 American Control Conference, Baltimore, Maryland, June 1994, pp. 1771–1775.

[96] D. G. Luenberger, *Linear and Nonlinear Programming*, Addison-Wesley, second ed., 1984.

[97] D. L. Lukes, *Optimal regulation of nonlinear dynamical systems*, SIAM J. Contr., 7 (1969), pp. 75–100.

[98] A. M. Lyapunov, *Problème general de la stabilite du mouvement*, Ann. Fac. Sci. Toulouse, 9 (1907), pp. 203–474. In French.

[99] I. M. Y. Mareels and D. J. Hill, *Monotone stability of nonlinear feedback systems*, J. Math. Syst. Estim. Contr., 2 (1992), pp. 275–291.

[100] R. Marino and P. Tomei, *Global adaptive output-feedback control of nonlinear systems, part I: Linear parameterization*, IEEE Trans. Automat. Contr., 38 (1993), pp. 17–32.

[101] ——, *Robust stabilization of feedback linearizable time-varying uncertain nonlinear systems*, Automatica, 29 (1993), pp. 181–189.

[102] ——, *Nonlinear Control Design: Geometric, Adaptive, and Robust*, Prentice Hall, London, 1995.

[103] P. MARTIN, S. DEVASIA, AND B. PADEN, *A different look at output tracking: Control of a VTOL aircraft*, in Proceedings of the 33rd IEEE Conference on Decision and Control, Lake Buena Vista, Florida, Dec. 1994, pp. 2376–2381.

[104] J. L. MASSERA, *Contributions to stability theory*, Annals of Mathematics, 64 (1956), pp. 182–206.

[105] A. I. MEES, *Dynamics of Feedback Systems*, John Wiley & Sons, Chichester, 1981.

[106] B. P. MOLINARI, *The stable regulator problem and its inverse*, IEEE Trans. Automat. Contr., 18 (1973), pp. 454–459.

[107] B. S. MORDUKHOVICH, *Generalized differential calculus for nonsmooth and set-valued mappings*, Journal of Mathematical Analysis and Applications, 183 (1994), pp. 250–288.

[108] ——, *Necessary optimality and controllability conditions for nonsmooth control systems*, in Proceedings of the 33rd IEEE Conference on Decision and Control, Lake Buena Vista, Florida, Dec. 1994, pp. 3992–3997.

[109] P. J. MOYLAN, *Implications of passivity in a class of nonlinear systems*, IEEE Trans. Automat. Contr., 19 (1974), pp. 373–381.

[110] P. J. MOYLAN AND B. D. O. ANDERSON, *Nonlinear regulator theory and an inverse optimal control problem*, IEEE Trans. Automat. Contr., 18 (1973), pp. 460–464.

[111] H. NIJMEIJER AND A. J. VAN DER SCHAFT, *Nonlinear Dynamical Control Systems*, Springer-Verlag, New York, 1990.

[112] P. C. PARKS, *Lyapunov redesign of model reference adaptive control systems*, IEEE Trans. Automat. Contr., 11 (1966), pp. 362–367.

[113] W. R. PERKINS AND J. B. CRUZ, JR., *The parameter variation problem in state feedback control systems*, Trans. ASME Ser. D: J. Basic Eng., 87 (1965), pp. 120–124.

[114] ——, *Feedback properties of linear regulators*, IEEE Trans. Automat. Contr., 16 (1971), pp. 659–664.

[115] M. M. POLYCARPOU AND P. A. IOANNOU, *A robust adaptive nonlinear control design*, in Proceedings of the 1993 American Control Conference, San Francisco, California, June 1993, pp. 1365–1369.

[116] J.-B. POMET, R. M. HIRSCHORN, AND W. A. CEBUHAR, *Dynamic output feedback regulation for a class of nonlinear systems*, Math. Control Signals Systems, 6 (1993), pp. 106–124.

[117] L. PRALY, *Lyapunov design of a dynamic output feedback for systems linear in their unmeasured state components*, in Proceedings of the IFAC Nonlinear Control Systems Design Symposium, Bordeaux, France, June 1992, pp. 31–36.

[118] L. PRALY, B. D'ANDRÉA NOVEL, AND J.-M. CORON, *Lyapunov design of stabilizing controllers for cascaded systems*, IEEE Trans. Automat. Contr., 36 (1991), pp. 1177–1181.

[119] L. PRALY AND Z.-P. JIANG, *Stabilization by output feedback for systems with ISS inverse dynamics*, Syst. Contr. Lett., 21 (1993), pp. 19–33.

[120] Z. QU, *Global stabilization of nonlinear systems with a class of unmatched uncertainties*, Syst. Contr. Lett., 18 (1992), pp. 301–307.

[121] ——, *Robust control of nonlinear uncertain systems under generalized matching conditions*, Automatica, 29 (1993), pp. 985–998.

[122] R. T. ROCKAFELLAR, *Convex Analysis*, Princeton University Press, Princeton, New Jersey, 1970.

[123] H. L. ROYDEN, *Real Analysis*, Macmillan, New York, third ed., 1988.

[124] A. SABERI, P. V. KOKOTOVIĆ, AND H. J. SUSSMANN, *Global stabilization of partially linear composite systems*, SIAM J. Contr. Optimiz., 28 (1990), pp. 1491–1503.

[125] M. SAFONOV, *Stability and Robustness of Multivariable Feedback Systems*, MIT Press, Cambridge, Massachusetts, 1980.

[126] I. W. SANDBERG, *On the L_2-boundedness of solutions of nonlinear functional equations*, Bell Sys. Tech. J., 43 (1964), pp. 1581–1599.

[127] J. S. SHAMMA, *Construction of nonlinear feedback for ℓ_1-optimal control*, in Proceedings of the 33rd IEEE Conference on Decision and Control, Lake Buena Vista, Florida, Dec. 1994, pp. 40–45.

[128] ——, *Optimization of the ℓ^∞-induced norm under full state feedback*, IEEE Trans. Automat. Contr., 41 (1996). To appear.

[129] J. J. E. SLOTINE AND K. HEDRICK, *Robust input-output feedback linearization*, Int. J. Contr., 57 (1993), pp. 1133–1139.

[130] E. D. SONTAG, *A Lyapunov-like characterization of asymptotic controllability*, SIAM J. Contr. Optimiz., 21 (1983), pp. 462–471.

[131] ——, *Smooth stabilization implies coprime factorization*, IEEE Trans. Automat. Contr., 34 (1989), pp. 435–443.

[132] ——, *A 'universal' construction of Artstein's theorem on nonlinear stabilization*, Syst. Contr. Lett., 13 (1989), pp. 117–123.

[133] ——, *Further facts about input to state stabilization*, IEEE Trans. Automat. Contr., 35 (1990), pp. 473–476.

[134] ——, *Input/output and state-space stability*, in New Trends in System Theory, Birkhäuser, Boston, 1991.

[135] ——, *State-space and I/O stability for nonlinear systems*, in Feedback Control, Nonlinear Systems, and Complexity, B. A. Francis and A. R. Tannenbaum, eds., Lecture Notes in Control and Information Sciences, Springer-Verlag, Berlin, 1995, pp. 215–235.

[136] E. D. SONTAG AND H. J. SUSSMANN, *Further comments on the stabilizability of the angular velocity of a rigid body*, Syst. Contr. Lett., 12 (1988), pp. 213–217.

[137] E. D. SONTAG AND Y. WANG, *On characterizations of the input-to-state stability property*, Syst. Contr. Lett., 24 (1995), pp. 351–359.

[138] ——, *New characterizations of input-to-state stability*, in Proceedings of the Conference on Information Sciences and Systems, Princeton, New Jersey, Mar. 1996. To appear.

[139] M. W. SPONG AND M. VIDYASAGAR, *Robot Dynamics and Control*, John Wiley & Sons, New York, 1989.

[140] H. J. SUSSMANN AND P. V. KOKOTOVIĆ, *The peaking phenomenon and the global stabilization of nonlinear systems*, IEEE Trans. Automat. Contr., 36 (1991), pp. 424–440.

[141] A. R. TEEL, *A nonlinear small gain theorem for the analysis of control systems with saturation*, IEEE Trans. Automat. Contr., (1994). To appear.

[142] A. R. TEEL, T. T. GEORGIOU, L. PRALY, AND E. D. SONTAG, *Input-output stability*, in The Control Handbook, W. S. Levine, ed., CRC Press, 1996, pp. 895–908.

[143] A. R. TEEL AND L. PRALY, *On output feedback stabilization for systems with ISS inverse dynamics and uncertainties*, in Proceedings of the 32nd IEEE Conference on Decision and Control, San Antonio, Texas, Dec. 1993, pp. 1942–1947.

[144] J. S. THORP AND B. R. BARMISH, *On guaranteed stability of uncertain linear systems via linear control*, J. Optimiz. Theory Appl., 35 (1981), pp. 559–579.

[145] A. F. TIMAN, *Theory of Approximation of Functions of a Real Variable*, Macmillan, New York, 1963.

[146] J. TSINIAS, *Sufficient Lyapunov-like conditions for stabilization*, Math. Control Signals Systems, 2 (1989), pp. 343–357.

[147] ——, *Optimal controllers and output feedback stabilization*, Syst. Contr. Lett., 15 (1990), pp. 277–284.

[148] J. TSINIAS AND N. KALOUPTSIDIS, *Output feedback stabilization*, IEEE Trans. Automat. Contr., 35 (1990), pp. 951–954.

[149] ——, *A correction note on "Output feedback stabilization"*, IEEE Trans. Automat. Contr., 39 (1994), p. 806.

[150] A. J. VAN DER SCHAFT, *On a state space approach to nonlinear H_∞ control*, Syst. Contr. Lett., 16 (1991), pp. 1–8.

[151] ——, *L_2-gain analysis of nonlinear systems and nonlinear state feedback H_∞ control*, IEEE Trans. Automat. Contr., 37 (1992), pp. 770–784.

[152] M. VIDYASAGAR, *Nonlinear Systems Analysis*, Prentice-Hall, Englewood Cliffs, New Jersey, second ed., 1993.

[153] T. WAZEWSKI, *On an optimal control problem*, in Proceedings of the Conference on Differential Equations and Their Applications, Prague, 1963, pp. 229–242.

[154] K. WEI, *Quadratic stabilizability of linear systems with structural independent time-varying uncertainties*, IEEE Trans. Automat. Contr., 35 (1990), pp. 268–277.

[155] J. C. WILLEMS, *Dissipative dynamical systems, Part I: General theory*, Arch. Rational Mech. Anal., 45 (1972), pp. 321–351.

[156] B. YAO AND M. TOMIZUKA, *Smooth robust adaptive sliding mode control of manipulators with guaranteed transient performance*, in Proceedings of the 1994 American Control Conference, Baltimore, Maryland, June 1994, pp. 1176–1180.

[157] ——, *Robust adaptive nonlinear control with guaranteed transient performance*, in Proceedings of the 1995 American Control Conference, Seattle, Washington, June 1995, pp. 2500–2504.

[158] T. YOSHIZAWA, *Stability Theory by Liapunov's Second Method*, The Mathematical Society of Japan, Gakujutsutosho Printing Co., Ltd., Tokyo, 1966.

[159] G. ZAMES, *On the input-output stability of time-varying nonlinear feedback systems. Part I: Conditions using concepts of loop gain, conicity, and positivity*, IEEE Trans. Automat. Contr., 11 (1966), pp. 228–238.

[160] ——, *On the input-output stability of time-varying nonlinear feedback systems. Part II: Conditions involving circles in the frequency plane and sector nonlinearities*, IEEE Trans. Automat. Contr., 11 (1966), pp. 465–476.

Index